Crop Chemicals

PUBLISHER

Fundamentals of Machine Operation (FMO) is a series of manuals created by Deere & Company. Each book in the series was conceived, researched, outlined, edited, and published by Deere & Company. Authors were selected to provide a basic technical manuscript which could be edited and rewritten, if necessary, by staff editors.

PUBLISHER: JOHN DEERE TECHNICAL SERVICES, Dept. F, John Deere Road, Moline, Illinois, 61265; Manager: John A. Conrads; Service Training Manager: John D. Spuller.

FUNDAMENTAL WRITING SERVICES EDITORIAL STAFF
Managing Editor: Louis R. Hathaway
Editor: Michael H. Johanning
Assisting Editor: Laurence T. Hammond
Promotions Clerk: Annette M. LaCour

AUTHOR: *Harold A. Hughes,* Associate Professor and Extension Project Leader for the Agricultural Engineering Department at Virginia Polytechnic Institute and State University, has been working with machines for 12 years. Dr. Hughes received his Ph.D. in System Studies of Energy Consumption at Michigan State University and is co-author of the FMO Agricultural Machinery Safety Manual published in 1974. A member of ASAE, Dr. Hughes has written over 30 technical papers and many extension publications that have appeared in a variety of magazines.

COPY EDITOR: *Kenneth H. Smith,* copy editor (retired), Deere & Company Advertising Department.

TECHNICAL CONSULTANTS: *E. L. Knake,* Professor of Weed Science at the University of Illinois, has been associated with the Department of Agronomy for 21 years and has over 31 years of teaching experience. In 1974, Dr. Knake received the Midwest Agricultural Chemicals Association Educator Award, and is past president of the Weed Science Society of America. Dr. Knake has been a committee member for the Northcentral Weed Control Conference and the author of numerous publications on weed control.

Malcolm Shurtleff, Extension Specialist in Plant Pathology at the University of Illinois, is the nation's outstanding Plant Pathologist. Dr. Shurtleff received his Ph.D. in Plant Pathology from the University of Minnesota and was the recipient of the Paul A. Funk Award for his Outstanding Work in the State in Agriculture. In 1970, Dr. Shurtleff was the first scientist to publicize the Southern Corn Leaf Blight and fully inform farmers of the disease. He is a member of the American Phydopathological Society and is the author of many books, publications and color-slide sets on plant diseases.

Richard S. Fawcett, Extension Weed Control Specialist at Iowa State University, has been involved in teaching and research in the field of Weed Science since 1974. Dr. Fawcett was one of ten recipients of the 1981 American Soybean Association Research Recognition Award. He is a member of several professional societies related to weed control and is the author of numerous papers, reports, and scientific journal articles on the subject.

D. E. Gates, Extension Entomologist, has been with the Kansas Extension Service at Kansas State University since 1949. Professor Gates is co-author of the publication, *Insects in Kansas.*

CONSULTING EDITORS: *Ronald F. Espenschied,* Ed.D., Professor of Agricultural Engineering and Vocational Agriculture Service, University of Illinois, has written many publications and prepared other types of audio-visual materials in agricultural mechanization during his 32 years as a teacher and teacher-educator. Professor Espenschied has spent 22 years specializing in agricultural mechanics in the Agricultural Egnineering Department of the University of Illinois following ten years of teaching vocational agriculture.

Thomas A. Hoerner, Ph.D., Professor and Teacher-Educator in Agricultural Engineering and Agricultural Education at Iowa State University. Dr. Hoerner has 24 years of high school and university teaching experience. He has authored numerous manuals and instructional materials in the agricultural mechanics area.

Keith R. Carlson has 13 years of experience as a high school vocational agricultural instructor. Mr. Carlson is the author of numerous instructor's guides. All instructor's kits for the FMO texts are being prepared by Mr. Carlson, who is president of Agri Education Inc., an agricultural consulting firm in Stratford, Iowa.

SPECIAL ACKNOWLEDGEMENTS: The authors and editors wish to thank the following Deere & Company people for their assistance: Robert A. Sohl, Director of Service; Arland W. Pauli, Product Planning Department; and a host of other John Deere people who gave extra assistance and advice on this project.

PUBLISHER: Fundamentals of Malchine Operation (FMO) texts and visuals are published by John Deere Service Training, Dept. F., John Deere Road, Moline, Illinois 61265.

FOR MORE INFORMATION: This text is part of a complete series of texts and visuals on agricultural malchinery called Fundamentals of Malchine Operation (FMO). For more information, request a free FMO Catalog of Manuals and Visuals. Send your request to John Deere Service Training, Dept. F., John Deere Road, Moline, Illinois 61265.

We have a long-range interest in good machine operation

ISBN 0-86691-024-7

CONTRIBUTORS: The publisher is grateful to the following individuals and companies who were helpful in providing some of the photographs used in this text:
Ag Chemicals and Commercial Fertilizer, Willoughby, Ohio — pg. 67, Fig. 2.
Allis-Chalmers, Milwaukee, Wisconsin — pg. 164, Fig. 19.
American Fruit Growers, Willoughby, Ohio — pg. 79, Fig. 24.
Aquatic Biology Section, Illinois Natural History Survey, Urbana, Illinois — pg. 9, Fig. 14; pg. 41, Fig. 8.
Asgrow Seed Co., Kalamazoo, Michigan — pg. 72, Fig. 13; pg. 73, Fig. 15; pg. 85, Fig. 3.
Automatic Equipment Manufacturing Co., Pender, Nebraska — pg. 9, Fig. 13 (lower left); pg. 122, Fig. 23.
Century Engineering Corp., Cedar Rapids, Iowa — pg. 116, Fig. 6; pg. 119, Fig. 12; pg. 130, Fig. 41; pg. 130, Fig. 43; pg. 135, Fig. 52; pg. 136, Fig. 53; pg. 137, Fig. 57.
Cessna Aircraft Co., Wichita, Kansas — pg. 9, Fig. 37 (lower right); pg. 160, Fig. 8.
Ciba-Geigy, Greensboro, North Carolina — Appendix, Insects of Agronomic Crops, Fig. 24.
Charles R. Drake. Associate Professor of Plant Pathology and Physiology, Virginia Polytechnic Institute and State University, Blacksburg, Virginia — pg. 68, Fig. 5; pg. 69, Fig. 6.
Clemson University Extension Service, Clemson, South Carolina, in cooperation with USDA — pg. 65, Intro; pg. 107, Intro; Appendix, Insect and Disease Illustrations.
Delavan Manufacturing Co., West Des Moines, Iowa — pg. 135, Fig. 51; pg. 136, Fig. 53; pg. 141, Fig. 68.
Dow Chemical Company, Midland, Michigan — pg. 111, Fig. 6.
Elanco Products Co., Indianapolis, Indiana — pg. 42, Fig. 10.
F.E. Myers & Bro. Co., Ashland, Ohio — pg. 101, Fig. 26.
Farmhand, Inc., Hopkins, Minnesota — pg. 28, Fig. 17; pg. 160, Fig. 7.
The Fertilizer Institute, Washington, D.C. — pg. 4, Fig. 5.
Florida Agricultural Experimental Station, Gainesville, Florida — pg. 72, Fig. 12; pg. 128, Figs. 36, 37, 38.
FMC Agricultural Machinery Division, Jonesboro, Arkansas — pg. 121, Fig. 18; pg. 121, Fig. 19; pg. 121, Fig. 20; pg. 122, Fig. 21; pg. 130, Fig. 42.
General Metals Inc., Greensboro, North Carolina — pg. 160, Fig. 6.
Grant Heilman Photography, Lititz, Pennsylvania — pg. 2, Fig. 1; pg. 41, Fig. 7; pg. 46, Fig. 15.
Highway Equipment Co., Cedar Rapids, Iowa — pg. 5, Fig. 9; pg. 8, Fig. 12 (left center and right center).
Hypro Division, Lear Sigler, Inc., St. Paul, Minnesota — pg. 135, Fig. 51 (right); pg. 137, Fig. 55; pg. 144, Figs. 73, 74; pg. 145, Fig. 76.
Irrigation Age, Webb Co., St. Paul, Minnesota — pg. 74, Fig. 16.
John Blue Co., Hunterville, Alabama — pg. 159, Fig. 4.
Lely, Wilson, North Carolina — pg. 159, Fig. 5.
Len Lindstrom, Deere & Co., Moline, Illinois — pg. 83, Intro.
Louisiana State University, Baton Rouge, Louisiana — pg. 126, Fig. 33.
Meister Publishing Co., Ag Consultant and Fieldman, 1982 Weed Control Manual, Willoughby, Ohio — Appendix, Weed Identification Photos.
Michigan State University Agricultural Experiment Station, East Lansing, Michigan — pg. 7, Fig. 11; pg. 133, Fig. 48.
Mobay Chemical Corp., Kansas City, Missouri — pg. 193, Intro.
Nalco Chemical Co., Oak Brook, Illinois — pg. 155, Fig. 89.
National Agricultural Chemical Association, Washington, D.C. — pg. 176, Fig. 4; pg. 181, Fig. 18.
National Fertilizer Solutions Association, Peoria, Illinois — pg. 18, Fig. 4.
Ohio State University, Columbus, Ohio — pg. 72, Fig. 11; pg. 85, Fig. 2; pg. 86, Fig. 4; pg. 86, Fig. 5; pg. 87, Fig. 6; pg. 88, Fig. 7; pg. 88, Fig. 8.
P-A-G Seeds, Minneapolis, Minnesota — pg. 43, Fig. 11.
Potash Institute, Atlanta, Georgia — pg. 17, Fig. 3; pg. 21, Fig. 8; pg. 23, Fig. 11.
Powell Manufacturing Co., Inc., Bennettsville, South Carolina — pg. 33, Fig. 26.
Rickel Manufacturing Corp., Salina, Kansas — pg. 8, Fig. 12 (lower right); pg. 120, Fig. 16.
Solo Inc., Newport News, Virginia — pg. 116, Fig. 5.
Stauffer Chemical Co., Westport Connecticut — pg. 37, Intro; pg. 38, Fig. 1; pg. 38, Fig. 2; pg. 47, Fig. 16; pg. 47, Fig. 17; pg. 50, Fig. 22; pg. 58, Fig. 32; pg. 104, Fig. 29.
Tennessee Valley Authority, Muscle Shoals, Alabama — pg. 31, Fig. 23; pg. 131, Figs. 44, 45; pg. 132, Fig. 47; pg. 166, Fig. 22; pg. 167, Fig. 23; pg. 169, Figs. 25, 26.
Uniroyal Ag Chemicals, Nougatuck, Connecticut — pg. 41, Fig. 6.
University of Arkansas Rice Branch Experiment Station (Bobby Huey), Stuttgart, Arkansas, Appendix, Insects of Agronomic Crops, Figs. 42-44.
University of Georgia Cooperative Extension Service, Athens, Georgia — Appendix, Weed Identification Photos.
University of Illinois, Urbana, Illinois — pg. 67, Figs. 3, 4; pg. 69, Fig. 7; pg. 71, Fig. 10; pg. 77, Fig. 21; pg. 108, Figs. 3, 4; Department of Illinois Plant Pathology, Appendix, Diseases of Agronomic Crops, Figs. 8, 9.
USDA Photograph, Washington, D.C. — pg. 40, Figs. 4, 5; pg. 73, Fig. 14; pg. 79, Fig. 24; pg. 91, Fig. 12; pg. 108, Fig. 2; Appendix, Diseases of Crops and Insect Identification.
USDA Soil Conservation Service, Washington, D.C. — pg. 2, Fig. 1.
U.S. Government Illustrations, Washington, D.C. — pg. 94, Fig. 18; pg. 96, Fig. 20; pg. 162, Fig. 13.
Virginia Polytechnic Institute and State University Cooperative Extension Service, Blacksburg, Virginia — pg. 76, Fig. 18; pg. 186, Fig. 28.
Wallaces Farmer, Des Moines, Iowa — pg. 8, Fig. 12 (lower left); pg. 126, Fig. 31.

Contents

1
INTRODUCTION

2
FERTILIZERS AND LIME

3
WEED CONTROL

4
DISEASE CONTROL

5
INSECT CONTROL

6
FUMIGATION

7
APPLICATION OF LIQUID CHEMICALS

8

APPLICATION OF DRY CHEMICALS

9

SAFETY

APPENDIX

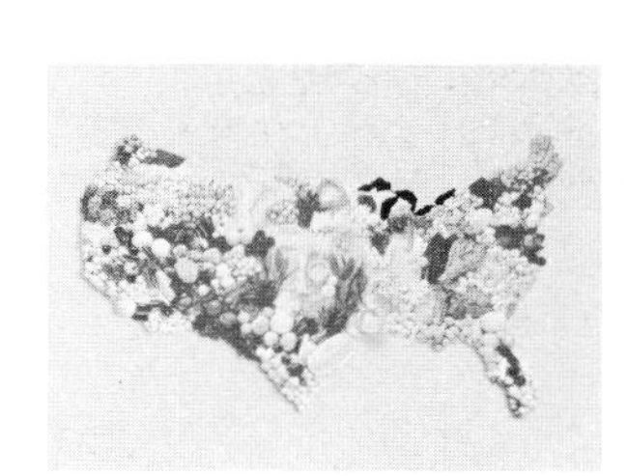

1
Introduction

Fig. 1—Chemicals Help Assure Good Harvest

AGRICULTURAL CHEMICALS

Chemicals are "tools" for a modern farmer. They are as important as mechanical tools for growing successful crops (Fig. 1). While chemicals perform many functions, the two most important are:

- **Improving plant growth**
- **Protecting against pests**

Chemicals can be used:

- *Before planting (Fig. 2)*
- *During growth (Fig. 3)*
- *After a crop has been harvested (Fig. 4)*

There are so many agricultural chemicals that it is impractical to discuss them all in detail in this book. Two classes of chemicals which will be emphasized are:

- **Fertilizers**
- **Pesticides**

FERTILIZERS

Plants, like animals, need food to help them grow and develop. Plant food is composed of parts called plant-food elements. At least 16 elements are needed for a plant to grow normally. If any one of the essential elements is missing, the plant will die. If an element is in short supply, the plant will be undernourished and will not grow properly (Fig. 5). Men and animals which eat undernourished plants may also be poorly nourished. Well-fed plants provide better nutrition for animals or people who consume the crop (Fig. 6).

Fertilizers are materials which supply one or more of the essential plant-food elements. Many materials have been used as fertilizers. The Pilgrims are credited with burying a fish under each hill of corn for fertilizer. Common natural materials that have been used as fertilizer include guano (manure from seabirds which accumulates in dry coastal areas, chiefly South America), leaves and other plant materials, tankage, blood and other animal materials, animal manure, and "night soil" (human waste).

Fig. 2—Seed Is Often Treated with Chemicals to Reduce Insect Damage

Fig. 3—Chemicals Applied during the Growing Season Help Control Weeds, Insects, and Crop Diseases

Fig. 4—Chemicals Applied to Harvested Products Help Prevent Spoilage and Insect Damage

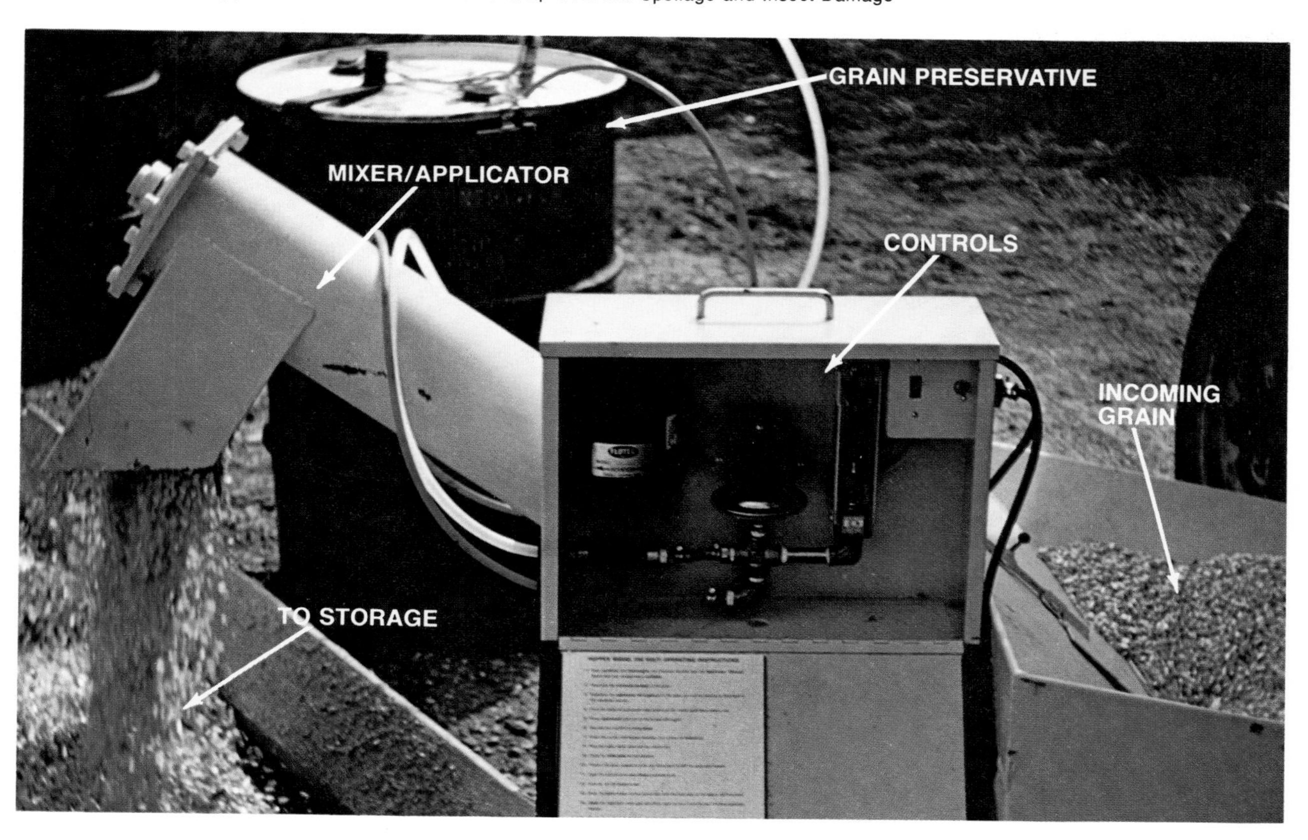

Fig. 5—A Shortage of Plant Food Results in Poor Crop Quality and Low Yields

Fig. 6—There Is a Definite Relationship between Healthy Plants and Healthy Animals

COMMERCIAL FERTILIZERS

The "natural" materials listed above are still widely used. But, as fertilizer use has expanded, emphasis has shifted to inorganic chemical fertilizers which are mixtures of plant-food elements with inert "carriers". Commerical fertilizers have several advantages when compared to natural materials.

1. Commercial fertilizers are more concentrated than most natural materials. Less weight and volume are needed to provide equivalent nutrients.

2. The plant food content of commercial fertilizers is known and is uniform throughout.

3. Commercial fertilizers can be tailor-made to suit a particular requirement.

4. Commercial fertilizers are usually easier to handle than many "natural" materials.

IMPORTANCE OF COMMERCIAL FERTILIZER

The quantity of fertilizer used on U.S. farms has increased annually for many years. Since 1900, when about 2 million tons were used, fertilizer use has climbed to more than 50 million tons (45 million metric tons) per year (Fig. 7). The increase is even more dramatic when you realize that modern fertilizers contain a higher percentage of plant food than those of the earlier era.

Fertilizer is used on virtually all crops grown in the U.S. It is also critically important to the agriculture of other countries. Recent innovations, such as the introduction of high-yielding grain varieties, give promise of the "Green Revolution" which could make some parts of the world self-sufficient in food production for the first time in many years. But without adequate fertilization programs, the "Green Revolution" will never reach its full potential.

Fig. 7—Increased Use of Commercial Fertilizer

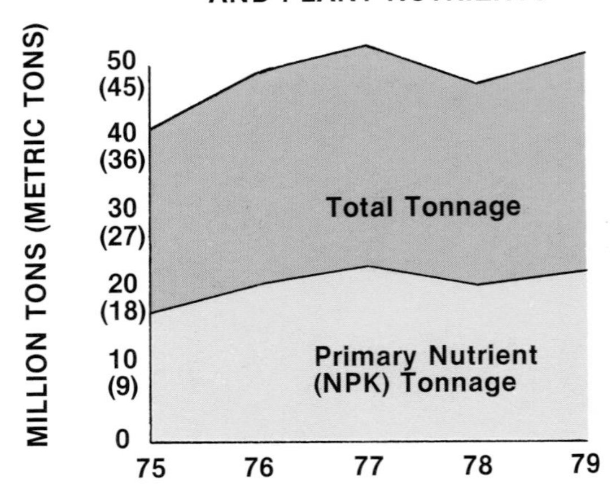

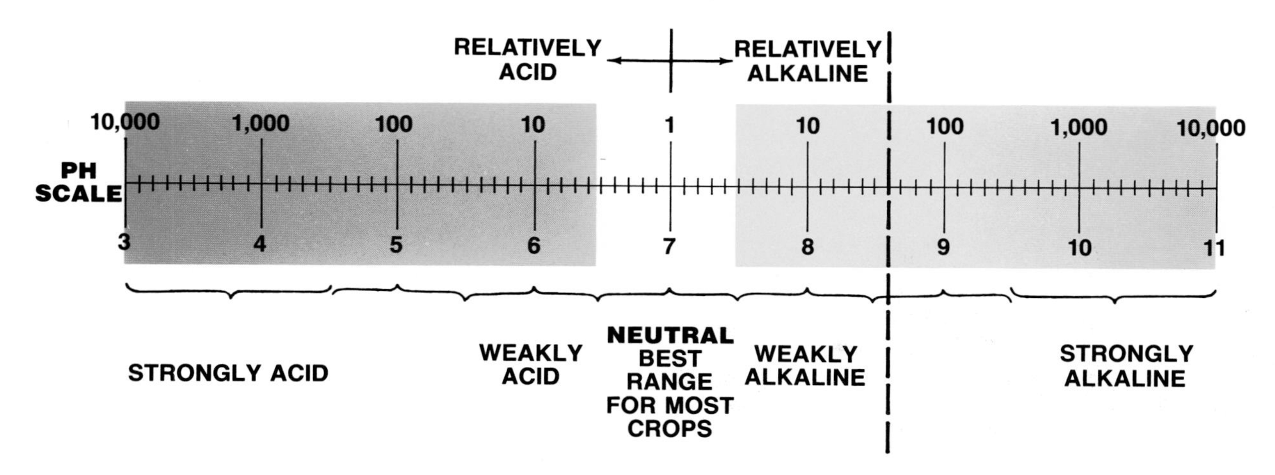

Fig. 8—The pH Scale

EFFECT OF SOIL pH

Most soils in humid regions are acid. The degree of acidity is expressed as soil pH. The pH scale has 14 units (Fig. 8). A pH of 7.0 is neutral. Lower pH indicates acid soil; higher pH indicates alkaline soil.

The pH scale is **logarithmic.** That means a pH of 5.0 is 10 times as acid as a pH of 6.0. Similarly, pH of 9.0 is 10 times as alkaline as pH 8.0 (top of scale, Fig. 8).

Most agricultural crops grow best in soil with pH near 7.0. However, there are exceptions. Crops which grow better with lower pH include blueberries and cranberries. Alfalfa and sweet clover thrive at pH levels above 7.0. Plant food availability is closely associated with soil acidity. Therefore, soil pH can affect plant response to fertilizer. If soil is too acid or too alkaline, the pH can be changed by adding the proper material.

Lime Corrects Soil Acidity

The standard way to correct soil acidity is to add crushed limestone to the soil (Fig. 9). Lime supplies some plant food elements, but is not generally thought of as a fertilizer. Lime raises the pH of acid soil so plants can extract food elements more easily. The relationship between lime and plant food availability has led to a commonly used expression, "Lime and fertilizer work as a team".

Gypsum and Sulfur Correct Soil Alkalinity

While acid soil is a common problem in humid areas, alkaline soil is a concern in arid areas. To lower soil pH, materials such as gypsum and sulfur are applied.

PESTICIDES

Pesticides are substances or combinations of substances used to kill or control pests. A pest is anything that:

- *Competes with man, his animals, or crops for food and feed*
- *Causes injury to man, his livestock, or crops*
- *Carries disease to man, his livestock, or crops*
- *Causes annoyance to man, his livestock, or crops*

The ending "cide" on the word pesticide means killer. Some common pesticides and the pests they are used on are:

Fig. 9—Correcting Soil pH — Lime for Acid Soil, Gypsum for Alkaline Soil.

Fig. 10—Some Pesticides Were Discovered by Accident

- *Insecticides – control insects*
- *Herbicides – control undesirable plants*
- *Rodenticides – control rats, mice and other rodents*
- *Nematicides – control nematodes*
- *Fungicides – control fungus diseases*
- *Acaricides – control mites and spiders*
- *Bactericides – control bacteria*

Primary emphasis in this book will be on application of insecticides and herbicides for protection of field crops.

PESTICIDE DEVELOPMENT

Use of toxic chemicals to control pests is not new; the Greek poet Homer mentioned the value of burning sulfur as a fumigant about 1000 B.C. Arsenic was being used as an insecticide a century before the birth of Christ.

The first so-called "natural" insecticide was tobacco. It was in use at least 300 years ago. Today, we know that tobacco was effective because of nicotine, which is still being used in a few insecticides.

Most early pesticides were discovered by accident. Bordeaux mixture, a weapon against plant disease, is an interesting example. The owner of a small vineyard near Bordeaux, France, mixed lime, copper sulfate, and water and splashed the bluish mixture on grape vines near a path which crossed his property. He was trying to discourage schoolboy "pests" from picking his fruit by making it look unappetizing (Fig. 10). Later, the vineyard was stricken by powdery mildew, but the vines which had been "treated" survived.

By the middle of the 19th Century, the first scientific studies of pesticides were appearing. Experimentation with arsenical compounds led to the introduction of Paris green which was used to check the spread of the Colorado Potato beetle. It was so successful that it was the most widely used agricultural insecticide for many years.

Near the end of the 19th Century, a vineyard operator, using Bordeaux mixture on his grapes, noticed that the fungicide caused the leaves of certain weeds to turn black. The idea of selective herbicides may have developed from this chance observation.

Another early herbicide was iron sulfate. If sprayed on a mixture of cereal plants and broad-leaved weeds, it kills the weeds but not the small grain.

Use of pesticides expanded shortly before World War II. That increased use was a factor contributing to the dramatic increase in farming efficiency at that time.

Further advances were made during the war. DDT was discovered in Switzerland, organophosphorus compounds were introduced in Germany, and dithiocarbamate fungicides were discovered in the U.S.

The discovery that organochlorine compounds made effective pesticides followed. The great potential of DDT was not appreciated until its use enabled a severe outbreak of Typhus to be controlled. Organophosphorous and organochlorine compounds of many kinds have since been used as pesticides.

Research also led to the introduction of herbicides which affected plants somewhat like natural growth regulators (hormones). Early herbicides of this type included 2,4-D and MCPA.

The development and use of pesticides expanded rapidly after World War II. The increased use led to concerns about the safety and long term effects of chemicals which were heightened by publication of the book "Silent Spring." More recently, there has been widespread concern and press discussion about exposure of Vietnam era veterans to "Agent Orange," a defoliant mixture which included 2,4,5-T. These concerns resulted in new laws and regulations controlling pesticide use and the removal of some pesticides, notably DDT, from the market. Such regulations will be discussed more fully in a later chapter.

PESTICIDE USE AND IMPORTANCE

In spite of the concerns about safety, large quantities of pesticides are used. For example, in 1980, 846 million pounds of pesticide active ingredients were used on U.S. farms (Table 1) — about .81 pounds per acre (0.9 kg/ha) of farmland. These materials cost $3.6 billion, approximately $16 for every person in the U.S.

Most fruit and vegetable crops would be more expensive, less available, and of lower quality without pesticides. Yields of fiber, cereal, and forage crops would drop. One estimate of crop losses that might occur if insecticides were not used is shown in Table 2. Crop losses from pests can be high enough to preclude production of some crops if pesticides are not used.

TABLE 1

QUANTITY AND VALUE OF PESTICIDES USED ON U.S. FARMS

Year	Quantity used (active ingredients) (million lbs.)	(million kg)	Farm expenditures (million dollar)
1966	353	160	561
1971	494	225	1,002
1976	660	300	1,934
1980	846	385	3,600

TABLE 2

LOSS IN VALUE WITHOUT INSECTICIDES

CROP	PERCENTAGE
Field Crop	10
Soybeans	1
Wheat	2
Alfalfa	60
Sugar Beets	5
Sweet Corn	40
Tomatoes	15

CHEMICAL APPLICATION EQUIPMENT

Choosing the best method for applying a chemical to a crop is as important as choosing the correct amount of the proper material. The growth stage and condition of the crop, local soil and climactic conditions, and the chemical formulation all influence the method of application. The chemical must be applied properly for maximum effectiveness. Proper placement is very important (Fig. 11).

Details for selecting and using equipment will be covered in later chapters. (Figs. 12, 13).

NONCHEMICAL ALTERNATIVES

The benefits that have resulted from the use of agricultural chemicals are generally recognized. Despite the benefits, some people advocate elimination of

Fig. 11—Correct Fertilizer Placement Pays Off

Fig. 12—Fertilizer Application Equipment

some chemical uses and strict control of others. Some even urge elimination of all chemical pesticide and commercial fertilizer use. Safety, ecological, and energy concerns are given as reasons.

ECOLOGY AND SAFETY

There are some valid concerns. Pesticides are poisons! Most can adversely affect people, pets, livestock, wildlife, and desirable plants, as well as the pests they are intended to control. If not used correctly, they may damage nearby desirable plants or contaminate water. If applied too heavily, herbicide residue in the soil can damage the next crop.

Fertilizers may also have undesirable side effects. Runoff from agricultural land is a major source of surface and groundwater pollution. Nutrients, attached to sediment or dissolved in water, can be carried into lakes and streams. This may result in excessive growth of aquatic plants (Fig. 14). When these plants die, the biological breakdown consumes oxygen in the water, leading to "dead" lakes. The lack of oxygen can also cause fish kills. Fish kills, however, are often blamed on pesticides until the real cause is identified.

Dissolved nutrients can also be carried downward through the soil into groundwater. When the concentration of certain nutrient forms exceed critical levels, the water may become toxic to fish, animals, or

Fig. 13—Pesticide Application Equipment

Fig. 14—Fertilizer Runoff Can Pollute Water and Cause Rapid Growth of Aquatic Weeds

Fig. 15—Natural Predators Attack Cereal Leaf Beetle

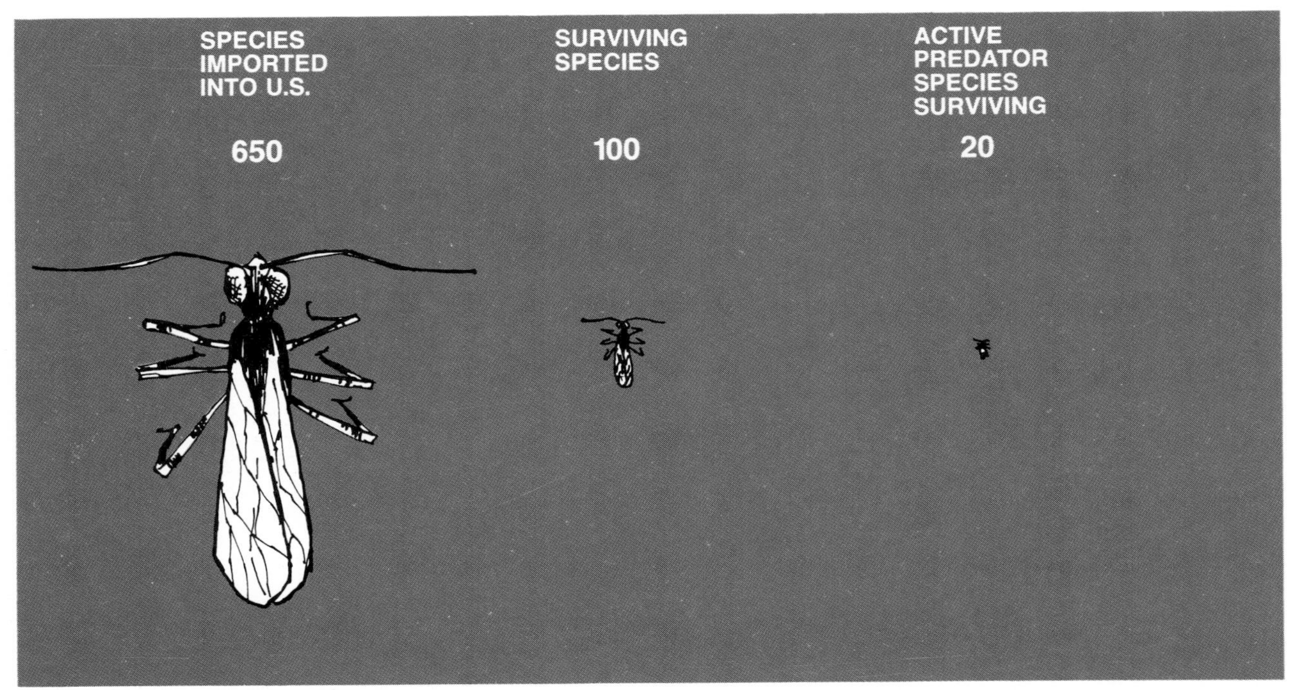

Fig. 16—Natural Enemies Are Hard To Find

even humans. The nitrate form of nitrogen is the most toxic. It interacts with other blood components and interferes with transportation of oxygen. However, most of the recorded occurances of nitrate poisoning have been associated with farm wells located too close to manure concentrations rather than runoff or leaching from fields.

The dilemma is that chemicals do so much good and yet may be harmful unless used wisely and carefully.

ENERGY

According to a USDA estimate, 652,532 billion Btus were used to manufacture fertilizers in 1978. This total is about 36 percent of total on-farm energy use. The energy is mainly used to manufacture nitrogen fertilizers. Nitrogen is abundantly available in the atmosphere, but energy is needed to convert it to the form needed by plants. Smaller amounts of energy are used to develop and process other nutrients and to manufacture pesticides.

CAN NONCHEMICAL METHODS DO THE JOB?

Nonchemical methods can sometimes be used instead of chemicals for pest control. But unfortunately, results with some nonchemical controls are often variable and not consistently effective. The research needed to develop a nonchemical technique is as difficult and complicated as developing a chemical pesticide. Also, caution must be used! We have discussed some of the possible side effects of using chemicals. Nonchemical methods can also give unexpected and undesirable results. For example, an insect introduced to control weeds might also attack some desirable plants.

PREDATORS AND PARASITES

Natural enemies, predators or parasites, can sometimes be used to control a pest (Fig. 15). Control of the citrus blackfly is an example. The citrus blackfly was discovered in Florida in 1934. It was quickly eradicated there, but not until it had spread to the citrus producing regions of Mexico. Four species of small wasps from India were found to be natural enemies of the pest. These wasps now keep the fly under control in Mexico, minimize the need for insecticides, and reduce the likelihood of the fly returning to the U.S.

It must be emphasized that the predator method has limited application. The search for such natural enemies is slow and difficult. Of an estimated 650 species of insects brought to this country to use as natural enemies against some of the 10,000 destructive insects, only about 100 have even survived. No more than 20 are helpful in controlling pests (Fig. 16).

Introducing predators also has dangers. The "friend" may turn out to be an enemy. Introducing the mongoose into Hawaii was such a case.

Rats had come to the island in ships. The abundance of food and lack of natural enemies caused a rodent population explosion. Mongooses, being natural enemies of rodents, were brought in to control the rats.

But the mongooses did not stop with rats. They also killed poultry, game birds, and pets. Several native duck species were almost exterminated. Eventually a new pest control program had to be started—to control the mongooses!

Some weeds also have natural enemies. St. Johnswort was a problem on western rangelands. Chemical control was impractical because the problem was so widespread. Control was achieved by using leaf-eating beetles that thrived on St. Johnswort, but did not seem to like other plants.

Use of vegetation such as desirable perennial grasses can provide considerable competition for weeds. These grasses help control weeds along roadsides, in fence rows, and on ditch banks.

Fig. 17—Scent of Insect Sex Attractant Confuses Males and Reduces Breeding

BAITS, ATTRACTANTS, AND REPELLENTS

Many insects are attracted to light. Moths and many other insects swarm to light. Light is an example of a bait or attractant — something to attract insects. If you can bait insects, it is relatively easy to trap or kill them.

Several other baits are being used. Some of the most effective are the sex attractants. The powerful sex attractant of the female gypsy moth has been isolated and can be used in traps for the males. By trapping the males, breeding frequency is reduced and the population declines. Releasing the scent of the female pink bollworm's sex attractant into the air confuses males and reduces mating and reproduction (Fig. 17).

Repellents work like attractants in reverse. They are materials which the insect dislikes. Common mosquito repellent is an example.

Another example is the use of aluminum foil spread on top of the soil to repel aphids which spread mosaic disease. Another interesting example was the method used to get rid of pigeons at Buckingham Palace. Strips of jelly were laid along the ledges in nesting areas. The birds disliked having their feet in the jelly and left. Excellent control was achieved.

Fig. 18—Insect-Resistant Soybean Plant (Right), Nonresistant Plant (Left)

Fig. 19—Mechanical Methods can Complement the Use of Herbicides (Rotary Hoe Shown)

STERILE-MALE RELEASE METHOD

The sterile-male technique is easy to understand. Large numbers of sexually sterile males of the species to be controlled are reared in a laboratory and released in the field. They compete successfully with normal males for mates, resulting in many infertile eggs being laid. The overall population thus declines. The process is repeated with each succeeding generation until, finally, only infertile eggs are laid and the species dies out. The method was used successfully to eradicate the screwworm (a cattle parasite) in parts of the United States.

Control by the sterile-male method works only if the conditions are advantageous. The insect population must be confined to a relatively small or isolated area, such as an island. A method to sterilize the males without other damage is needed. The problem must be important to justify the expense.

Even if these conditions are met, control may not be effective. Many insects mate several times. Thus, many females may be fertilized even if most of the males are sterile.

RESISTANT VARIETIES

Disease- and insect-resistant crop varieties can sometimes eliminate the need for a pesticide (Fig. 18). A good example is the resistance of some wheat varieties to the Hessian fly and such diseases as rusts, powdery mildew, and soil-borne mosaic. Until resistant varieties were available, these pests often destroyed the entire crop of Midwestern wheat growers. Other examples are corn varieties resistant to European corn borers, northern leaf blight, and diplodia stalk rot, and alfalfa varieties resistant to spotted alfalfa aphids, pea aphids, and bacterial wilt.

This approach takes a lot of time and effort. First, resistance must be developed or isolated. Then, the resistance must be bred into varieties that have other desirable characteristics. Plant breeding for disease and insect resistance is never ending. New races of disease-causing organisms and insect strains are constantly arising to attack the crop varieties that were previously resistant.

CULTURAL PRACTICES FOR WEED CONTROL

Despite the success of herbicides, most weed control nonchemical methods are still important for weed control (Fig. 19). Common methods include:

- **Fallowing**
- **Preplant tillage**
- **Rotary hoeing**
- **Row cultivation**
- **Mulching**
- **Cutting or mowing**
- **Use of weed-free seed**
- **Crop rotation**

Nonchemical methods are constantly being evaluated and improved to improve their effectiveness.

SUMMARY

Chemicals are "tools" for a modern farmer. They can be used at various stages of crop production.

The two most important classes of agricultural chemicals are:

- **Fertilizers**
- **Pesticides**

Fertilizers supply food for plant growth.

Pesticides control pests of various kinds . . . weeds, insects, diseases.

There is a wide variety of chemical application equipment. Accurate placement is very important.

Because of ecological and safety concerns, some people would like to see chemical use limited. They advocate the use of non-chemical methods of pest control. Non-chemical methods include:

- *Predators and parasites*
- *Baits, attractants and repellents*
- *Release of sterile-male*
- *Resistant varieties*
- *Cultural practices*

CHAPTER QUIZ

1. The two major classes of agricultural chemicals are __________ and __________.

2. List four "natural" foods for plants.

a. ____________________

b. ____________________

c. ____________________

d. ____________________

3. (True or false) The pH of a "neutral" soil is 6½.

4. __________ was in use at least 300 years ago as a "natural" insecticide.

5. Name five methods of non-chemical weed control.

6. Name four approaches that have been used to control pests without chemicals.

2
Fertilizers and Lime

Fig. 1—Adequate Nutrients Help Ensure Healthy Productive Crops

INTRODUCTION

The terms "plant nutrients", "plant foods", and "fertilizers" are more or less interchangeable. They identify the chemical elements and compounds present in or applied to soil or to plants to ensure good plant growth (Fig. 1). At present, 16 plant nutrients essential for normal vegetative and reproductive phases of plant growth have been identified. They are:

- **Boron**
- **Calcium**
- **Carbon**
- **Chlorine**
- **Copper**
- **Hydrogen**
- **Iron**
- **Manganese**
- **Magnesium**
- **Molybdenum**
- **Nitrogen**
- **Oxygen**
- **Phosphorus**
- **Potassium**
- **Sulfur**
- **Zinc**

SOURCES OF PLANT NUTRIENTS

Plants can obtain nutrients from three sources: air, water and soil (Fig. 2). Oxygen and hydrogen are taken from water. Carbon and oxygen come from air. Some plants (alfalfa and other legumes) are also able to remove nitrogen directly from the air. Other nutrients come from the soil. A crop uses large quantities of nutrients (Table 1).

PRIMARY NUTRIENTS

Three nutrients, nitrogen, phosphorus and potassium, usually are applied in large quantities to many soils to maintain and improve crop growth. These three nutrients are present in the plant in larger quantities than other nutrients and are called the major or primary nutrients. Other nutrients are known as secondary and micronutrients (Table 2).

Fig. 2—Plants Have Three Sources of Nutrients

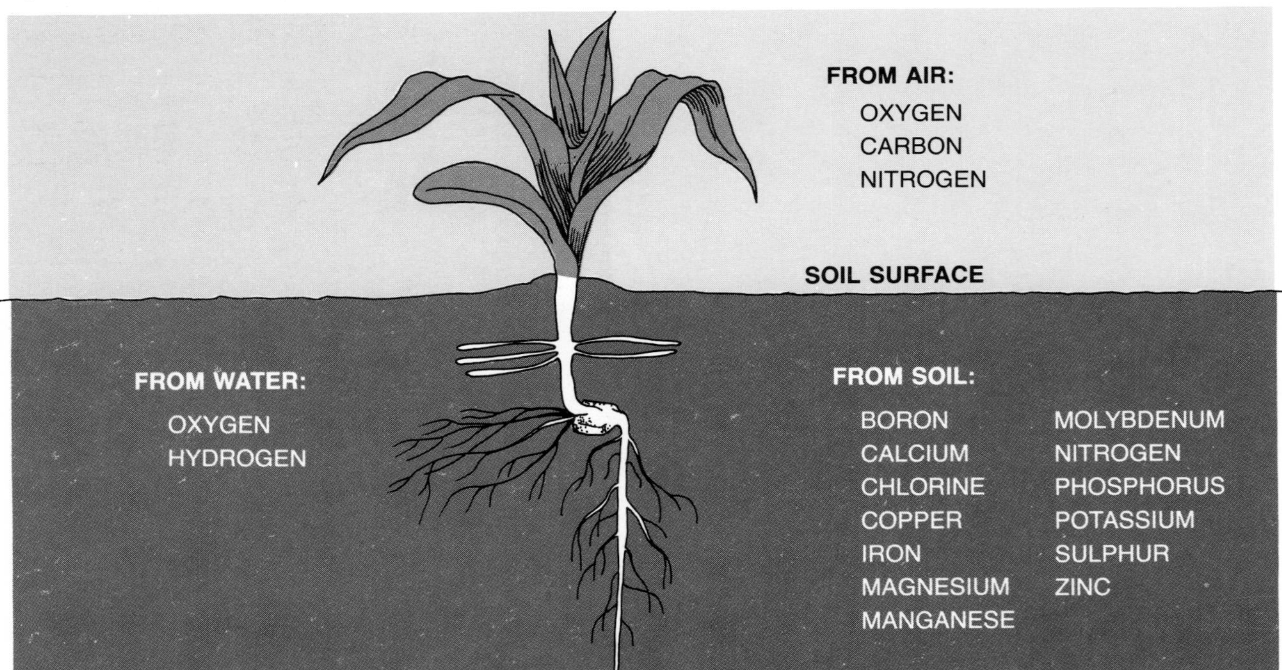

TABLE 1*

PLANT NUTRIENTS NEEDED TO PRODUCE 100 BU/ACRE CORN

Substance	Pounds per Acre	Approx. Equiv.
Water	6 to 8,000,000	19 to 24 in. of rain
Oxygen	10,100	
Carbon	7,800	4 tons of coal
Nitrogen	195	
Potassium	165	
Phosphorus	33	
Calcium	55	140 lb. of limestone
Magnesium	50	400 lb. of Epsom salts
Sulfur	33	
Iron	3	
Manganese	0.45	
Boron	0.09	3/8 lb. of borax
Chlorine	trace	
Zinc	trace	1 dry-cell-battery case
Copper	trace	57 ft. of No. 9 wire

*For equivalent table with metric units, see p. 201 Table 1.

TABLE 2

THREE CLASSES OF PLANT NUTRIENTS

Primary Nutrients	Secondary Nutrients	Micro-nutrients
Nitrogen	Calcium	Boron
Phosphorus	Magnesium	Chlorine
Potassium	Sulfur	Copper
		Iron
		Manganese
		Molybdenum
		Zinc

NITROGEN

Crops use large amounts of nitrogen (Table 3). In fact, more nitrogen is used as plant food than any other two elements combined.

Nitrogen is essential for cell division and plant growth. It is an element in proteins, amino-acids, sugars, and chlorophyll, and is involved in the manufacture, by the plant, of sugars and starches which do not contain nitrogen.

Nitrogen promotes rapid leaf and stem growth and gives plants succulence and a dark-green color. If nitrogen is limited, plants are small and usually have poor color; corn ears may have unfilled tips (Fig. 3).

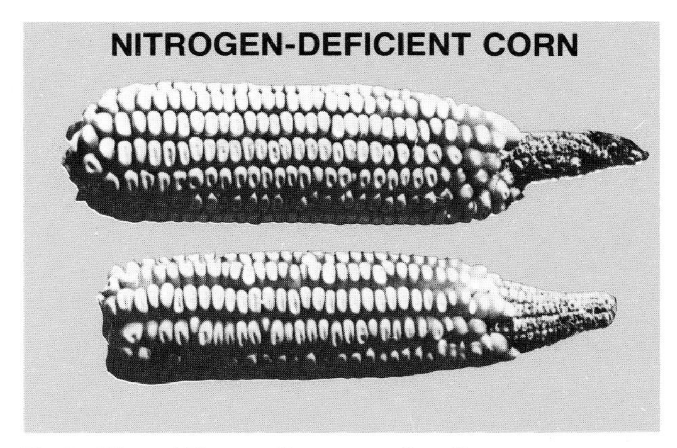

Fig. 3—Effect of Nitrogen Shortage on Corn Ears

Atmospheric Nitrogen

Approximately 80 percent of the atmosphere is nitrogen gas. There are 35,000 tons of nitrogen in the air above every acre (78,400 metric tons above each hectare). Nitrogen is constantly being removed from and returned to the atmosphere in a group of processes collectively known as the *Nitrogen Cycle* (Fig. 4).

Obviously, the supply of nitrogen in the atmosphere is adequate for any crop needs we can foresee. However, atmospheric nitrogen is a relatively inert gas that most plants cannot use directly. These plants must rely on nitrogen that has been "fixed" by a biological or industrial process. Fixed nitrogen is combined with other elements to form compounds which the plant can use. Legumes are plants that can use atmospheric nitrogen which is fixed by certain types of soil bacteria. Research is underway to transfer the nitrogen fixing

TABLE 3*

POUNDS OF NITROGEN IN CERTAIN CROPS

Crop	Yield	Nitrogen (lb.)
Alfalfa	6 tons	335
Apples	600 bu.	50
Barley	100 bu.	110
Corn	150 bu.	135
Cotton	3200 lb. of lint and seed	75
Oats	100 bu.	65
Oranges	800 boxes	85
Peaches	600 bu.	35
Potatoes	400 bu.	110
Soybeans	50 bu.	135
Sugar Beets	30 tons	125
Timothy	3 tons	85
Tobacco	2800 lb. of stems and leaves	95
Tomatoes	30 tons	170
Wheat	60 bu.	75

*For equivalent table with metric units, see page 201 Table 2.

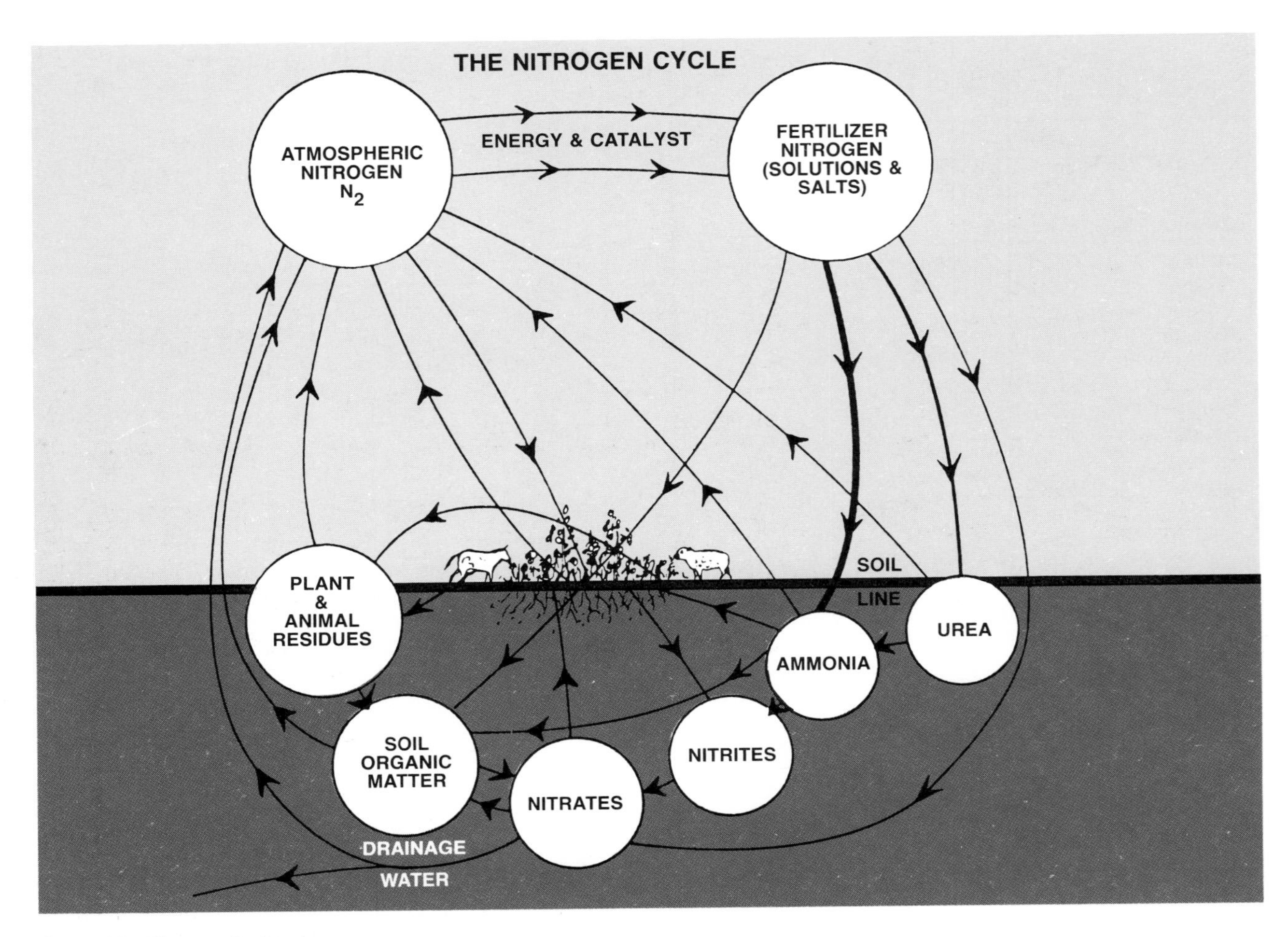

Fig. 4—The Nitrogen Cycle

capability to corn and other non-legumes. However, the amount of nitrogen that can be fixed biologically is currently not sufficient to provide for maximum growth of most crops. Consequently, most of the nitrogen used by crops is manufactured and applied as commercial fertilizer.

NOTE: An over abundance of nitrogen may cause a deficiency of potassium or other elements; stunted growth; loss of green color (chlorosis); lack of or delay in flower and fruit development; bud drop of certain plants; and possible winter injury.

Plants can use two forms of nitrogen (Fig. 5):

- *Ammonium–nitrogen combined with hydrogen*
- *Nitrates–nitrogen combined with oxygen*

Forms of Soil Nitrogen

The three principal forms of soil nitrogen are:

- **Organic nitrogen**
- **Ammonium nitrogen**
- **Nitrate nitrogen**

The largest portion of nitrogen in most soils is **organic nitrogen.** The top 7 inches (18 cm) of mineral soils contain 2,000 to 20,000 pounds of nitrogen per acre (2240 to 22 420 kilograms per hectare). Peat soils, which are up to 95 percent organic matter, may hold up to 50,000 pounds of nitrogen per acre (56 050 kg/ha). However, only 1 to 2 percent of this nitrogen is released for crop use during a normal growing season. This amounts to 30 to 40 pounds per acre (34 to 45 kg/ha) for typical Corn Belt soil. The remaining crop needs are met by application of nitrogen fertilizers.

Organic nitrogen exists in many forms. Much of it is undergoing a constant process of change from one form to another. The action of soil microorganisms can release ammonia. The ammonia is then held in the soil until it is taken up by plants or lost to the atmosphere.

Research has shown that plants respond identically to nitrogen whether the source is organic or inorganic. However, nitrogen from organic matter has several important advantages over other nitrogen fertilizers:

- *It is released slowly*
- *It rarely exists in toxic concentrations*

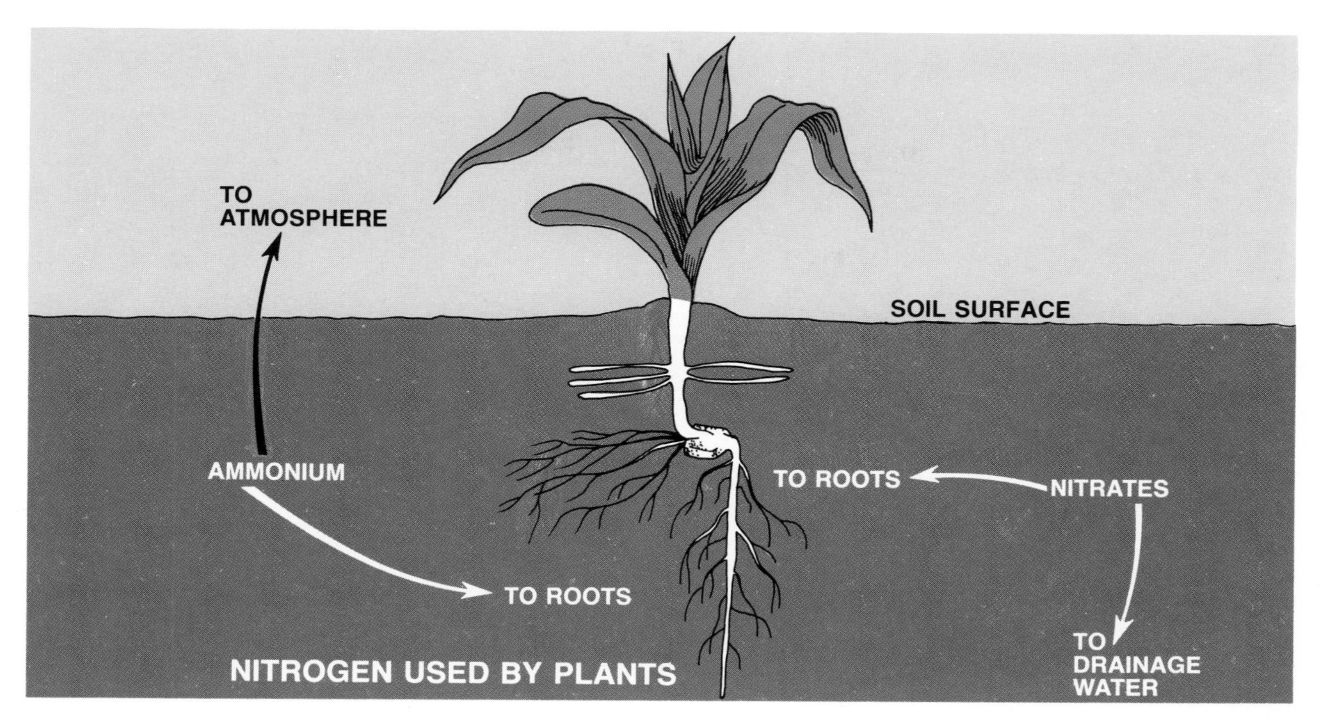

Fig. 5—Forms of Nitrogen Used by Plants

- *It is not lost by leaching*
- *It is released continuously*

Ammonium nitrogen may either be applied to the soil or be produced by soil bacteria. Soils will absorb and hold ammonium nitrogen for only a short time. The holding ability depends on soil type and ranges up to 2,000 pounds per acre (2240 kg/ha) for clay. Ammonia is held weakly on soil particles and can be "knocked loose" or released by soil microorganisms. Then the ammonium nitrogen is reabsorbed by soil, used by plants, changed to nitrates, or is lost to the air. The activity of soil bacteria speeds up as the temperature rises above 50°F (10°C). Unless there is a growing crop to use the displaced or released nitrogen, much is lost.

Nitrate nitrogen is supplied in fertilizer (Table 4) or produced in the soil by transformation from the ammonia or ammonium forms. The transformation from the ammonium form, called nitrification, occurs in two steps (Fig. 6):

- *Ammonium nitrogen changes to nitrites*
- *Nitrites change to nitrates*

Fortunately—because nitrites are toxic to plants—the second step in the process occurs quickly.

TABLE 4

COMPOUNDS USED FOR COMMERCIAL FERTILIZER

Nitrogen	Phosphorus	Potassium
Ammonium Chloride	Ammonium Phosphates	Kainite
Ammonium Nitrate	Basic Slag	Potassium Carbonate
Ammonium Phosphates	Bone Meal	Potassium Chloride
Ammonium Sulfate	Dicalcium Phosphate	Potassium Hydroxide
Anhydrous Ammonia	Nitric Phosphates	Potassium Nitrate
Aqua Ammonia	Phosphoric Acid	Potassium Phosphate
Nitric Phosphates	Rock Phosphate	Potassium Sulfate
Sodium Nitrate	Superphosphate	Tobacco By-Products
Urea		

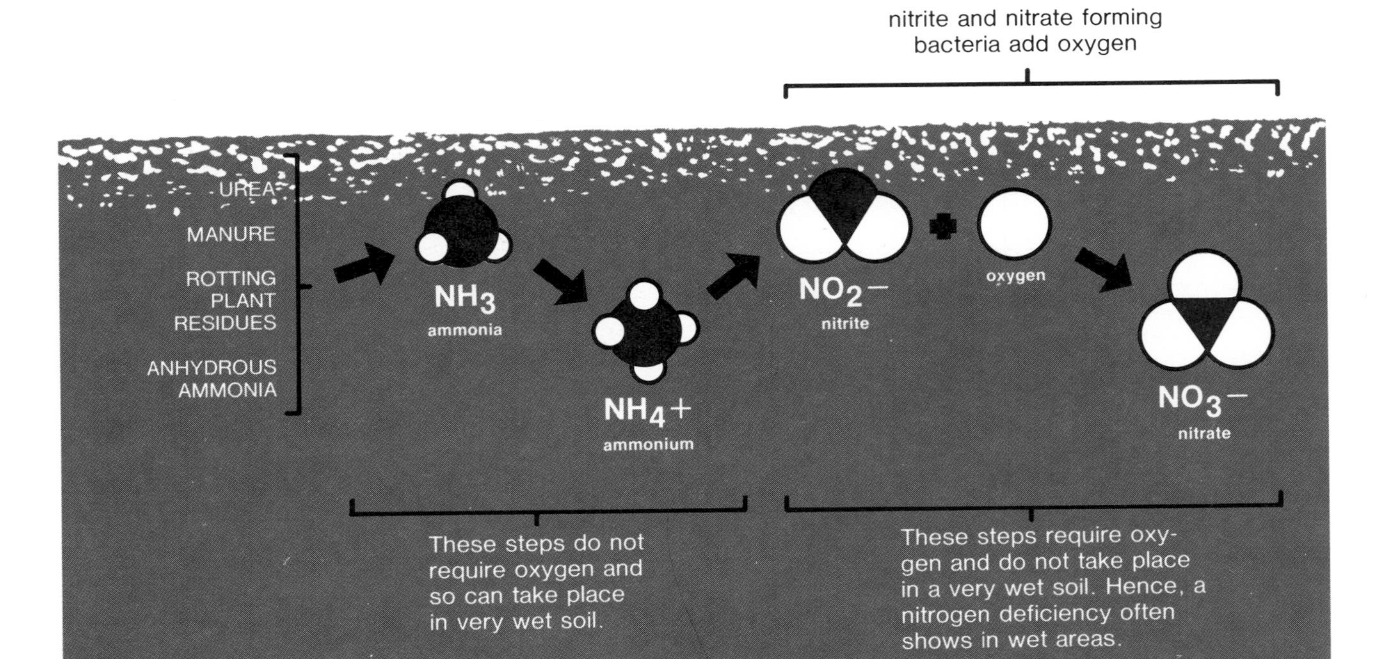

Fig. 6—Nitrification process

Nitrates are held in the soil solution. Thus, they are more readily available to the plants than nitrogen in the ammonium form. However, nitrates may also be lost easily in drainage water.

Nitrates can also be lost if the process of denitrification occurs (Fig. 7). The N_20 and N_2 gases, formed in the volatilization process, can escape through the soil surface.

Urea

Urea is a synthetic form of organic nitrogen. It is quickly broken down into the ammonium form which can be used by the plant. Urea may also be taken into the plant through the leaves, if used as foliar application (applied to leaves and stems of the plant).

Fig. 7—Denitrification process

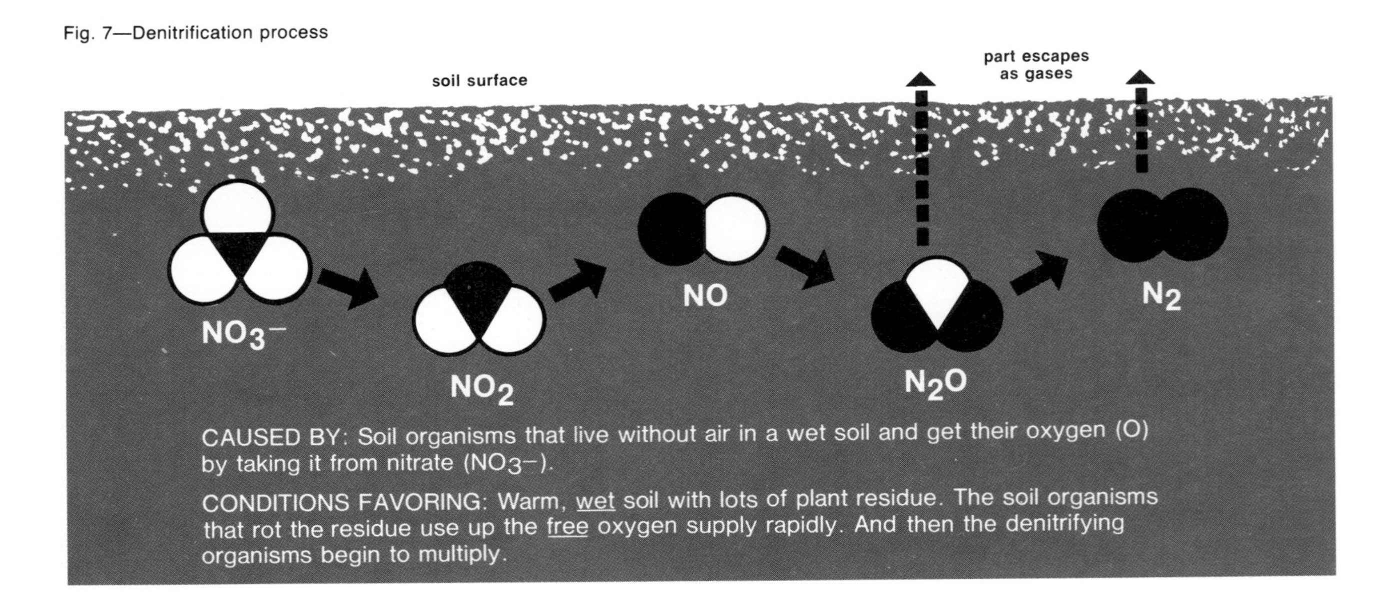

PHOSPHORUS

Phosphorus, as rock phosphate, was one of the earliest chemical fertilizers. Though the use of nitrogen is now greater, phosphorus is still an important fertilizer (Table 5).

TABLE 5*

POUNDS OF PHOSPHORUS IN CROPS

Crop	Yield	Phosphoric Acid (lb.)
Alfalfa	6 tons	70
Apples	600 bu.	15
Barley	100 bu.	40
Corn	150 bu.	50
Cotton	3200 lb. of lint and seed	30
Oats	100 bu.	25
Oranges	800 boxes	30
Peaches	600 bu.	10
Potatoes	400 bu.	35
Soybeans	50 bu.	40
Sugar Beets	30 tons	50
Timothy	3 tons	30
Tobacco	2800 lb. of stems and leaves	25
Tomatoes	30 tons	55
Wheat	60 bu.	35

*For equivalent table with metric units, see page 201 Table 3.

Phosphorus stimulates early root growth and blooming, and hastens crop maturity. It contributes to strong plant structures, improves seed quality, and increases resistance and winter hardiness in some plants.

Phosphorus-deficient plants are unable to use other nutrients properly. A common symptom of a phosphorus shortage is a buildup of nitrogen in leaves, which then take on a bluish or dark green color. When a plant develops a phosphorus deficiency, it usually is stunted and does not recover completely. Animals fed phosphorus-deficient plants may also have symptoms of phosphorus deficiency, such as low vitality, poor reproductive efficiency, and weak bones.

Typical phosphorus-deficiency symptoms in corn include small twisted ears with crooked and missing rows of kernels (Fig. 8).

Plants obtain phosphorus from the soil solution. The rate of phosphorus consumption depends on the stage of growth (Table 6). It has long been known that an application of phosphorus fertilizer at planting time helps get a crop off to a rapid start (Fig. 9).

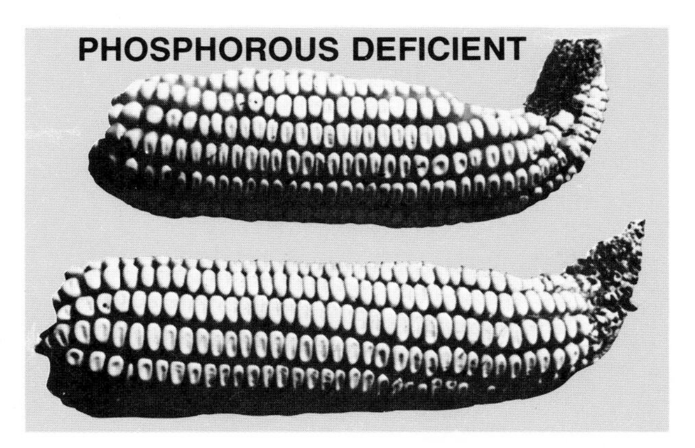

Fig. 8—Effect of Phosphorus Shortage on Corn Ears

Phosphorus Fertilizers

As a pure element, phosphorus is a gray metal which bursts into flame when exposed to air. So, obviously, the pure element cannot be used for fertilizing. It must be combined with other elements to make it safe to handle and useful to crops.

In the fertilizer trade, the phosphorus content of fertilizers is expressed in terms of phosphoric acid (P_2O_5).

Many materials containing phosphorus may be used as fertilizers. Details of particular materials are not included here, but a list of the most common phosphorus-bearing materials is shown in Table 4. The basic material is rock phosphate which is mined at locations around the world. Various materials are then combined to make fertilizers (Table 4).

Water Solubility

Water solubility of phosphorus determines its availability to plants. Solubility of phosphorus-bearing materials can vary widely (Table 7) and this factor has a marked effect on plant response (Fig. 10). Too much phosphorus in the soil (over-fertilization) can cause iron and zinc deficiencies in many crop plants.

TABLE 6

PHOSPHORUS USE BY PLANTS

Crop Growth Stage	Percent of Total Phosphorus Uptake
First month	1 - 15
Second month	30 - 60
Third month	10 - 30
Fourth month	5 - 10

Fig. 9—Phosphorus Applied at Planting Time Stimulates Early Growth

Loss of Water Solubility

Phosphates tend to change into insoluble forms and become unavailable to plants. This process is called **phosphate reversion.** The first crop seldom is able to take up more than 30 percent of a phosphorus application. Each year, the percentage of recovery decreases. Consequently, regular applications of phosphate fertilizers are needed to help ensure adequate phosphorus for each crop.

TABLE 7

WATER-SOLUBILITY OF PHOSPHORUS MATERIALS

Material Analysis	Percent Water Solubility
Ammonium Phosphate(11-48-0)	89
Ammonium Phosphate(16-20-0)	86
20% Superphosphate(0-20-0)	78
Triple Superphosphate(0-47-0)	84
Dicalcium Phosphate(0-40-0)	5
Calcium Phosphate(0-62-0)	4

POTASSIUM

Potassium is one of the essential elements required by all plants and animals. It is absorbed by plants in greater quantities than any other nutrient except nitrogen (Table 8). It is found in cell fluids throughout the plant, but is not built into the plant structure. Symptoms of potassium deficiency include curling, yellowing,

Fig. 10—Phosphate Solubility Affects Corn Yield

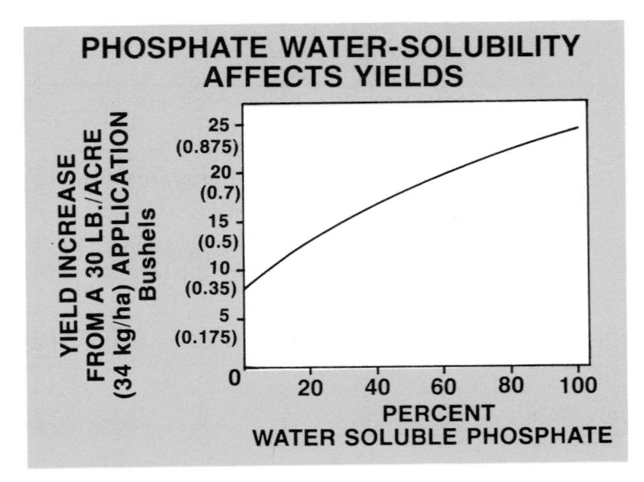

TABLE 8 *

POTASSIUM CONTENT OF CROPS

Crop	Yield	Potassium
Alfalfa	6 tons	270
Apples	600 bu.	70
Barley	100 bu.	35
Corn	150 bu.	35
Cotton	3200 lb. of lint and seed	40
Oats	100 bu.	20
Oranges	800 boxes	140
Peaches	600 bu.	65
Potatoes	400 bu.	200
Soybeans	50 bu.	70
Sugar Beets	30 tons	200
Timothy	3 tons	90
Tobacco	2800 lb. of stems and leaves	190
Tomatoes	30 tons	360
Wheat	60 bu.	15

*For equivalent table with metric units, see page 201, Table 4.

"scorching" or bronzing of leaf margins and tips; stems are weak; roots and tubers (such as potatoes) are undeveloped; and plants are often stunted.

Potassium is sometimes described as a "chemical policeman" which keeps things moving. The functions of potassium include:

- *Increases root and stem growth*
- *Improves drought resistance*
- *Improves product quality*
- *Helps retard diseases*
- *Builds cellulose, adds stem or stalk strength, and helps prevent lodging*
- *Aids formation and movement of starches and sugars*
- *Enhances protein production*
- *Aids in photosynthesis*
- *Reduces water loss and wilting*

Potassium Fertilizers

Pure potassium, like phosphorus, is a light gray metal that burns when exposed to air. Like nitrogen and phosphorus, potassium must be combined with other elements before plants can use it.

In the fertilizer trade, the potassium content of fertilizer is expressed in terms of potash (K_2O).

Many potassium-bearing materials may be used as fertilizers. A list of common potassium-fertilizer materials is included in Table 4.

Potassium Availability

Most U.S. soils contain large quantities of potash — up to 40,000 pounds per acre (44 800 kg/ha) in the plow layer. However, much of it is not in an available form for plant use. Most crops place a fairly heavy drain on the supply of potash in the soil. When crops are produced successively on the same land, the soil potash is gradually depleted and crop yields decline unless additional amounts are made available. Potassium shortage can cause small chaffy ears of corn with dull colored, loose kernels and unfilled tips (Fig. 11).

SECONDARY NUTRIENTS

Secondary nutrients are essential to plant growth, but are usually required in smaller amounts than the three primary nutrients (Table 1). The amount of each secondary nutrient needed depends on:

- **Plant species**
- **Soil conditions**
- **Climate**
- **Availability**

CALCIUM

Calcium stimulates root, stem, and leaf growth, and improves general vigor and disease resistance. It increases the uptake of other essential nutrients and is needed to assure crop seed production and maturity.

Calcium improves soil structure by causing aggregation—forming granules of soil. Granular soil pro-

Fig. 11—Potassium Shortage Causes Small, Chaffy Corn Ears

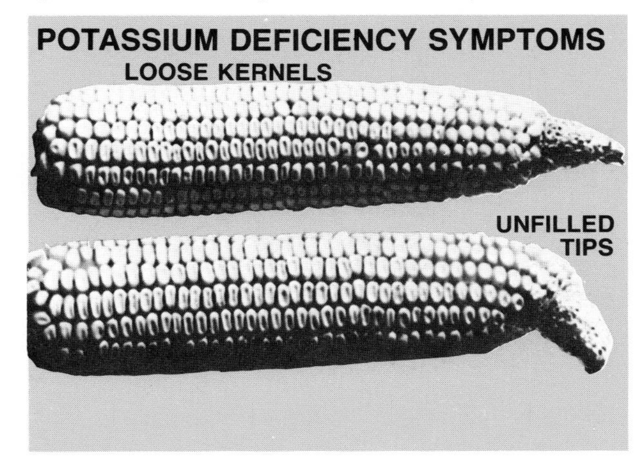

vides a good root environment by allowing easy penetration of air and water.

Calcium is also used to correct soil acidity. This need becomes more evident as soils become older. Crops have absorbed a greal deal of calcium from the soil and large amounts have been removed by leaching due to rainfall. The tremendous increase in the use of acid-forming nitrogen fertilizers has also intensified the need for calcium.

NOTE: Excessive calcium in the soil may cause deficiencies of boron, iron, magnesium, manganese, potassium, and zinc.

Ground and burned limestone and gypsum are the most widely used calcium sources.

MAGNESIUM

Magnesium is essential for production of chlorophyll. It aids in formation of many plant compounds such as sugars, starches, oils, and fats. It also has a role in movement of phosphorus and other plant foods to various parts of a plant. The most common source of magnesium is dolomitic limestone, which contains both magnesium and calcium, although other magnesium compounds are also used.

SULFUR

Sulfur is found in all living plant cells as part of the protoplasm. Plant seeds have a higher concentration of sulfur than other plant parts. Sulfur aids in formation of proteins, encourages vigorous growth, and helps plants withstand cold temperatures.

Sulfur is released into the atmosphere when sulfur-bearing fuels (coal and oil) are burned. Sulfur is then added to the soil by snow and rainfall which washes it out of the air. Enough sulfur is deposited this way to satisfy the needs of most crops on land near sulfur-releasing industries (Fig. 10).

However, supplying the sulfur for fertilization is only one effect. Less desirable effects include air pollution and acid rain which can be produced hundreds of miles away from the power plant or factory which burns the fuel. Consequently, much effort is being expended to eliminate this source of sulfur for plants.

Soil in rural areas, away from industrial concen-

Fig. 12—Sulfur is Carried by Wind and Returned to Soil by Rainfall

TABLE 9

MICRONUTRIENT DEFICIENCIES AFFECT MANY CROPS

Micronutrient	Soil Conditions	Crops
Boron	Sands, overlimed acid soil, organic soil	Cotton, tomatoes citrus, sweet potatoes, leafy vegetables, tree fruits, legumes with heavy lime.
Copper	Sands, organic soils, high pH	Small grains, vegetables, tree fruits
Iron	High pH soils, high phosphate	Blueberries, corn, sorghum, soybeans, ornamentals
Manganese	Sands, overlimed soils, organic soils	Soybeans, small grains, tree fruits, cotton, sweet potatoes, leafy vegetables
Zinc	Sands, high pH soils, high phosphate	Soybeans, corn, citrus, rice, sorghum, pecans, tree fruits, some vegetables
Molybdenum	Highly weathered acid soils	Legumes, citrus, cauliflower, cabbage, and similar plants

trations, may require additional sulfur for good crop yields. It may be added by using fertilizer materials which contain sulfur such as ammonium sulfate and superphosphate. Irrigation water and decomposition of plant matter also supply some sulfur.

MICRONUTRIENTS

Interest in and use of micronutrients, which are applied in very small amounts, has expanded greatly in the last 25 years (Table 1). Micronutrient deficiencies have been found in all parts of the country and affect many different crops grown under various soil conditions (Table 9).

Most micronutrient-deficient plants exhibit two symptoms:

- **Stunted growth**
- **Discoloration of foliage**

Diagnosis of particular deficiencies requires experience as well as knowledge of the symptoms. Tissue and soil tests can confirm visual evaluation. These tests can also be used to detect deficiencies before the symptoms become visible. Corrective action can then be taken before serious growth impairment results. In tests at Louisiana State University, addition of the equivalent of 1.7 pounds of actual zinc per acre (1.9 kg/ha) increased rice yield as much as 2,736 pounds per acre (3065 kg/ha) (Fig. 13).

SOIL pH

Soil pH exerts a tremendous influence on the availability of plant nutrients. For example, nitrogen is most available when the soil pH is 6.0 to 8.0. It is much less available at pH 5.0 (Fig. 14). Thus, plants grown in acid soil may display symptoms of nitrogen and calcium deficiency even if adequate nitrogen is in the soil, because it is not available to the plants.

As mentioned previously (Chapter 1), crops also have pH preferences. Most crops prefer a neutral soil, but some respond better to more acid or alkaline conditions (Table 10).

INDICATIONS OF ACID SOIL

Observing plant growth can provide indications of soil acidity (Fig. 15). Acid-sensitive plants will experience partial or complete failure. At the same time, there may be an encroachment of acid-tolerant plants such as:

- **Wild daisies**
- **Red sorrel**
- **Redtop**
- **Broomsedge**
- **Bentgrass**
- **Plantain**

Fig. 13—Micronutrient Deficiencies May Cause Complete Crop Failures Such as in this Rice Field

Visual observations can indicate a need for lime. But, don't rely on appearance alone! Take a soil sample (Fig. 16) and have a pH test made. Local county extension offices can provide instructions and location of the nearest lab. There may be a nominal fee for the service.

CORRECTING ACID SOIL

Soil acidity can be corrected by adding lime. There are five commonly used commercial forms of lime:

Ground limestone is pulverized limestone rock. Dolomitic limestone can supply the nutrient magnesium in addition to correcting acidity.

TABLE 10

RANGE OF pH FOR SOME CROPS

Soil pH	Crops
6.5-7.5	Alfalfa, Sweet Clover
6.0-7.0	Red Clover, White Clover
5.5-7.0	Alsike Clover, Barley, Birds-Foot Trefoil, Corn, Cowpeas, Crimson Clover, Lespedeza, Millet, Oats, Rye, Sorghum, Soybeans, Sundangrass, Wheat, Grasses
5.5-6.5	Buckwheat, Cotton, Vetch
5.3-5.7	Tobacco
5.0-5.4	Potatoes

Fig. 14—Soil pH Affects Nutrient Availability

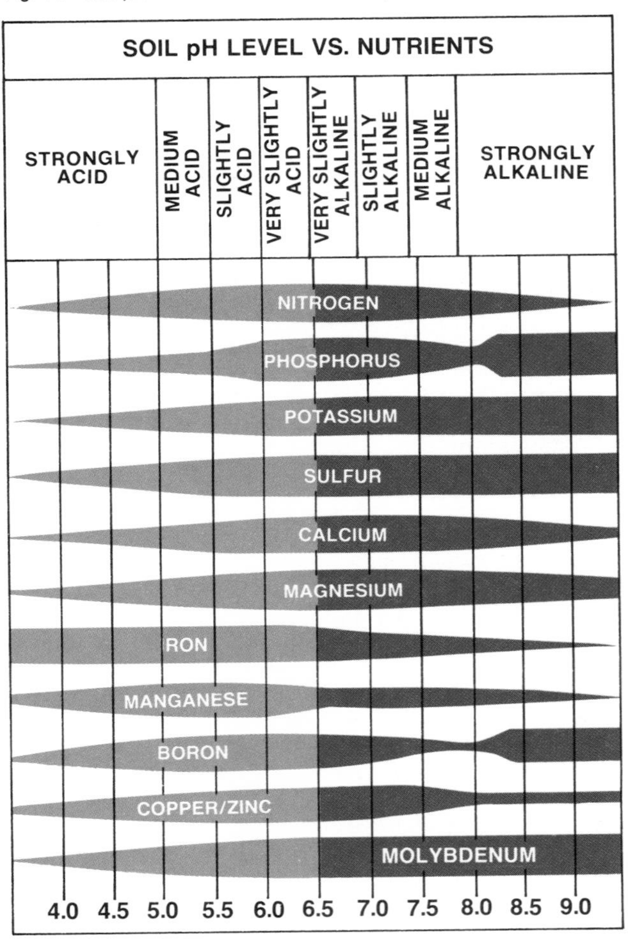

Fig. 15—Iron Chlorosis of Soybeans Resulting from Low pH

Burnt lime is produced by burning ground limestone to drive off carbon dioxide. It is the most concentrated form of lime.

Hydrated lime is made by adding water to burnt lime.

Marl is a naturally-occurring material formed by the decomposition of marine animals or by removal of calcium carbonate from water.

Oyster shells are sometimes ground and marketed for agricultural use.

Calcium Carbonate Equivalent

The acid-neutralizing ability of the liming materials is expressed as calcium carbonate equivalent (Table 11).

TABLE 11*

LIME NEEDED FOR EQUAL ACID-NEUTRALIZATION

Material	Pounds needed to equal one pound of ground limestone
Ground limestone	1.0
Burnt lime	0.6
Hydrated lime	0.67
Marl	1.0-1.5
Oyster shells	1.0-1.5

*For equivalent table with metric units, see page 202, Table 5.

Soil Conditions

Soil type and texture and amount of organic matter affect the amount of lime that should be applied. Generally, fine-textured soil requires more lime to achieve an equivalent pH modification than coarse soil; the application rates (Table 12) would have to be increased up to 100 percent for the same effect in soils with high-organic-matter content.

TABLE 12**

LIME TO RAISE pH ONE UNIT

Soil type	*Ground limestone	*Burnt lime	*Hydrated lime
Sandy	1500	840	1110
Sandy loams	2000	1120	1480
Loams	3000	1680	2220
Silt loams and clay loams	3500	1960	2590

*Application rate in pounds per acre
**For equivalent table with metric units, see page 202, Table 6.

Fineness

Fineness is expressed in mesh sizes. The higher the mesh number, the finer (smaller) the particles that pass through sieves used to measure fineness. Effectiveness of liming materials is higher for fine particles (Table 13). It takes about 5 tons (4.5 tonnes) of coarse lime with a 20 percent effectiveness rating to produce the same effect as one ton of fine material at 100 percent effectiveness.

Time and Method of Application

Subsoils are usually less acid than topsoils, so the best place for lime is in the upper few inches (centimeters) of soil. To be most effective, it should be mixed into the plow layer.

Fig. 16—Take Enough Soil Samples to Represent each Entire Field

Lime may be applied at any time, but the most convenient time is usually before plowing.

Lime may be applied by spreaders of several kinds. Either truck-mounted or tractor-powered equipment may be used (Fig. 17). Lime is often applied by custom operators with high-capacity equipment.

In some areas, acid subsoils have recently been identified. Yield limitations caused by the acid subsoil were not identified until the combination of adequate fertilization, pest control, irrigation and other modern production practices failed to produce the anticipated yield increases. The best method for correcting acid subsoil has not yet been identified. Researchers are testing several techniques, including: chiseling — in, deep tillage, and leaching with heavy applications of irrigation water, to get lime to the subsoil.

TABLE 13

SMALLER LIME PARTICLES ARE MORE EFFECTIVE

Particle Size	Effectiveness
Material passing a 60-mesh sieve	100
Material passing a 20-mesh sieve but not a 60-mesh sieve	60
Material passing an 8-mesh sieve but not a 20-mesh sieve	20

Benefits from Lime

Besides correcting pH, lime also produces changes that make soil more suitable for growing plants. In summary, the benefits of lime use are:

- *pH correction*
- *Improved physical condition of soil*
- *Calcium added to soil*
- *Accelerated decomposition of organic matter for nitrogen release*
- *Increased fertilizer efficiency*
- *Increased nutrient availability*

Fig. 17—Pull-type Lime Applicator

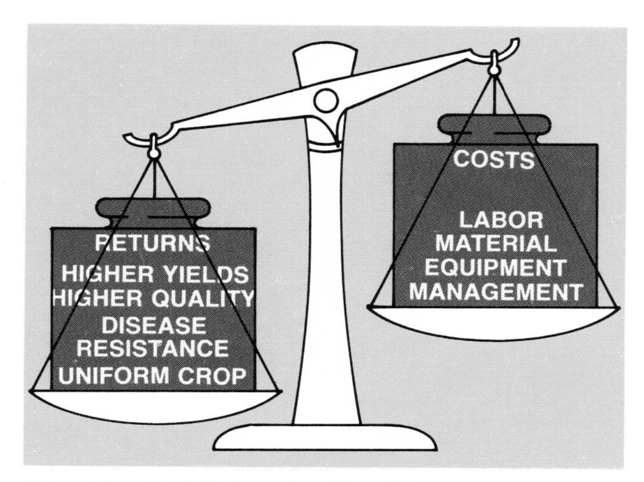

Fig. 18—Costs and Benefits of Fertilizer Use

FERTILIZER MANAGEMENT

Farmers spend money on fertilizer because they expect to make a profit on the investment. To be worthwhile, benefits have to outweigh costs (Fig. 18).

BENEFITS

Higher yield is the most frequent benefit of fertilizer use. But, benefits such as earlier maturity and improved quality are also important. This is particularly true for high-value, short-season crops such as strawberries. Growers who reach the market early almost always enjoy higher than average prices for their produce. High quality, uniform products also command higher prices.

Another important benefit of fertilizer use is reduction of the risk of crop failure. A well nourished crop is better able to resist disease, insect, and weed infestations. A high level of nutrients also helps ensure that most crops will be less sensitive to stresses such as winter injury or moisture shortages.

Costs

Costs of fertilizer use are easy to identify. They are:

- *Cost to purchase fertilizer*
- *Cost of owning and operating or hiring equipment for handling and applying fertilizer*
- *Cost of labor for operating equipment*
- *Cost of needed management*

Management

In the past, benefits from fertilizer use almost always exceeded the cost. Reasonable fertilizer use readily produced a profit.

This is not always true today. The costs of fertilizer use have in many cases risen faster than the value of the crops produced. Good management is needed to keep costs as low as possible and to help ensure that expected benefits actually occur. Some points the manager should consider are outlined below. They will be discussed more fully elsewhere.

Formulation and analysis of the fertilizer affect the purchase price and application costs. Also, the kind of fertilizer should be matched to the specific crop needs and expected benefits.

Time of application influences reaction of the crop. Also, fertilizer application should be planned to avoid interference with other farm operations.

Fertilizer placement is critical if optimum results are to be achieved.

Equipment must be matched to the kind of fertilizer being applied and the placement requirements.

FERTILIZER FORMS

Fertilizers are mixtures of materials which contain plant nutrients. Complete fertilizers contain all three primary nutrients. Fertilizers are available in several forms.

PRE-MIXED DRY FERTILIZERS

Many varieties of pre-mixed dry fertilizers are on the market. Most are bagged for sale. The mixtures, or analyses, are described by numbers such as 4-16-8 (Fig. 19):

- *The numbers indicate the fertilizer analysis (three major plant food elements)*

Fig. 19—Fertilizer Mixtures Described by Numbers

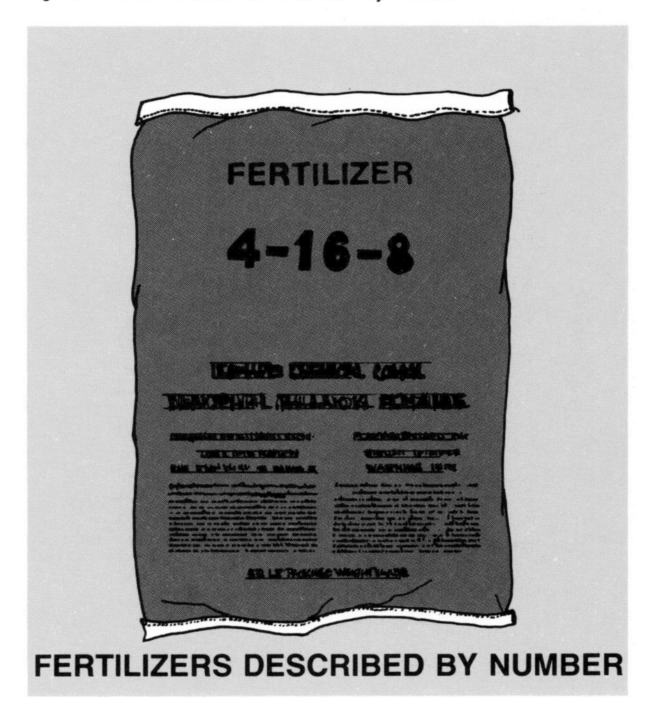

- *The analysis is always listed in whole numbers on the bag*
- *The numbers, in the example, indicate that the mixture contains at least 4 percent nitrogen (N), 16 percent phosphoric acid (P_2O_5) and 8 percent potash (K_2O). Thus, one ton of 4-16-8 contains at least:*

80 pounds of nitrogen (0.04 X 2,000 lbs.)
(40 kilograms or 0.04 X 1,000 kg)

320 pounds of phosphoric acid (0.16 X 2,000 lbs.)
(160 kg or 0.16 X 1,000 kg)

160 pounds of potash (0.08 X 2,000 lbs.)
(80 kg or 0.08 X 1,000)

The remaining 1440 pounds (720 kg) may be sand, ground limestone or other carrier materials (Fig. 20).

There is a trend, especially among soil and plant scientists, to express fertilizer analysis on an elemental basis. This has already happened with nitrogen. Formerly, the analysis indicated the percentage of ammonia (NH_3), but it now indicates nitrogen percentage. This change, when it occurs, will eliminate the confusion about terminology which sometimes happens.

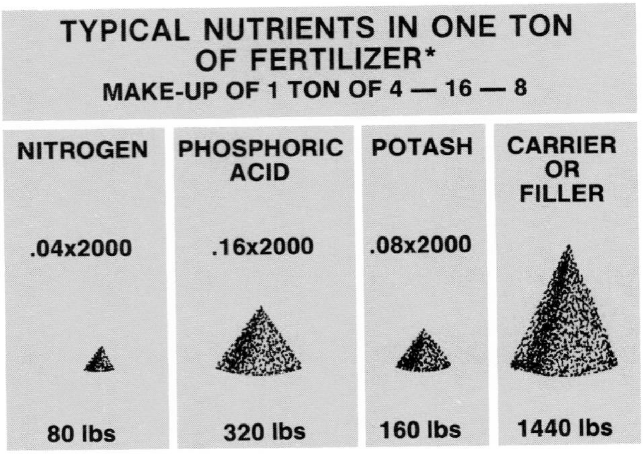

TYPICAL NUTRIENTS IN ONE TON OF FERTILIZER*

MAKE-UP OF 1 TON OF 4 — 16 — 8

NITROGEN	PHOSPHORIC ACID	POTASH	CARRIER OR FILLER
.04x2000	.16x2000	.08x2000	
80 lbs	320 lbs	160 lbs	1440 lbs

*For equivalent table in metric units, see page 203, Table 10.

Fig. 20—Typical Fertilizer Analysis

BULK-BLENDED DRY FERTILIZER

Bulk-blended fertilizers are mixed to a customer's specifications just before delivery. A bulk-blending plant essentially consists of bins of fertilizer ingredients and a mixer for combining them. Bulk-blending plants have been established in all parts of the country.

A wide variety of ingredients can be used for bulk blending (Table 14). Most are compatible, but some, such as urea and ammonium nitrate, should not be mixed.

TABLE 14

COMMON BULK-FERTILIZER INGREDIENTS

Material	Nitrogen (Percent)	Phosphoric Acid (P_2O_5) (Percent)	Potash (K_2O) (Percent)
Ammonium Nitrate	33.5-34	--	--
Ammonium Nitrate Sulfate	26-30	--	--
Ammonium Sulfate	20.5-21	--	--
Urea	45-46	--	--
Sodium Nitrate	16	--	--
Monoammonium Phosphate	11	48	--
Diammonium Phosphate	16-21	48-53	--
Ammonium Phosphate Sulfate	16	20	--
Muriate of Potash	--	--	60-62
Sulfate of Potash	--	--	--
Nitrate of Potash	13	--	44
Single Superphosphate	--	45-46	--
Nitric Phosphate	10-22	10-22	0-16
Nitrate Phosphate	20-27	12-14	--

Fig. 21—Liquid Fertilizers can be Handled Mechanically

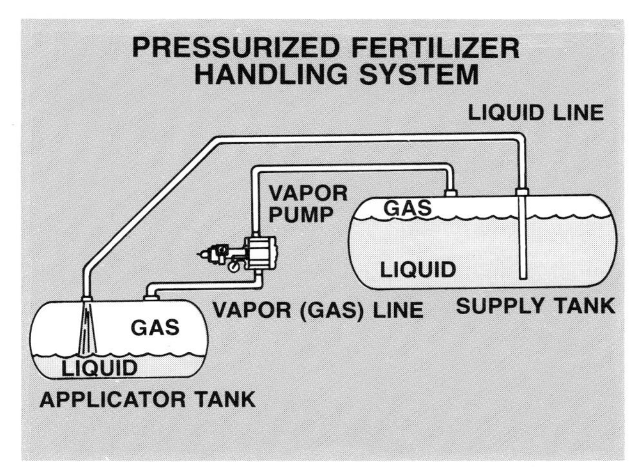

Fig. 22—Volatile Fertilizers Are a Combination of Vapor and Liquid Under Pressure

LIQUID FERTILIZERS

Liquid fertilizers are time and labor savers. Handling is done mechanically or by gravity flow (Fig. 21). When properly applied, they are as effective as solid nutrients for promoting plant growth. Volatile materials, such as anhydrous ammonia, must be handled under pressure to prevent loss (Fig. 22).

There are two common types of liquid fertilizers:

- **Liquid-mixed**
- **Slurry**

A **liquid-mixed** fertilizer is a solution of the correct proportions of solid-plant-food carriers, such as ammonium phosphates or urea in water.

A **slurry** fertilizer is a suspension of the chemical materials in a liquid carrier. Slurries may be prepared with higher analysis than the liquid mixed type.

However, a slurry must be agitated continually (Fig. 23) to prevent the solids from settling out. The concentration of solid material in a slurry also causes added wear on pumps, meters, and other equipment.

Marketing channels for liquid fertilizer are similar to those for bulk-blended dry fertilizer — sales are direct from the blender to the consumer. Application service is frequently provided by the blending plant or by custom operators. For those customers who have their own application equipment, liquid fertilizer may be transported to the farm in a nurse tank. It is then transferred to the field equipment for application.

Liquid-Nitrogen Fertilizer

Liquid nitrogen is available in several forms. Anhydrous ammonia, which has to be transported and handled under pressure, is the most popular. Aqua-ammonia and solutions of other nitrogen compounds may be handled under low pressure or unpressurized.

Fig. 23—Agitation of Slurries Prevents Solids from Settling Out

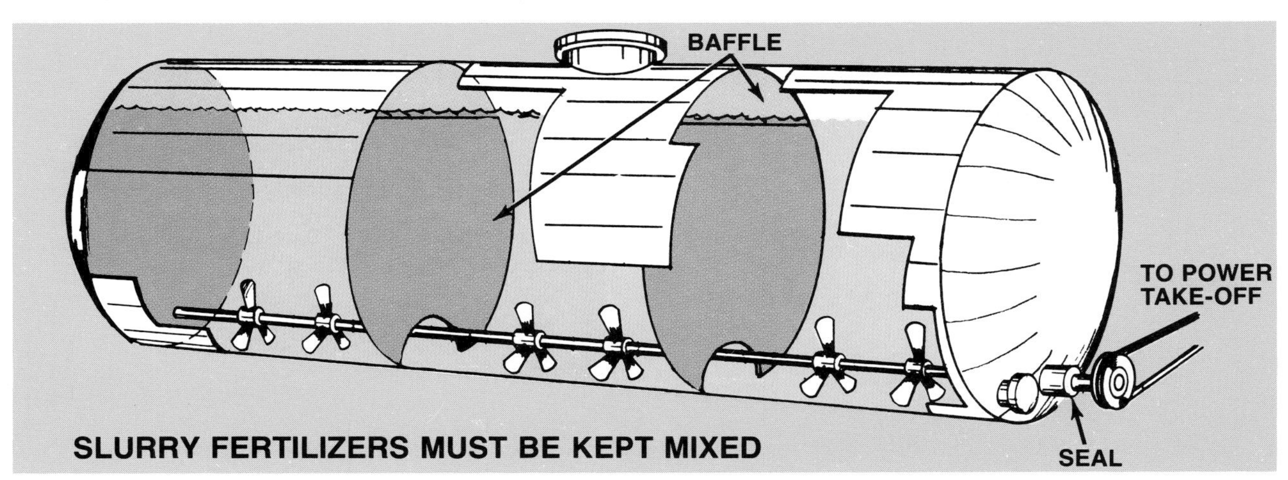

FERTILIZER APPLICATION RATES

Application rates are normally selected which will:

- *Replace nutrients used by a previous crop*
- *Replace nutrients lost by leaching, runoff, or volatilization*
- *Add nutrients which a crop is expected to remove*
- *Provide additional nutrients for future crops*

Soil Tests

Soil tests determine the existing fertility level of the soil. Most tests are performed in commercial laboratories or laboratories supervised or operated by State Extension Services. Many fertilizer companies and crop consultants provide soil sampling and testing services. A fertilizer application recommendation is usually provided along with the results of each soil test. These recommendations are based on the cropping history of the field tested, the soil test results, and the nutrient needs of the next crop. The accuracy of these recommendations is improved by providing accurate cropping information and by collecting a representative soil sample. The soil sample should include plow layer soil from representative parts of the field. The soil should be mixed thoroughly before a sample is taken for analysis (Fig. 24).

Tissue Tests

Tissue tests provide a chemical analysis of the plant. They measure the nutrient status of the plant and are an excellent indication of the soil's ability to furnish plant nutrients. The results are often used to correct for nitrogen and potash deficiencies, if they are found early in the growing period.

Tissue tests can be made on the farm with a kit or they may be made in testing laboratories.

FERTILIZER PLACEMENT

Nutrients must be dissolved by soil mositure and in general must be close enough to roots to be readily

Fig. 24—Taking a Soil Sample

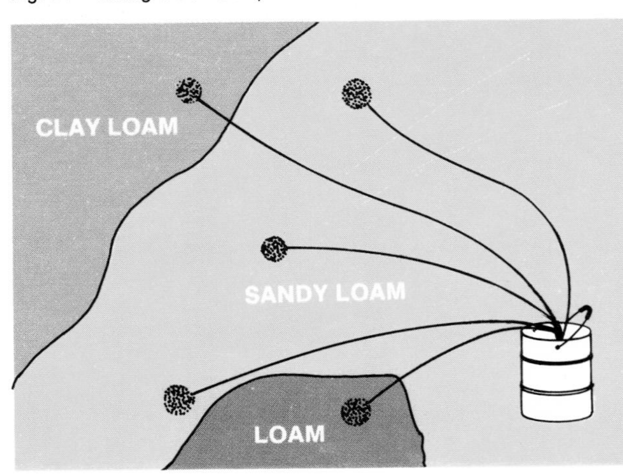

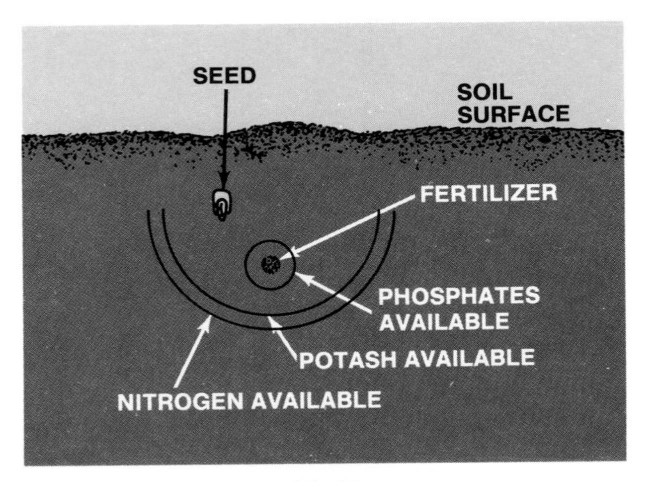

Fig. 25—Relative Movement of Fertilizer

taken up by plants. Nitrogen placement is not critical because it dissolves easily and moves readily in the soil. Phosphorus does not move readily in soil and should be placed close to the seed or to plant roots. Potash moves more than phosphorus but less than nitrogen (Fig. 25). Placing heavy applications of these materials too close to roots can cause "burning," particularly if a dry period follows fertilizer application. Obviously, a compromise is needed to determine the best placement. Banding the fertilizer below and to one side of row crops is a frequent solution. Specific placement recommendations for common crops can be obtained from local county extension agents or other fertilizer specialists.

Banding

Fertilizer placement in a band near the roots of row crops is desirable for three reasons:

- *Reduced contact between soil and fertilizer means less loss of phosphorus and potash availability because of fixation to soil particles*
- *Nutrients are within easy reach of plant roots*
- *Fewer nutrients are available to weeds growing between rows*

Pop-Up Fertilizer

Pop-up fertilizers are used in small amounts at planting time. They are placed in direct contact with the seed and are an immediate nutrient source for the seedling. This method is used to supplement broadcast applications made prior to planting to ensure fast starting of the crop. Only suitable fertilizer materials should be used for pop-up applications.

Starter Solutions

A starter solution is made by dissolving fertilizer in water. Such solutions are often used to hasten recovery of transplanted seedlings or for rapid establishment of a crop stand. Normal benefits are more rapid early growth and earlier maturity.

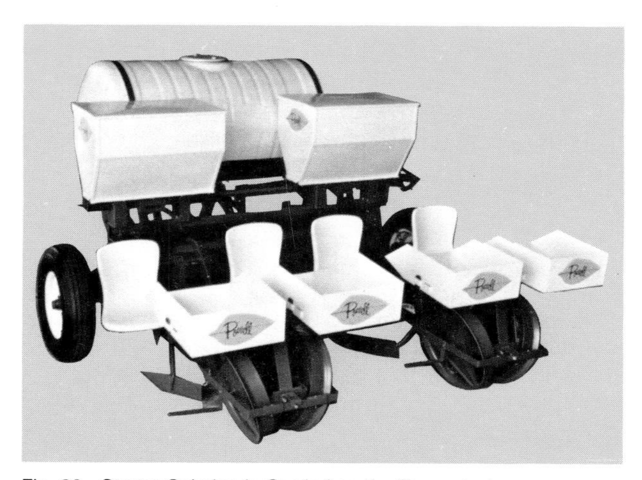

Fig. 26—Starter Solution is Carried on the Transplanter

Most starter solutions are high in phosphorus. In large-scale transplanting operation, starter solution is carried in a tank on the transplanter and is applied automatically as each plant is set (Fig. 26).

Foliar Application

Some fertilizers may be dissolved in water and sprayed on plant foliage (Fig. 27). Foliar applications are usually used to correct particular nutrient deficiencies which could seriously impair crop growth and production. Complete fertilizers for foliar application are also available. Fertilizer sprayed on the plant is usually absorbed and used faster than when applied to soil. However, under most conditions, the spray should be supplemented with a soil application. Fertilizers in the soil become effective later and last longer so additional foliar applications are not usually required.

Foliar sprays should be dilute solutions to avoid plant injury. All fertilizers are toxic and can cause crop damage if applied in too high concentrations. Hard water reduces the risk of toxic effects.

FERTILIZER-PESTICIDE MIXTURES

Applying mixtures of fertilizers and pesticides in one operation is not a new practice. It started before World War II. The big advantage of using such mixtures, of course, is that it permits doing two jobs at one time:

- **Fertilize the crop**
- **Control pests—either insects or weeds**

Chlorinated hydrocarbon and organic phosphate soil insecticides are commonly used in mixtures. Insecticide-fertilizer mixes are usually solid materials.

Mixtures must be treated like pesticides because they contain toxic materials and can be dangerous if not properly handled. Three basic safety rules (Fig. 28) are:

- *Wear a protective respirator*

Fig. 27—Foliar Application of Fertilizer with Irrigation Water

Fig. 28—Treat Pesticide-Fertilizer Mixtures Like Pesticides

- *Wear gloves and long sleeves and keep the dust off your hands and arms*
- *Wash frequently to remove dust from skin*

Herbicides are often applied with liquid fertilizers. The materials used must be compatible to ensure that the combination does not deteriorate. Soil-applied herbicides that can be put on before or at planting time are the ones most commonly mixed with liquid fertilizers. However, use of post-emergence herbicides with liquid fertilizer may be satisfactory under some conditions. Use of fertilizer-herbicide mixtures has become quite popular because there are many materials that work well together. Some herbicides may be mixed with certain dry fertilizers. Uniform mixing and application is necessary.

Simultaneous Application

An alternative to fertilizer-pesticide mixtures is the use of equipment which meters and handles the materials separately, but at the same time (Fig. 29).

Planters can be equipped to apply either solid or liquid fertilizer, herbicide, and insecticide while the crop is being planted. This method has several advantages over using mixtures of fertilizer and pesticide:

- *Application rates of each material are controlled separately and can be changed when variety, field, or soil type changes*
- *Lower cost because "standard" materials are used*
- *Fewer compatibility and storage problems*
- *Avoids possible "label restrictions" on mixing different materials*

Mixing of fertilizer and pesticide can save trips, time, and costs and can be especially helpful where availability of water for spraying is limited.

SUMMARY

Plants require sixteen nutrients. These are divided into three groups:

- **Primary nutrients**
- **Secondary nutrients**
- **Micronutrients**

Soil pH is very important because nutrient availability is affected by pH. Most crops prefer pH of about 6.5 to 7.0, but some grow better under more acid or more alkaline conditions.

A soil test provides the most reliable method of evaluating fertilizer needs and soil pH. There are also visual

Fig. 29—Simultaneous Application of Fertilizers and Pesticides While Planting

clues which an experienced observer can use to diagnose crop-nutrient problems. Tissue tests can give the nutrient status of a plant and thus indicate a possible need for fertilizer.

Fertilizers are available in several forms, including:

- **Pre-mixed dry fertilizers**
- **Bulk-blended dry fertilizers**
- **Liquid fertilizers**

Fertilizer application is important. If the fertilizer is too close to a seed, the seedling may be damaged. If too far away, the young plant may be undernourished during early growth stages. Banding below and to the side of the seed is a popular application technique. Special fertilizer application methods are sometimes used. These include foliar application, starter solutions, and pop-up fertilizers.

Mixtures of pesticides and fertilizers can be used to save time and labor in the field. These materials may also be applied simultaneously, but by separate parts of a machine (such as fertilizer and pesticide attachments on planters). Treat mixtures with the same respect you would give a pesticide alone.

CHAPTER QUIZ

1. What are plant nutrients?

2. ________, ________ and ________ are the primary nutrients used to improve and maintain crop growth.

3. Name the three secondary nutrients.

4. Name four micro-nutrients.

5. (True or false.) Soil pH influences the availability of nutrients to the plant.

6. What are the three principal forms of soil nitrogen?

7. What is the nitrogen cycle?

8. (True or false.) Visual observations along with soil samples determine if lime is needed.

9. What are three benefits of fertilizing crops?

10. (True or false.) Fertilizer-pesticide mixtures do not need to be treated carefully.

11. How much nitrogen, phosphoric acid and potash is contained in 5000 lbs. (2270 kg) of 5-20-10 fertilizer?

12. Two types of liquid fertilizers are ________ and ________.

13. What are two reasons for banding fertilizer?

3
Weed Control

Fig. 1—Highest Crop Yields are Achieved when Weeds are Controlled.

INTRODUCTION

Weeds are unwanted plants. Any plant can be a weed in a particular situation (Fig. 2). Johnsongrass may be used as a forage in some areas but is considered a serious weed in row crops. Even corn can be a weed. When corn grows in a soybean field, the two crops compete and the corn kernels are foreign material in the harvested beans. Volunteer corn in a soybean field can also negate the advantages of a crop rotation for controlling corn rootworm. Multiflora rose, once planted because it was considered desirable, is now often considered a weed because it has spread from fence rows to pastures.

Most people, when they speak of weeds, are referring to a group of plant species that are nearly always undesirable. Weeds may be either broadleaf or grass plants. It is incorrect to think of weeds as only broadleaf plants and say "weeds and grasses." Weeds are a factor in the management of all kinds of land and water. The economic costs of weeds are especially significant in agriculture.

Fig. 2—Weeds Compete with Crops for Nutrients and Moisture

WEEDS IN AGRICULTURE

No one knows, with certainty, the worldwide costs or economic loss caused by weeds. But, it is generally conceded that *losses from weeds exceed the losses from any other category of agricultural pest.*

In developed countries with temperate climates, weed losses are often 10-15 percent of the total value of agricultural and forest products (Table 1, 2). According to one estimate, weeds in the United States cost at least $15 billion per year. That is an average of more than $6000 per year per farm.

TABLE I

FOXTAIL REDUCES CORN YIELDS

Yield	Foxtail
91.4 Bu/Ac. (7.9 m³/ha)	No Foxtail
86.4 (7.5)	1Plant/24"(61cm)
84.6 (7.3)	1 Plant/12" (30)
83.2 (7.2)	1 Plant/4" (10)
81.0 (7)	1 Plant/2" (5)
76.0 (6.6)	1 Plant/1" (2.5)
68.2 (5.9)	Unthinned

TABLE 2
PIGWEED REDUCES SOYBEAN YIELDS

Pigweed plants per foot (meter) of row	Pigweed yield (dry wt.) (lb/ac)	(kg/ha)	Soybean yield (bu/ac)	(kg/ha)
0 (0)	0	0	50	(3363)
1 (3)	1704	1908	35	(2355)
2 (7)	4065	4550	25	(1680)
4 (13)	4882	5470	22	(1480)
8 (26)	3395	3800	27	(1815)
16 (52)	4580	5130	22	(1480)
32 (105)	6485	7265	14	(940)
uncontrolled	8072	9040	10	(675)

Losses in tropical areas are often much higher. For example it has been estimated that effective weed control could double or triple rice yields in some areas.

EFFECTS OF WEEDS

Weeds reduce profits in three ways:

- *By reducing crop yields*
- *By lowering product quality*
- *Because of cost of control activities*

Weeds have an economic affect on nearly everyone.

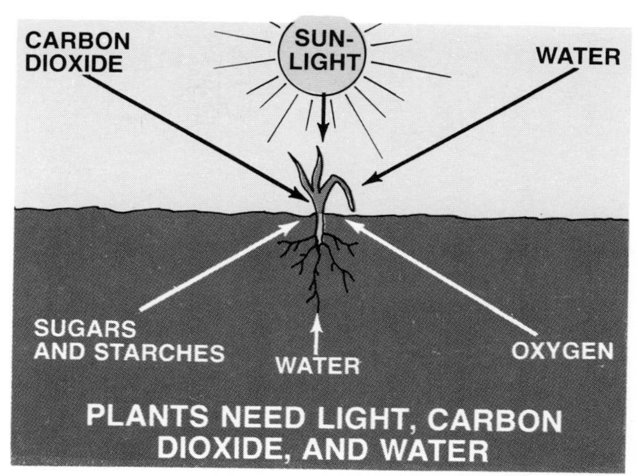

Fig. 3—Photosynthesis Requires Light, Water and Air

Weeds Compete With Crops for Nutrients, Moisture, and Light

Competition by weeds for these essential elements causes considerable crop loss. All three are needed by every plant and those used by weeds are simply not available for the crop.

Weeds use as much water as crops per pound of dry matter produced (Table 3). Obviously, the more weeds in a field, the higher the water loss.

Weeds, especially large ones, can shade crop plants and limit the light reaching them. Since light is essential for photosynthesis and growth to take place (Fig. 3),

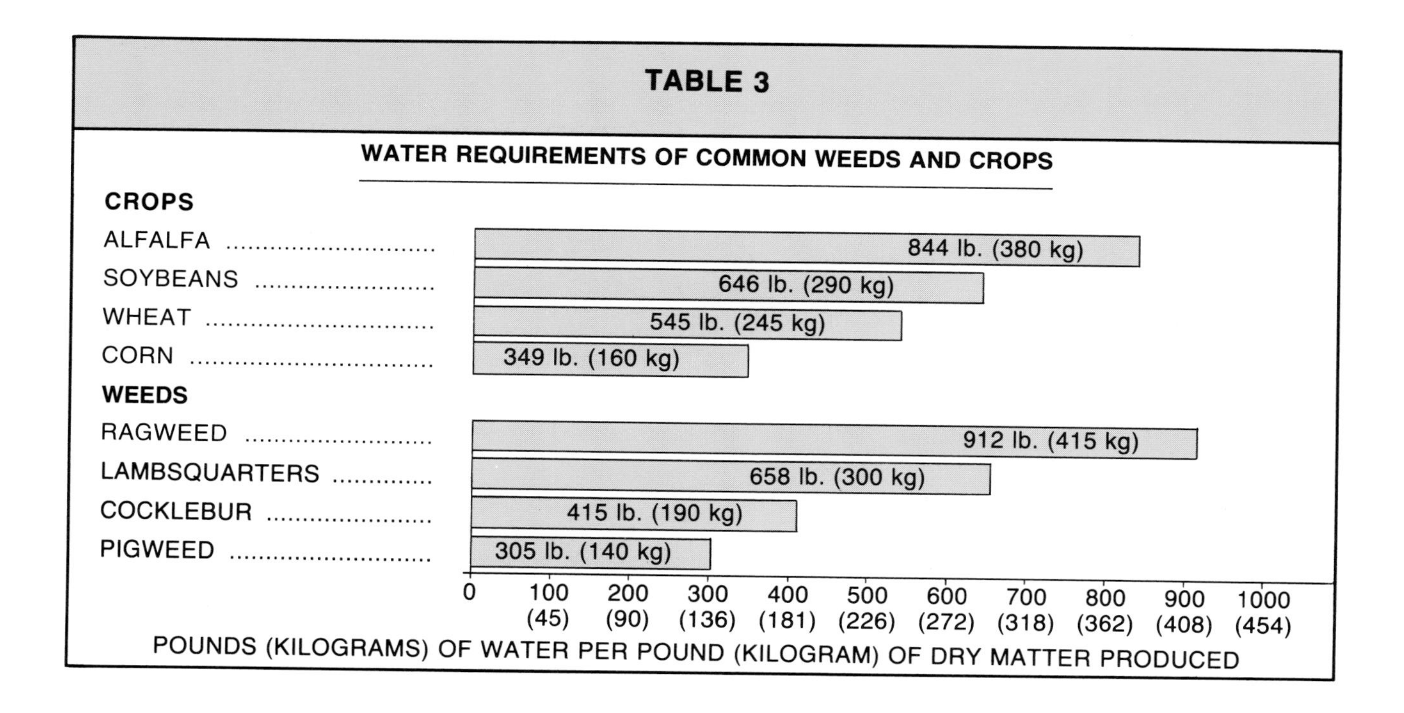

Fig. 4—Dodder is a Parasitic Weed

shading reduces crop growth. Shading affects all kinds of crops. The seriousness depends on the degree of shading which, in turn, depends on the number and kinds of weeds present.

Most weeds are vigorous plants which use large amounts of nutrients. The timing of nutrient uptake, as well as the total quantity used by weeds is important.

For example, pigweed takes up and stores nitrates during the early growth period. These nitrates are not available for the crop to use. This is one reason pigweed is such a successful competitor with most crops. The stored nitrates can also make pigweed hazardous to cattle because of possible nitrate poisoning.

Some weeds have the capacity to obtain water and food directly from crop plants rather than from the soil. An example of such a parasitic weed is dodder (Fig. 4). It has no true leaves or roots and is unable to manufacture its own food. After the seeds germinate in the spring, long slender stems wrap around the host plant and small suckers penetrate the host to get food and water. If the dodder cannot find a suitable host it dies. Alfalfa and clover are favorite host plants of dodder.

Other Weed-crop Competition

Other plants, such as field bindweed, vine tightly about crop plants and restrict normal growth and development and also compete for nutrients, water, and light (Fig. 5).

There are also bio-chemical interactions between plant species. Such interactions occur when a plant of one species secretes a substance which alters the growth pattern of a nearby sensitive plant of another species. Effects can include:

- *Inhibition of seed germination*
- *Development of abnormal seedlings*
- *Restriction of root growth*

An example of such interaction is the phytotoxic chemical produced by black walnut which injures or kills plants, such as tomatoes and apples, planted near the walnut tree.

Fig. 5—Field Bindweed Vines Around Crop Plants

Fig. 6—Weedy Crops can cause Serious Harvesting Problems

Weeds Lower Crop Quality

Weed contamination reduces the harvestability, quality and marketability of many crops (Fig. 6). Weeds reduce the nutritional value and palatability of hay and pasture. Consequently, most people are willing to pay more for weed-free hay forages than they would for a weedy crop.

All states have seed laws which may prohibit the sale of agricultural seeds containing certain noxious weed seeds or limit the amount of other kinds of weed seeds that may be present.

Fig. 8—Water Weeds Completely Cover Parts of this Pond

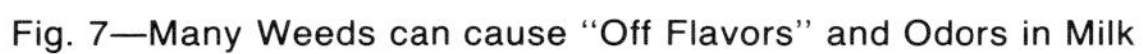

Fig. 7—Many Weeds can cause "Off Flavors" and Odors in Milk

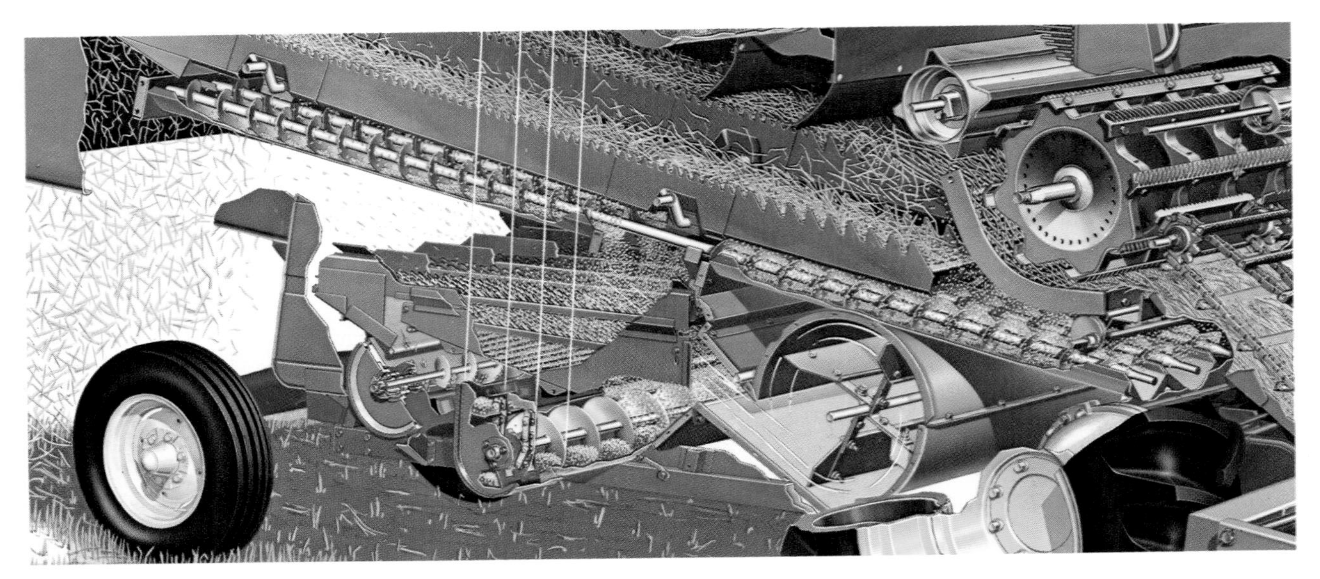

Fig. 9—Cleaning Unit on a Grain Combine

Weed seeds in stored grain may retard drying and contribute to mold growth. Small quantities of certain weed seeds such as wild garlic and mustard impart objectionable odors and flavors to flour when the wheat is ground.

Weeds Lower Quality of Animal Products

Cows eating some weeds, such as wild garlic and onion weed may produce milk with an undesirable flavor (Fig. 7). Also, some weed species, such as white snakeroot, are poisonous to grazing animals.

Fig. 10—Many Allergy Problems are Caused by Plants such as Common Ragweed

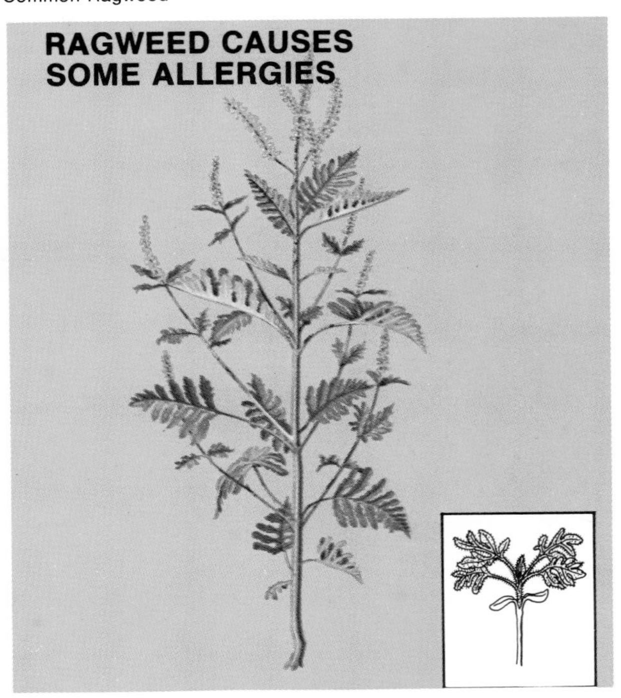

Weeds Can Poison Animals

Each year, many animals are killed or injured by feeding on poisonous plants. Livestock poisonings are most common during late fall and early spring on over-grazed pasture when grass is sparse and livestock consume plants that would ordinarily be avoided.

Some plants contain naturally occuring toxic substances. jimsonweed, for example, contains an alkaloid which acts as a halucinogin (Table 4). In small quantities, it may not be harmful. However, it can be poisonous if the animal ingests a large enough quantity.

Certain weeds contain photo active chemicals which travel through the blood stream to all parts of the animal's body. In areas of skin not protected by hair or dark pigmentation, the chemical is activated by light and produces an effect similar to severe sunburn. For example, klamath weed contains a substance which makes the mouths of livestock so tender they cannot eat properly. Spring parsley makes ewes' udders so sensitive that the mothers refuse to let the lambs nurse.

Other plants accumulate toxic substances from the surrounding environment. Lambsquarters, mustard, and pigweed are examples of rapidly growing plants that can accumulate nitrates at levels high enough to be poisonous to livestock. Nitrate concentrations about 1.5 percent (expressed as KNO_3, dry weight) is generally lethal to livestock. Ruminants are more susceptible than animals with simple stomachs, such as horses.

Other plants can absorb and accumulate toxic minerals, such as selenium, cadmium, copper, lead and manganese. Selenium, which is stored in locoweed accounts for many instances of livestock poisoning.

Weeds Harbor Insects and Diseases of Crop Plants

Some weeds harbor fungal or bacterial disease organisms and insect pests. Black stem rust of wheat spends a part of its cycle on the common European barberry. Besides being a strong competitor for water and nutrients in a corn field, johnsongrass harbors maize dwarf Mosaic virus which attacks corn.

Clogged Channels

Water hyacinth and other aquatic weeds grow in irrigation or drainage ditches, ponds, and lakes. Such weeds reduce the water flow, interfere with drainage and can require considerable money and effort to keep channels clean (Fig. 8). In addition, these plants are heavy water users. In arid areas, water used by aquatic weeds is lost and cannot be used for irrigating crops.

Ditches and natural waterways also frequently have a heavy growth of willows and other weeds on their banks.

Weeds Reduce Land Values

Most farmers are not willing to pay as much for land with serious weed infestations as they would for "clean" land. In some areas, banks are reluctant to lend money on land badly infested with weeds.

Fig. 11—Typical Seed Tag

A

PAG SEEDS

P.O. BOX 9480, MINNEAPOLIS, MINNESOTA 55440

HYBRID FIELD CORN SEED NET WT. 50 LBS.

NOTICE: See reverse side of tag for treated seed information and limitation of warranty.

VARIETY: SX397 LOT: 266708
ORIGIN: ILL GERM: 95%
TESTED: JAN. 81 SINGLE CROSS
KERNAL COUNT APPROX. 80,700
WE SUGGEST YOU TRY THESE PLATES
FIRST: B7-24X/B7-16/C7-24X/C7-16X

PURE SEED -	98%
INERT MATTER -	1.5%
OTHER CROP SEED-	.5%
WEED SEED -	NONE
NOXIOUS WEED SEEDS -	NONE

B

LIMITATION OF WARRANTY

NOTICE TO BUYER: P A G warrants that the seeds are described as hereon, within recognized tolerances. THIS EXPRESS WARRANTY EXCLUDES OTHER WARRANTIES, EXPRESSED OR IMPLIED, AND EXCLUDES THE WARRANTY OF MERCHANTABILITY OTHER THAN THAT THE SEED CONFORMS TO THE DESCRIPTIVE STATEMENTS HEREON. THERE ARE NO WARRANTIES WHICH EXTEND BEYOND THE DESCRIPTION HEREON. By acceptance of the seed, buyer expressly agrees that the company's liability and the buyer's exclusive remedy for any loss, shall be limited in all events to a return of the purchase price of the seed.

No liability hereunder shall be asserted unless the buyer or user reports to the warrantor within a reasonable period after discovery (not to exceed 30 days), any conditions that might lead to a complaint.

— CAUTION —

This seed has been treated with EVERSHIELD seed protectants: .65 oz. active CAPTAN-THIRAM/ 50 lbs., plus 3.9 ppm MALATHION and 23 ppm METHOXYCHLOR.
Do not use for FOOD, FEED, or OIL purposes. Avoid prolonged and repeated contact with skin. In case of contact, flush with water, For eyes get medical attention.

MISSOURI PERMIT NO. W13750 FORM XPAG-59

Fig. 12—Ensiling Can Destroy the Viability of Some Weed Seeds

Bankers have learned that the high cost of weed control programs and lower yields may not produce adequate income to repay a mortgage.

Weeds Increase Production Costs

The presence of weeds requires additional cultivations and expenditures for herbicides for control. Weeds also increase machinery wear and slow harvesting.

In addition, extra time, labor and equipment is needed to clean weed seeds from harvested grain (Fig. 9).

Weeds May Cause Human Discomfort

There are many weeds that can cause health problems to humans and animals. Some of these problems are:

- *Pollen from ragweeds contribute to asthma, hayfever and other respiratory problems (Fig. 10).*
- *Skin irritation can result from contact with poison ivy or poison sumac.*
- *Stems, leaves and fruit of some weeds are poisonous, especially to children.*
- *Poisonous plants may be eaten by animals.*

CHANGING WEED PROBLEMS

Weed problems change as cultural practices, crop selection, and weed control practices are modified. In general, specific weeds are a problem because they thrive and grow under the same conditions being produced for the crop. The advent of herbicides has changed but not eliminated weed problems. For example, when broadleaved weeds, which were the major problem in corn were brought under control by the use of 2,4-D, grasses became a major problem. In a similar way, prickly sida became a serious weed in cotton after trifluralin was used to control other, more aggressive weeds.

The widespread use of herbicides has not led to the development of weed resistance in most cases, but has caused shifts to new weed species which always were tolerant to the herbicide. However, triazine herbicide resistant weeds have developed in certain isolated areas of the United States and Canada.

WEED VALUE

In spite of being generally undesirable, some weeds do have value. For example, an African weed was the early source of the drug cortisone. Researchers are constantly studying weeds and other plants in an effort to identify new and useful products.

COMMON WEEDS

Before attempting to control weeds, they should be identified. Items to observe include leaf shape, leaf arrangement, type of growth, flower color and flower shape. There are thousands of weeds. You don't need to know them all. The most practical approach is to become familiar with the weeds common to your area. Some common and serious weeds have a wide range,

Fig. 13—Weed Seeds are Carried by Livestock

and appear in many parts of the world. If a weed is found which cannot be identified, contact your local County Extension Agent. He should be able to help you. For names and illustrations of many different weeds see the Appendix.

WEED PREVENTION

The best way to avoid weed problems is to prevent weeds from spreading to clean fields from infested areas. Prevention is not easy because of the many ways that weed seeds can be transported. Some precautions are discussed below.

Use Weed-Free Seed

Many weeds are introduced to new areas by being planted along with crop seeds. The majority of plant species known as weeds were introduced to America from other parts of the world, mainly Europe. Some of them got a "free ride" with imported crop seeds.

All states have laws and regulations requiring strict labeling of seed, including the kind and quantity of weed seeds, if any (Fig. 11). There are many state and area "crop improvement associations" which help to assure the availability of high quality crop seed with little or no weed seed.

Avoid Spreading Weed Seeds With Feed

Weed seeds are often viable (able to germinate) even after having passed through an animal's digestive tract. When the manure is spread, such seeds are "planted" perhaps where they never grew before. Grinding may not destroy all weed seeds either. Feeding material such as screenings should be discouraged.

Ensiling (Fig. 12) can destroy the viability of many weed seeds. However, some such as field bindweed may remain viable in silage. In one test, some seeds were still viable after storage in a silo for 4½ years.

Ferment Manure Before Spreading

Farm products such as hay, grain, and other feeds often contain large numbers of weed seeds. It is impractical to destroy them all before feeding. Thus, if viable seeds are to be kept from fields, an attempt can be made to reduce the viability by storing in manure. Fermenting manure reaches temperatures of 150 to 200°F (66 to 93°C). Sixty days of such temperatures can kill most weed seeds.

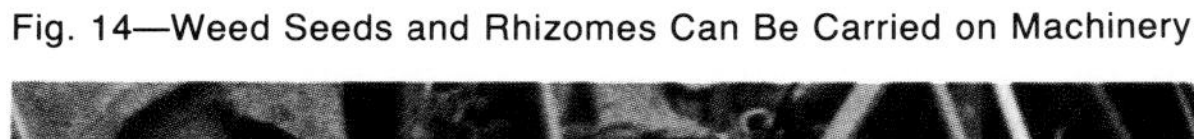

Fig. 14—Weed Seeds and Rhizomes Can Be Carried on Machinery

Anaerobic digesters (methane gas generators) which operate at the thermophilic temperature range (140°F, 60°C) should effectively kill weed seeds. However, the effect of the more common mesophilic temperatures (95°F, 35°C) are largely unknown.

Do Not Move Livestock Directly From an Infested Area to a Clean Area

Livestock frequently give weed seeds a ride when they are moved. Seeds stick to wool, hair, and muddy feet and are carried in digestive tracts and bedding (Fig. 13). Weeds can be introduced to new communities when cattle are unloaded from trucks or railroad cars. Cattle are frequently transported hundreds of miles. If possible, hold new cattle in a "decontamination area" for a few days.

Clean Tractors and Equipment Before Leaving a Weedy Area

All kinds of equipment, particularly harvesting machinery can carry weed seeds from field to field or farm to farm. Thoroughly clean machinery before leaving a weedy field so that seeds do not infest additional fields (Fig. 14).

Inspect Nursery Stock for Weed Seeds or Rhizomes

Nursery stock is regularly inspected for diseases and insects, seldom for weeds. The packing around plants and the soil ball on tree roots can carry seeds or rhizomes. The orchardist or homeowner should be aware of these sources of possible infestation. Watch for weeds in and around new plantings and destroy any that get started.

Fig. 15—Windborne Seeds

Control Weeds on Ditch Banks

Conditions are usually good for profuse weed growth along ditch banks. Many weed seeds float. They can be carried for miles in irrigation or drainage water. Many irrigation and drainage districts try to keep ditch banks clean to prevent spread of weeds as well as to conserve water. A desirable perennial grass can provide effective weed control and also protect the soil from erosion.

Prevent Production of Windborne Seeds

Many weed species produce windborne seeds (Fig. 15). It is very difficult to prevent them from spreading. No matter how careful a grower is on his own land, he has no way to keep seeds from blowing in from a neighbor's land. Canada thistle and milkweed are typical weeds which spread this way.

In spite of all prevention and control efforts, unless all farmers in an area cooperate weeds producing windborne seeds continue to spread.

Most states in the U.S.A. have weed laws designed to prevent the spread of specific noxious weeds. Landowners must prevent noxious weeds from producing seed. Weed laws in many states provide that if the landowner fails to control noxious weeds, county personnel may enter the land, destroy the weeds, and add the cost onto the landowner's taxes. Recently, the Federal Noxious Weed Act was passed which is intended to prevent new weed problems from being imported into the United States.

ERADICATION OR CONTROL

Eradication is the complete destruction of weeds. It can be done mechanically or with chemicals.

Control procedures do not attempt to kill all of the weeds, but aim to keep weeds at a level that does not seriously interfere with crop production or yield.

Eradication is more difficult, and usually more expensive than control. Eradication may be advisable and practical if a small area is being treated, the weed species is new, or if it is a noxious species.

Control is usually lower in cost and feasible. It is often the only practical approach if the weeds are wide spread.

Control efforts must be repeated each year, because weeds always seem to be present. The next section discusses some special characteristics of weeds that enable them to persist and survive in spite of efforts to kill them.

SURVIVAL MECHANISMS OF COMMON WEEDS

The key factor in the survival of weeds is the reserve of seeds and vegetative plant parts in the soil. Crop fields sometimes contain 25,000 weed seeds per square

TABLE 4	
WEED-SEED PRODUCTION	
Weed Plant	**Number of Seeds Per Plant**
Lambsquarters	72,000
Mustard	13,500
Ragweed	3,400
Pigweed	118,000
Curly dock	30,000

foot in the plow depth; over 1 billion seeds per acre.

Seeds are the primary survival mechanism of annual weeds. Perennial weeds also produce seeds. In addition, perennials can reproduce vegetatively, by roots, bulbs, buds, tubers, rhizomes and other means.

Seed Production

Most weeds are prolific seed producers (Table 4). Many weed seeds mature before crop plants ripen. They can

TABLE 5	
Weed-Seed Dormancy	
Name of Weed	**Years Some Seed Remain Viable (Alive) in the Soil**
Common Chickweed	30
Shepherdspurse	35
Pigweed	40
Black Mustard	50
Curled Dock	more than 80

thus complete their reproductive cycle before being disturbed by crop harvesting.

Seed Spreading

Seed dormancy is a characteristic that enables weeds to persist and continue infesting an area even though the soil is tilled frequently. Some weed seeds can remain dormant for many years (Table 5). Unlike crop plants, which are bred and selected to produce under a wide range of conditions, some weeds require rather specific conditions for seeds to break their dormancy and germinate. Thus seeds remain dormant in the soil until all conditions are right, then they germinate.

Fig. 16—Rhizomes of Johnson Grass (Chopped Up by a Disk Harrow)

Plowing and other tillage operations help to ensure that some weed seeds are always in the proper location and ready to grow.

Fig. 17—Tillage may not Control Perennial Weeds Completely

Fig. 18—Mow before Annual Weeds Produce Seed

Vegetative Reproduction

Vegetative or asexual reproduction is important for perennial weeds. Such characteristics as deep root systems, large numbers of rhizomes or bulbs and a reserve of food in the roots of some species help perennial weeds to survive and spread. The underground parts are usually difficult to destroy completely (Fig. 16).

Quackgrass is a good example of a weed that can vigorously reproduce vegetatively. Most quackgrass rhizomes are located in the upper layers of soil and are frequently damaged by tillage. However, quackgrass compensates by being able to grow a new plant from a small piece of rhizome tissue. A thorough disking often results in a heavier stand of quackgrass because of the new tops that grow from the pieces of rhizomes cut up while tilling the soil (Fig. 17).

PRINCIPLES OF WEED CONTROL

Weed prevention is better than control, but most of our land was infested with weeds long before we understood the need for prevention. Consequently, weed control (including prevention of new infestation) is necessary on all kinds of crop land.

The best weed control results come from "custom designed" efforts for each specific situation. Often a variety of control techniques are used. The life span of weeds is an important characteristic to consider when planning a weed control program. Weeds fall into three life span groups (Table 6).

- **Annual**
- **Biennial**
- **Perennial**

Here is a brief discussion of each.

Annuals

Annuals are plants that complete their life cycle in one year. During that time, they grow from seed, produce new seed and then die. Summer annuals germinate in the spring, grow and produce seed during the summer and die in the fall. Winter annuals germinate in the fall, then complete flowering and seed production and die in the spring.

Since they can only reproduce by reseeding themselves, annual weeds can be controlled by preventing seed production. Stopping seed production every year (Fig. 18) causes the supply of seeds in the soil to be gradually reduced or depleted. Weed seed depletion can be accelerated by encouraging seeds already in the ground to germinate, then using tillage or chemicals to kill the seedlings. Frequent tillage and control of soil moisture by irrigation can cause many weed seeds to germinate.

Biennials

Biennials have a life span longer than one year, but not more than two years. Growth is entirely vegetative during the first year. Seed production, during the second year leads to the death of the plant. Control of biennials should be aimed at preventing seed production. Control efforts should be aimed at both vegetative and seed-producing stages for effective biennial weed control. Control of biennial weeds with herbicides is most effective during the vegetative stage.

TABLE 6

Life-Cycle Duration of Common Weeds

Annuals	Biennials	Perennials
Barnyardgrass	Burdock	Plantain
Black Nightshade	Bull Thistle	Canada Thistle
Chickweed	Mullein	Curled Dock
Ragweed	Poison Hemlock	Dandelion
Crabgrass	Wild Carrot	Field Bindweed
Dodder	Wild Parsnip	Johnsongrass
Foxtail	Yellow Goatsbeard	Leafy Spurge
Lambsquarters		Sow Thistle
Pigweed		Quackgrass
Purslane		Sheep Sorrel
Shepherdspurse		Wild Garlic
Wild Oats		Wild Onion

Perennials

Perennials live for more than two years. Seed is usually

Fig. 19—Cultivation Controls Shallow-Rooted Weeds

produced during the second year and in each succeeding year of the plant's life. Simple perennials reproduce mostly by seed and established plants are usually killed by tillage. Bulbous perennials may reproduce by both seeds and bulbs. Some species, such as wild garlic can produce bulblets on the above-ground portion of the plant. Creeping perennials spread by stems above ground (stolons) or underground (rhizomes) and by seeds. They regrow from buds on roots or rhizomes after tillage. Some species have most of the root system in the upper foot (0.3 meter) of soil (Fig. 19). However, roots of others, such as bindweed may grow as deep as 30 feet (9 meters).

To control perennial weeds, prevent vegetative spreading, stop seed production and kill existing plants. Weed identification photographs begin on page 209 of this manual.

WEED CONTROL METHODS

Many methods of weed control have been developed. They fall into three groups:

- **Mechanical and cultural**
- **Biological**
- **Chemical**

MECHANICAL AND CULTURAL WEED CONTROL

Physical weed control methods have been used for centuries in all parts of the world. Many cultural practices are based on the need for tillage operations to control weeds.

Although herbicides have become more important in weed control, mechanical and cultural methods are still widely used. Some physical methods are discussed in the next section.

Hand Pulling

Hand pulling is probably the oldest method of controlling weeds. Large scale use is limited, but it is still widely used in home gardening and developing countries. It is a practical way to remove a few scattered weeds in a field to prevent seeding. Pulling is most effective on annual or biennial weeds. It is less useful for perennials, because they grow back from root parts left in the soil.

Fig. 20—Using a Rotary Hoe for Blind Tillage

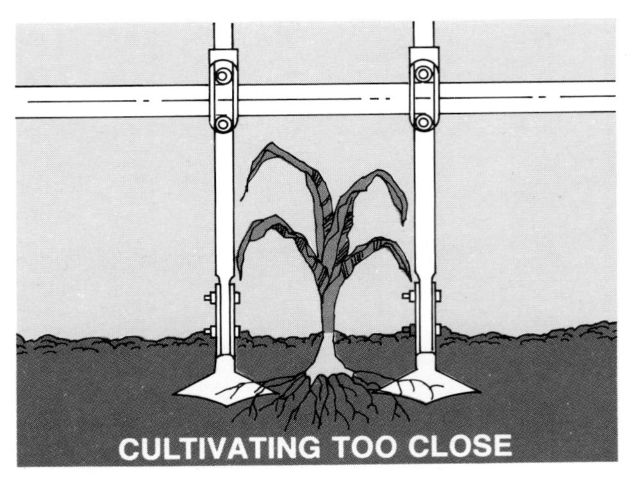

Fig. 21—Cultivating too Close or too Deep Cuts Crop Roots

Hand Hoeing

Hoeing is designed to cut off the tops of weeds. Hoeing has declined as herbicides have become available and as people have become less willing to work with a hoe. Hoeing is faster than hand pulling. It is most effective on annuals or biennials.

Tillage

Tillage is a common practice. Its primary function is weed control. Tillage is used before planting, while the crop is growing and after harvest.

Fig. 22—Flooding Helps Control Weeds in Rice Fields

Blind tillage (Fig. 20) is tilling the entire soil surface after a crop has been planted. Rotary hoes, finger weeders and spike tooth harrows are used.

Row crop cultivation has decreased since the advent of effective chemical weed controls. On soils in good tilth, weed control is often the only benefit of row crop cultivating, and improperly adjusted cultivators can damage a crop (Fig. 21).

The rotary hoe can be very effective for early control soon after the weed seeds have germinated and before or soon after the weed seedlings emerge. Even where a herbicide has been used, the rotary hoe can be very effective, especially if delayed rainfall decreases effectiveness of the herbicide. The rotary hoe can help to control the first flush of weeds and may even give very slight incorporation of the herbicide to increase effectiveness for the next flush of weeds.

Mowing and Grazing

Weeds can be mowed to prevent seed production and to remove unsightly growth. To be most effective, mowing must be close to the ground to ensure that all branches, tillers or stems are cut. Close mowing, however, is not always possible and weeds often regrow quickly. Grazing has sometimes been used to control weeds in pastures, and along ditches, fencerows, and roadsides. Sheep and geese are used for weeding some crops.

Flooding

Flooding controls weeds by excluding air that is essential for plant growth. It is only effective if the entire plant is covered for a fairly long period of time. It is used in conjunction with crops like rice that can grow in the flooded area (Fig. 22).

Burning

Heat is sometimes used to control weeds. Burning rangelands removes the existing plant cover, including weeds, to make room for more desirable grasses. Other than removing some of the plant parts, burning is not usually a very effective weed control method because many seeds remain viable on the soil surface even after a fire. Valuable organic matter is also destroyed by burning.

Selective weed killing in a crop field is possible with specialized flaming equipment. Temperatures as high as 2000°F (1090°C) can be reached by burning fuel oil or propane. Flaming is rather expensive, but effective on cotton. Smaller, more tender weeds are destroyed by the heat, but the woody cotton stems are not harmed significantly by the flame.

Flaming is used on a very limited basis today. The current cost of fuel makes this practice less economical than other weed-control methods.

Smothering

Weeds can often be smothered by competitive crops or by various kinds of mulches. Planting patterns that quickly produce a complete ground cover enable some crops to compete effectively by shading weeds (Fig. 23).

Fig. 23—Shading Can Help Control Weeds

Mulches are often used to control weeds around ornamental plants or in high-value crops such as strawberries and melons. Plastic, paper, straw, woodchips and manure are just a few of the materials used.

BIOLOGICAL CONTROL

Biological weed control involves the use of living organisms such as insects or fungi which attack certain weed species. An example is the control of St. Johnswort by the Chrysolina beetle in the Western United States.

To obtain effective biological control, several conditions are necessary.

- *The organism must be specific to the weed to be controlled, otherwise it may spread to other species, possibly crops and become a pest itself.*

- *The organism must have no natural enemies that interfere with their activity.*

It should be pointed out that biological control is seldom complete. The degree of control fluctuates and there is always a time lag between weed growth and the response of the insect or disease which is the control agent. Biological techniques usually can't be used to gain eradication, but under the proper conditions, biological controls can be effective enough to remove the threat posed by the weed pest.

Certain temperature and moisture conditions may not be conducive to survival of disease organisms for controlling weeds. However, there have been some successes as with control of Northern joint vetch in rice.

Biological control is particularly well suited to situations where other control methods, such as an effective and economical herbicide, are not available. Biological control can be a low cost method to use which is particularly well suited to low value land and large areas. However, the initial cost for locating, evaluating, developing, and applying biological controls can be quite high. Consequently, Federal Government support is usually required.

CHEMICAL CONTROL

Although herbicides are sometimes used to eliminate tillage, the more common procedure is to combine chemicals, tillage, and other cultural practices in a carefully designed, well coordinated weed control plan. When compared to other control methods, herbicides have the following advantages:

- *Appropriate herbicides may be applied to control weeds in crops where row spacing is too narrow for cultivation (Fig. 24).*

- *Preemergence application of herbicides provides early season weed control, allowing the crop to grow without competition during the critical early part of the growing season (Fig. 25).*

- *Selective herbicides reduce the need for cultivation which can injure crop roots as well as weeds (Fig. 21).*

Fig. 24—Herbicides can Control Weeds in Crop Rows

Fig. 25—Preemergence Herbicides Allow Crop Plants to Begin Growth Rapidly, Free of Weeds

- *Herbicides help maintain soil structure by reducing the need for tillage (Fig. 26).*
- *Erosion can frequently be reduced by vegetation or cover of crop residue while herbicides are used for weed control.*
- *Many weeds, especially perennials which cannot be effectively controlled by other means, are susceptible to herbicides.*
- *Although herbicides require significant amounts of energy to produce, transport, and apply, they can reduce the total amount of energy required for weed control when compared to the mechanical approach (Table 7).*

HERBICIDES

Many materials with varied chemical properties are used for weed control. These chemicals can be described or grouped in various ways, as discussed in the following sections. Such groupings are useful as a means of comparing and selecting materials.

Fig. 26—Herbicides Permit No-Till Planting

TABLE 7				
Method of Controlling Weeds	Energy Input for Controlling Weeds-Thousand Kilocalories/Acre (ha)	Yield of Corn Grain/Acre (ha)		Net Profit in Energy Due to Weed Control-Million Kilocalories Acre (ha)
		Bushels (kg)	Million Kilocalories	
None	0	54 (3390)	5.4 (13.3)	—
Cultivation	56 (138.3)	81 (5085)	8.2 (20.3)	2.7 (6.7)
Herbicide	37.9 (93.6)	90 (5650)	9.1 (22.5)	3.6 (8.9)
Hand labor	32.7 (80.7)	92 (5775)	9.3 (23)	3.8 (9.4)

MODE OF ACTION

Herbicides affect plants by either contact or systemic action. **Contact** herbicides kill only the plant parts to which the material is applied. **Systemic** herbicides are absorbed by plant roots or foliage and translocated within the plant to other tissues.

Systemics are particularly advantageous against perennial weeds since uniform coverage is not necessary for the chemical to reach all parts of the plant. Systemic herbicides may be applied to the foliage of perennial weeds and be translocated to reach deep root systems.

Contact herbicides are usually targeted against annuals since killing the foliage either kills the weed or damages it to the point that the crop can compete successfully. There are selective and nonselective herbicides of both types.

HERBICIDE SELECTIVITY

The selectivity of a herbicide depends on the concentration of the dosage applied. All herbicides are phytotoxic and, if applied at a high enough rate, any plant will die. Conversely, if the dosage is low enough at dosages between these extremes, some plants will be uninjured and other plants will be killed.

Several properties are involved in the selective effects of herbicides. Some examples include:

• *Early herbicides, such as aqueous solutions of iron and copper salts, were selective because of differential wetting. The droplets of spray bounced off the leaves of cereal plants but wet the broad leaves of other plants.*

• *Compounds, such as simazine, become strongly fixed in soil and kill shallow rooted weeds without harming deep rooted crops like orchard trees.*

• *Trifluralin inhibits secondary root growth. Weed seedlings die because they can't overcome the root damage. However, deep rooted plants, such as cotton, can penetrate the treated soil layer and grow normally.*

• *Atrazine inhibits photosynthesis in the leaves of most plants. Corn and certain other grasses take up the herbicide like susceptible weeds, but rapidly break down the atrazine to harmless compounds. Susceptible weeds lack the ability to break down the herbicide and are killed.*

Fig. 27—Selective Herbicides Allow Improved Weed Control

The selectivity of a chemical can often be changed by the addition of some type of adjuvant (adjuvants are discussed in a later section). For example, if a wetting agent is added to a solution of copper or iron salts, the differential wetting is prevented and all plants are injured.

The process of selectivity are not well understood for many herbicides. Even for 2,4-D, a widely used material, the mechanism of selectivity has not been fully explained. It is likely that as detailed research is carried out on the many new herbicides that have been developed, that new selectivity mechanism will be found.

Both selective and nonselective treatments are important. Selective herbicides are used to control weeds and reduce competition with crops (Fig. 27), lawns, nurseries, and ornamental plantings. Nonselective applications are used where no plants are wanted, such as fence rows, ditch banks, roadsides, railroad right-of-ways, driveways, and parking areas.

Fig. 28—Band Application of Herbicide

HERBICIDE APPLICATION PATTERNS

There are four common herbicide application patterns. **Band** application consists of a continuous restricted area, such as along a row (Fig. 28). Band treatments reduce costs because less material is used. **Broadcast** application spreads the herbicide uniformly over the entire area, including weeds and crop plants (Fig. 29). **Spot** treatments are applied to limited areas such as a patch of Canada thistle. **Directed sprays** are applied to weeds or soil in a controlled manner to minimize contact with the crop. This technique may allow a nonselective material to be used in a selective manner (Fig. 30).

HERBICIDE APPLICATION TIMING

Proper timing of herbicide application is dependent on many factors, including kind of herbicide, weed species, cultural practices, climate, and soil conditions.

Three major categories of timing are used: preplant, preemergence and postemergence.

Preplant treatments are applied before the crop is planted. The application may be either band or broadcast. Preplant treatments are often mechanically incorporated into the top one to three inches (2.5 to 7.6 cm) of soil. Some soil herbicides must be incorporated to prevent loss through evaporation or photodecomposition (breakdown in sunlight). Other herbicides can either be applied to the soil surface or incorporated.

Preemergence treatments are made after the crop is planted, but before it emerges. This may also include application after the crop emerges, but before weeds emerge.

Postemergence herbicides are usually applied after both the crop and weeds emerge. However, the application could be postemergence to either the crop or weeds but prior to emergence of the other.

HERBICIDE ACTION

The effectiveness of some herbicides can be modified by adding adjuvants, altering the dosage or changing the method of application. However, the mode of action, selectivity, and proper application pattern and

Fig. 29—Broadcast Application

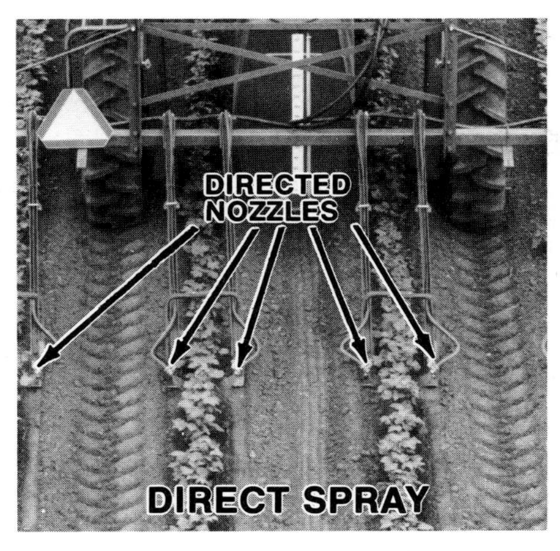

Fig. 30—Directed Spray Application Reduces Chance of Crop Injury

timing are basically related to the chemistry of the herbicide. A detailed discussion of the chemistry of all of the currently available herbicides is clearly beyond the scope of this publication. However, herbicides with similar characteristics can be grouped into classes. Table 8 presents one system which has been devised for grouping herbicides into classes and briefly lists some of the common features of each class.

Penetration Into Plants

A series of barriers must be overcome between the application of a herbicide and its ultimate effect on the plants. Any one of the barriers may stop or reduce herbicidal action. In crop plants, the barriers provide desired resistance to the herbicide. For weeds to be killed, the barriers have to be absent or nonfunctioning or the herbicide must be able to penetrate them. The surface of the plant presents the first barriers for most herbicides.

TABLE 8
HERBICIDES

TABLE 8
HERBICIDES

Herbicide family	Example materials common name (trade name)	Characteristics and uses
Aliphatics	Dalapon (Dowpon) TCA (TCA)	For control of annual and perennial grasses, such as johnsongrass, quackgrass, and bermudagrass. Usually applied in a water spray with conventional equipment. Translocates readily throughout the plant. Absorbed by roots or leaves. Easily washed off foliage. Dalapon breaks down rapidly and completely in soil; TCA breaks down more slowly.
Acetanilides	Alachlor (Lasso) CDAA (Randox) Naptalam (Alanap) Propachlor (Ramrod, Bexton) Metolachlor (Dual)	Inhibits root and shoot growth to kill weed seedlings. Must be present at early stages of germination and growth, so usually applied near planting time. Shoot uptake is important. Controls most annual grasses and some broadleaf weeds. Translocated throughout plant. Granular forms of some materials available.
Biphridyliums	Paraquat Diquat	Used postemergence. Causes rapid desiccation (drying) of plant tissue — primarily contact action. Not considered very selective. Used as desiccant to aid harvest. Paraquat used for no-till to kill existing vegetation. Inactivated rapidly in soil. **Use caution when handling.**
Benzoics	Chloramben (Amiben) Dicamba (Banvel)	Causes abnormal growth of roots and shoots by upsetting plant hormone balance. Dicamba is translocated readily after application to leaves or soil.

TABLE 8
HERBICIDES (CON'T.)

Herbicide family	Example materials common name (trade name)	Characteristics and uses
Benzoics (Con't.)		Chloramben usually applied as a preemergence treatment and taken up by seedlings at germination. Major uses are Chloramben on soybeans and Dicamba on corn.
Carbamates	Chlorpropham (Furloe) Propham (Betanol) Phenmedipham (Betanal) Barban (Carbyne)	Used on several crops, depending on the herbicide. Inhibits normal cell division. Applied as spray or granules with ground or air equipment. Taken up by roots or foliage with some translocation. Controls many annual grasses and broadleaf weeds.
Dinitroanilines	Benefin (Balan) Oryzalin (Surflan) Trifluralin (Treflan) Pendimethalin (Prowl) Fluchloralin (Basalin)	Most must be incorporated. Uptake through shoots of emerging seedlings is important but can also affect roots. Very little translocation. Material inhibits cell division and inhibits root growth. Controls grasses and certain broadleaf weeds, such as pigweed and lambsquarters. Usually applied as a spray but granular also available.
Diphenyl — ethers	Bifenox (Modown) Oxyfluorfen (Goal) Acifluorfen (Blazer)	Bifenox and Oxyfluorfen are used preemergence on soybeans or as directed sprays in certain crops. Acifluorfen is used postemergence in soybeans. Controls most annual broadleaf weeds. Very little translocation from either foliage or roots.
Phenols	Dinoseb (several)	Contact material, destroy cell membranes to kill by desiccation. Very little translocation.
Phenoxys	MCPA (several) Mecoprop (several) Silvex (several) 2,4-D (several) 2,4,5-T 2,4-DB (Butyrac)	Cause abnormal growth by changing plant hormone balance. Broadleaf weeds much more susceptible while grasses are tolerant. Rapid translocation in actively growing plants. Can be taken up by roots or foliage. Usually used for postemergence application. Application by ground or aerial equipment usually as a spray.
Thiocarbamates	Butylate (Sutan) CDEC (Vegadex) EPTC (Eptam, Eradicane) Metham (Vapam) Vernolate (Vernam)	Used on a variety of crops, depending on the herbicide. Volatile and must be incorporated. Alters distribution of plant hormones and inhibits cell division. Most effective on grass weeds.

TABLE 8 HERBICIDES (CON'T.)		
Herbicide family	**Example materials common name (trade name)**	**Characteristics and uses**
Thiocarbamates (Con't.)	Cycloate (Ro-Neet) Pebulate (Tillam) Diallate (Avadex)	Usually sprayed onto soil. Metham is a temporary soil fumigant.
Trianzines	Atrazine (several) Cyanazine (Bladex) Simazine (Princep) Propazine (Milogard) Metribuzin (Sencor/Lexone) Ametryn (Evik) Prometone (Pramitol) Prometryn (Caparol) Terbutryn (Igran)	Photosynthesis inhibitors causing chlorosis and death of leaf tissue. Materials can be taken up through roots or leaves. Moves from roots to leaves. Used on several crops, depending on the specific herbicide. Most effective on broadleaf weeds, but also certain grasses. Certain of these herbicides are persistent and may last in the soil from one season to the next. Most are either soil applied although certain of these herbicides may be applied postemergence.
Triazoles	Amitrole (Amitrole-T)	Used for general weed control, both broadleaf and grass, in noncropped areas and ornamental orchards. Used on some aquatic weeds. Usually applied as a spray with water. Slowly absorbed, rapidly translocated.
Uracils	Bromacil (Hyvar-X) Terbacil (Sinbar)	Controls a wide range of annual and perennial grasses and broadleaf weeds. Selective materials used to control weeds in citrus plantations. Applied to soil. Most effective when weeds are growing rapidly. Absorbed primarily through roots.
Ureas	Diuron (Karmex) Fluometuron (Cotoran or Lanex) Linuron (Lorox) Siduron (Tupersan) Tebuthiuron (Spike)	Used on a variety of crops and noncrop uses, depending on the herbicide. Controls a wide range of annual broadleaf and grasses, depending on the herbicide. Photosynthesis inhibitors causing chlorosis and death of leaf tissue. Usually applied to soil. Taken up by roots and translocated to leaves.
Unclassified	Chlorfluorenol (Maintain) Fenac (Fenac) Picloram (Tordon) Glyphosate (Roundup)	Characteristics depend on particular material.

Leaf Absorption

Leaves may absorb herbicides on both upper and lower surfaces, but entry is usually easier on the lower surface.

A wide variety of surfactants are used to improve penetration. Wetting agents, spreaders, and stickers are frequently used. Wetting agents reduce surface tension so that the herbicide can wet more of the plant surface and penetrate smaller openings. Spreaders help the herbicide spread over the surface and stickers help it stay on the leaf surface rather than runoff.

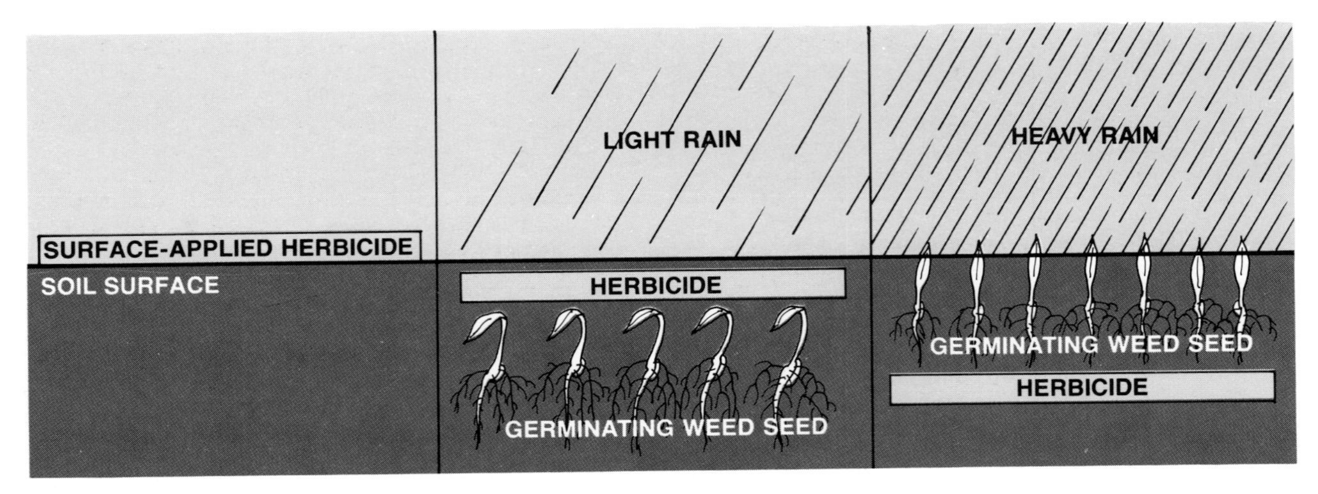

Fig. 31—The Amount of Rain Affects Depth of Herbicide Penetration

The **environment** affects foliar absorption. High relative humidity favors herbicide absorption. Warm, not excessively hot, temperatures also favor absorption, but high temperatures and low humidity reduce herbicide effectiveness.

Rain, soon after application, can wash spray deposits off the foliage. However, rate of entry into leaves is usually fairly rapid. Rain, several hours after application, may have little effect on absorption. However, rate of entry will vary with such factors as the specific herbicide, formulation, plant species, growing conditions, temperature, and humidity.

Light also affects herbicide penetration. Light actuates the photosynthetic processes to enhance herbicide movement through the plant away from the point of entry.

Action of Soil-Applied Herbicides

Placement of herbicide in the soil is extremely important in governing the effectiveness and selectivity. The top 6 inches of soil on an acre (15 cm of soil a hectare) weighs about 2 million pounds (5 million kg). The amount of herbicide incorporated into the soil may be only a few pounds per acre (kg/ha). The herbicide must be carefully concentrated in the "growth zone" of the weeds to be effective.

Most of the annual weeds to be controlled by a soil-applied herbicide will emerge from seeds that germinate in the upper two inches (5 cm) of soil. Although some herbicides may affect plant roots, the primary consideration for most common annual weeds and commonly used soil-applied herbicides is uptake of the herbicide by the emerging shoot. For adequate absorption of the herbicide by the shoot, moisture is generally required. This moisture may be from rainfall that moves the herbicide into the top inch or two (2 to 5 cm) of the soil. However, rain may not always come soon enough after herbicide application. Under dry conditions, the weed seedlings may emerge through

Fig. 32—Incorporating a Herbicide into Soil Kills Weeds as they Sprout

the herbicide on the dry soil surface without adequately absorbing it (Fig. 31).

Herbicide — Soil Interactions

Soil-applied herbicides are often affected by soil properties such as organic matter and clay content and sometimes by pH. Most herbicides are held to the surface of clay and organic matter. Generally, increasing organic matter and/or clay content reduces herbicide activity since tightly absorbed herbicides are not available for weed uptake. Often herbicide rates must be adjusted to match soil type, with higher rates needed on higher organic matter or clay soils. Crop injury due to excessive rates or poor weed control due to low rates can result if herbicide rates are not accurately matched to soil type.

Surface Application versus Incorporation

Soil-applied herbicides may be applied to the soil surface or incorporated. They are usually broadcast and incorporated into the soil prior to planting (Fig. 32). This is referred to as a preplant incorporated treatment (PPI).

It is sometimes indicated that rainfall is needed for "activation" of herbicides. Actually, the herbicide is quite active as it comes from the container and it is not essentially changed chemically or physically as the term "activation" falsely implies. The key is simply having moisture available from rainfall or present as soil moisture to enhance absorption of the herbicide by the emerging weed seedlings.

Incorporation of herbicides allows less dependence on rainfall and more consistent performance of the herbicide. This is because the herbicide can be mechanically incorporated into the soil where there is likely to be moisture available for absorption by the emerging weed seedlings. Generally, most herbicides can act on the shoots of the weed seedlings as they emerge through the top inch or two (2 to 5 cm) of the soil surface. There is usually no good reason to place the herbicide deep. It does not usually need to be in the root zone. Placing the herbicide deeper simply dilutes it, decreasing effectiveness.

Thus, the primary objective for incorporation is to mechanically move the herbicide into approximately the top two inches (5 cm) of soil where there is likely to be moisture available for absorption of the herbicide by the shoots of emerging weed seedlings. Since the top inch (2 cm) of soil can dry out relatively fast, incorporation into the top two inches (5 cm) is generally best. Also, it is difficult to limit incorporation to the top inch (2 cm) with most equipment.

Fig. 33—Field Cultivator for Herbicide Incorporation

Placement of Soil-Applied Herbicides

Surface-applied preemergence herbicides should be uniformly distributed to avoid skips and streaks of untreated soil where weeds will grow. Uniformity is also important to avoid overlaps and excessive amounts that may injure crops.

Distribution of an incorporated treatment will not be as uniform as a surface application, however you should aim for as much uniformity in distribution as possible. Many kinds of equipment are available for incorporating herbicides. Equipment already readily available such as disks and field cultivators (Fig. 33) have commonly been used. Generally, such equipment will incorporate the majority of the herbicide to about one-half the depth to which it is operated. Thus, if a disk or field cultivator is set to work the soil about four inches deep (10.2 cm), most of the herbicide will be placed in the upper two inches (5 cm).

Since only one-pass with a disk or field cultivator will streak the herbicide, a second pass should be made, preferably at a cross angle to first pass. Some type of harrowing device behind a disk or field cultivator can help further to achieve uniform distribution.

Many factors will affect incorporation: the moisture content and texture of the soil, amount of crop residue, type of equipment used, and speed. If soil is too wet or is fine textured it is more difficult to obtain good mixing action. Crop residues may interfere with good mixing. While a herbicide might be applied directly to soybean stubble and then incorporated, some prior tillage may be advisable for corn stubble. Smaller diameter and closer spacing of disk blades provides better mixing than large diameters and wide spacing. Wider sweeps and closer spacing of shanks on field cultivators is preferable to straight shanks and wide spacing. Speeds of about six miles per hour can help to improve mixing action for incorporation. Success will depend on the amount of mixing action and the degree of uniformity that results.

How Soon to Incorporate

Some herbicides can be lost from the soil surface by volatilization or photodecomposition. Most of the thiocarbonate herbicides are rather volatile and should be incorporated as they are applied in the same operation or very soon afterward. There is a little more flexibility with some of the dinitroanilines but incorporation within a few hours is generally best. For most of the triazines and amides, incorporation is usually optional and there is no urgency.

How soon you should incorporate will also depend on soil conditions and the type of day. If soil is warm and moist and it is a sunny, hot, windy day, the herbicide may be lost faster than if the soil is cool and dry and the day is cloudy, cool and calm. If a delay is unavoidable, the loss might be compensated for somewhat by using a higher rate of herbicide within the allowable limits. However, this increases cost. To obtain best results most economically, it is usually best to incorporate those herbicides that require it soon after application or in the same operation. The first pass moves the herbicide into the soil to avoid surface loss. There is usually no rush for the second pass and just before planting is generally a good time.

Time of Application

There is some flexibility in time of application for preplant herbicides. How long before planting the herbicide may be applied will depend on what herbicide is being used and its persistence. Most herbicides for preplant application should be applied within a week or two before planting. However, a more persistent one like trifluralin may be applied within a few months before planting. Refer to product labels for specifics on time of application and time of incorporation.

HERBICIDE FORMULATIONS

A wide variety of herbicide formulations have been developed for particular weed control problems. The active ingredients in the herbicide often dictate the most suitable type of formulation. Or, manufacturer may develop formulations for more convenient handling and mixing.

Typical formulations include:

- **Emulsifiable concentrates**
- **Solutions**
- **Soluble salts**
- **Wettable powders**
- **Water-dispersible liquids**
- **Water-dispersible granules**
- **Granules**

Emulsifiable Concentrates

Some herbicides are sold as emulsifiable concentrates which contain the active herbicide ingredient and an emulsifying agent. When the concentrate is mixed with water it usually forms a "milky-looking" emulsion. An emulsion is one liquid suspended in another. The emulsifying agent helps keep the two liquids in suspension. A stable emulsion does not need agitation.

Solutions

Soluble herbicides which mix with water are the easiest to handle in a sprayer. No agitation is required. When clean water is used, there are few nozzle-plugging problems. Both soluble liquids and soluble powders are used.

Soluble Salts

Soluble salts are dry materials that dissolve in water.

Wettable Powders

A wettable powder is a finely ground dry solid that can be suspended in water. Wettable powders are used when a herbicide is so insoluble that solutions and emulsions cannot be formed. Most wettable powders require constant agitation to stay in suspension.

Water-dispersible Liquid

This material is in a solid form but is suspended in a liquid. They pour easily from their container but still require agitation in the spray tank.

Water-dispersible Granules

Dry flowable or water dispersible granule formulations are made by forming wettable powders into small granules. This eliminates most dust and allows the material to be poured and measured almost like a liquid. The granules quickly break apart and go in suspension when added to water. Like wettable powders, dry flowables require constant agitation.

Flowable formulations or water-dispersible suspensions are made from wettable powders suspended in a petroleum or water-based liquid. Flowables may be more convenient to measure and pour than wettable powders. Flowable formulations require constant agitation.

Granules

Granular herbicide formulations are made by putting the active ingredients on an inert carrier such as clay. Granular herbicides are used primarily for preemergence soil-applied treatments and are not generally well adapted to postemergence foliar applications. Some advantages of granulars over sprays are:

- *No water is required*
- *Reduced drift*
- *Less retention on crop residue*
- *Convenient for band application*

Granular formulations are usually more expensive to manufacture and transport than liquids or wettable powders. Some of the more insoluble chemicals which do not move far in the soil do not perform as well in a granular form as in a spray.

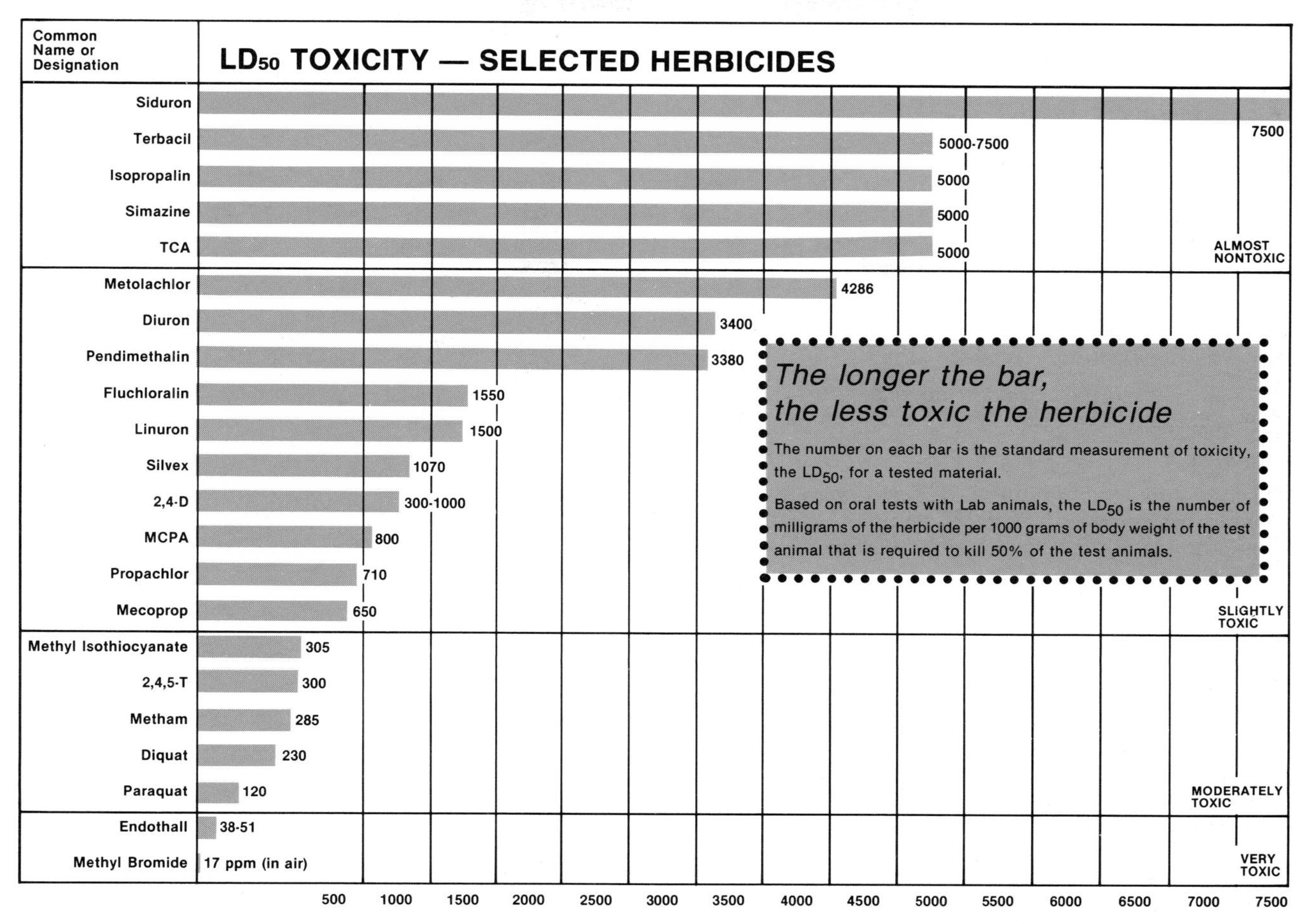

Fig. 34—Oral Toxicities of Various Herbicides

HERBICIDE COMBINATIONS

A herbicide will usually control some weed species better than other species. For example, the dinitroanilines, amides and thiocarbonates generally control annual grass weeds but are not very effective on most broadleaves. Some of the triazines generally give better control of broadleaf weeds but are weak on grass. Thus, by combining herbicides, the spectrum of weed control can be broadened — more different weed species can be controlled.

Using herbicide combinations also may allow use of reduced rates of certain herbicides to reduce risk of injury to the present crop. A reduced rate may also help to avoid injury to subsequent crops sensitive to herbicide residue.

Some herbicide combinations may be purchased as prepackaged mixtures. Some combinations may be tank-mixed by the applicator. Use only those combinations that have been tested and approved. Unapproved concoctions may result in crop injury, poor weed control, tank mixing problems or illegal use.

HERBICIDE HAZARDS

In general, when compared to some other pesticides, herbicides are not as dangerous. However, it must be emphasized that some can be quite hazardous and must be used with proper precautions.

Figure 34 gives the oral toxicities of many herbicides. More complete discussions of toxicities and pesticide hazards are given in Chapters 5 and 9.

ADJUVANTS

Adjuvants are materials added to a herbicide or other pesticide to improve performance. Such substances may increase toxicity, assist emulsification, improve spreading, promote retention on the leaf or increase penetration. The term surfactant is also used to include some of the following.

Wetting Agents

Waxy leaf surfaces tend to repel water and herbicide. To be effective herbicides must be brought into intimate contact with the plant. A wetting agent mixed with the herbicide helps the liquid spread over more of the leaf surface and promotes more effective action.

Spreader-Sticker Agents

Spreader-stickers reduce surface tension, increase adherence of spray to leaves and reduce bounce-off or runoff of spray droplets during spraying.

Emulsifiers

Emulsifying agents are used to maintain the stability of an emulsion.

Anticaking Agents

Anticaking materials prevent solid or granular herbicide products from forming hard aggregates.

Activators

Certain chemicals, though not toxic to plants, can increase the toxicity or effectiveness of a herbicide. Some adjuvants may also be activators.

Crop Oils

Certain oils can be used to improve performance of postemergence herbicides. Non-phytotoxic oils have various amounts of surfactant added. Those with 80 percent oil and 20 percent surfactant are referred to as crop oil concentrates.

Compatibility Agents

Compatibility agents are sometimes added to mixtures of agricultural chemicals used as herbicides mixed with fertilizer solutions. They help solve mixing problems.

SELECTING A HERBICIDE

How do you select a herbicide? What features make one material better for certain conditions than some other material? Consider these points.

Effectiveness

The herbicide selected should be effective against the weeds present. To determine which herbicides are effective, consult state weed control guides, farm publications, extension service personnel, chemical salesmen and neighbors.

Registration

Use a material which is registered for the specific crop to be treated. It is illegal to use herbicides not registered for the intended use. For instance, a herbicide might be approved for use on corn grown for grain, but not for silage. Registration regulations will be discussed further in Chapter 9.

Availability

Select a material which is available. To ensure availability, place orders early.

Cost

In addition to cost, consider crop tolerance, record of performance, suitability for the weeds to be controlled, and convenience. The least costly herbicide may not be the most economical.

Safe for Crop

Make sure that the materials selected are safe for the crops on which they will be used. Time and method of application must also be considered.

Soil Type

Select soil herbicides that can be used safely and effectively on your soil type. Some herbicides cannot be used on sandy or low organic matter soils. Others may be ineffective on very high organic matter soils.

Subsequent Operations

Determine whether any toxic residues may remain which could affect people entering the field for subsequent operations or activities. Consider the residue carryover potential of the herbicide. Certain herbicides can persist in the soil from one season to the next to harm sensitive rotational crops.

Toxicity

Herbicides are toxic chemicals (See: FMO-FARM MACHINERY SAFETY). If you have a choice, select herbicides which present the least risk of irritation or toxicity to the user, animals in the area, or other people who might come in contact with the product during or after application.

Application Method

Select a material with a formulation matched to available equipment, or choose equipment capable of handling the herbicide to be used. For example, do not purchase a granular herbicide if only spray application equipment is available.

Environment

Consider the environment. Select materials with low volatility which are less likely to drift. Persistent herbicides remain in the soil longer and are more likely to be washed into a stream or lake. Avoid herbicides or application methods that present significant danger to nearby desirable plants, people, animals or subsequent crops. Most herbicides are not highly toxic to humans or animals, but appropriate care must be taken in selection, storage and application.

Herbicides can improve the environment by controlling undesirable weeds which adversely affect the life, health or well-being of humans, livestock or wildlife. For instance, controlling such plants as ragweed can reduce the irritation of allergies, asthma and hay fever; control of poisonous plants can reduce livestock losses, possible death or injury to small children or the discomfort of poison ivy exposure.

WEED CONTROL PROGRAM

Prepare a complete weed control program before ap-

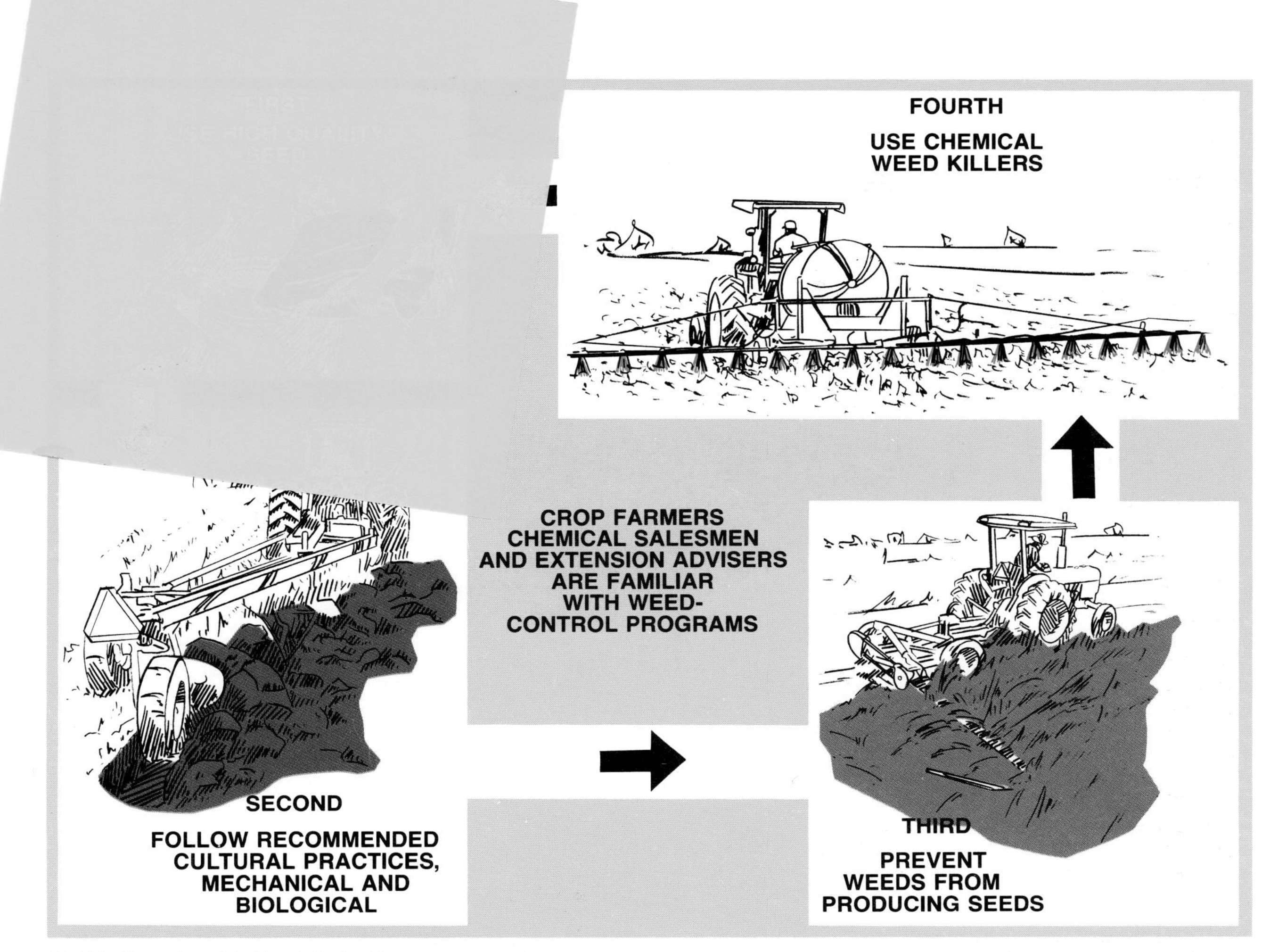

Fig. 35—Plan a Complete Weed Control Program

plying any herbicides. Consider control methods for all areas, not just in row crops. Weeds should be controlled in small grain, forage crops, pastures, fence rows, ditch banks, roads and driveways, around buildings and in other non-crop land to prevent production and spread of weed seed to uninfested areas. For economy and convenience, plan to use eradication measures in only limited situations.

Knowledge is the key to an effective weed control program. Know how to recognize weeds and how and when to attack them for most effective control.

Become familiar with herbicides and the weeds each material will and will not control. Stay informed about new materials and changes in registration of existing materials.

Consider all available weed control methods and use those methods most effective and economical for each weed infestation or crop (Fig. 35). Tillage and other cultural practices provide effective weed control when properly integrated into a weed control program. Competitive vegetation such as desirable perennial grasses can provide very effective and economical weed control in areas such as fence rows, ditch banks and around buildings after initial weed control has been obtained and desirable plants established.

Promote cooperation among neighbors to solve common weed problems. Very little is gained in attempting to control weeds which are immediately reseeded from adjoining land.

The final step is to carry out the plan! But be prepared to change weed control activities from time to time. As cropping methods are changed or control of some species is achieved, it is likely that different weed species will increase.

WEED CONTROL FOR CONSERVATION TILLAGE PROGRAMS

When tillage is reduced, weed control can be more of a challenge and usually greater reliance is placed on herbicides. Today's variety of tillage equipment and arsenal of herbicides offers many opportunities for conservation tillage, but careful planning, good management, frequent observation, quick decisions and timely operations are essential to success.

Zero-tillage can be practiced by using herbicides such as paraquat or glyphosate to control existing vegetation and also using a residual preemergence herbicide to control annual weeds that will be emerging. Paraquat is primarily a contact herbicide and "burns" the tops of plants. Deep rooted perennials such as alfalfa may regrow from the roots if only paraquat is used. For such perennials, a translocated herbicide such a glyphosate, 2,4-D or dicamba may be needed. In some grass sod with shallow roots, paraquat plus atrazine may be adequate.

We have been using crop rotations for many years. More recently, herbicide rotations have been promoted as a means of controlling more different weed species — using different herbicides to achieve broader spectrum control. It is also feasible to consider tillage rotations. Use more tillage for some parts of the cropping sequence and less tillage at other times. For example, in parts of the midwest, acreages of corn and soybeans are about equal. Following soybeans, there is little crop residue and tillage can be reduced considerably. However, following corn there is considerable crop residues and more tillage may be convenient and practical.

Preplant incorporated herbicides can be used rather easily on soybean stubble. However, where there is corn stubble, either more tillage may be needed prior to herbicide application or greater reliance will be needed on surface-applied or postemergence treatments. Each grower should determine the tillage and weed control program that best fits his farm and select equipment and develop a weed control program accordingly.

DOUBLE CROPPING

In some areas there is considerable opportunity for growing two crops in one season. For example, winter wheat can be planted in the fall and harvested in late June or early July. Soybeans can then be planted immediately to produce a second crop. This procedure takes advantage of the available nutrients, moisture and light that might otherwise "go to waste" for much of the growing season.

Since it is important to plant the second crop as soon as possible, no-till planters or drills are frequently used in the small grain stubble. A combination of herbicides is generally used to control existing vegetation and to also provide residual control of weeds that would be expected to emerge.

SUMMARY

Weeds cost money. They reduce yield and quality of crops and have many other characteristics that make them generally undesirable.

Weeds persist because they have well developed survival mechanisms. The most important are: prolific seed production, and the ability of seeds to stay dormant in the soil for years.

Prevention of weeds is the best policy, but weeds are difficult to prevent. As a result, control programs are necessary. Weeds can be controlled by three general methods:

- **Mechanical and cultural**
- **Biological**
- **Chemical**

Herbicides are selective or non-selective. They can be contact or systemic in action. Herbicides can be applied before a crop is planted, before it emerges or after it emerges.

Application patterns include: banding, broadcast, spot treatment and directed spray.

Most herbicides are applied as sprays or granules. Adjuvants are sometimes added to herbicides to improve their action.

Select a herbicide suited to each particular situation. Consider all aspects: crop, weeds, soils, available herbicides, and equipment, and seek help from competent sources.

Plan weed control programs carefully to include all appropriate control methods. Cooperate with neighbors to combat common weed problems.

CHAPTER QUIZ

1. What is a weed?
2. What are the differences between annual, biennial and perennial weeds?
3. List three factors which allow weeds to be troublesome year after year.
4. What are three methods of weed control? Give two examples of each.
5. Why isn't biological control used more widely?
6. List advantages of granules over spray formulations.
7. What are "parasitic" weeds?
8. Give five common ways that weeds can spread.
9. What basic technique is used to control annual weeds?
10. Give types of herbicide action.
11. List the three major times for herbicide application.
12. List three kinds of application.
13. List and give the use of three adjuvants.

4
Disease Control

Fig. 1—Plant Diseases must be Controlled to Grow Healthy Crops

INTRODUCTION

Plant diseases and microscopic worms called nematodes are among the greatest hazards in man's continuing struggle to produce food. Many diseases can spread explosively (Fig. 2) and cause wholesale destruction of crops. Because of the heavy damage they can inflict, diseases often limit the kinds or varieties of crops that can be produced in a particular region.

THE CONCEPT OF PLANT DISEASE

What is a disease? Diseases are different than other plant enemies like insects and weeds and are not as easy to define.

Disease affects the plant's physiological processes directly. More specifically, disease is a change in one or more of the physiological processes that enable plants to utilize energy. Usually, it causes a progressive and continuous disturbance of cellular activities which is eventually manifested as "symptoms" (Fig. 3).

Disease is any extreme variation in the metabolism of the plant. It can be caused by virtually any environmental factor—biological, physical, or chemical. The most common disease agents are biological.

Pathogens

The disease-producing agent is called a **pathogen.** There are three important groups of biological pathogens.

- **Fungi**
- **Bacteria**
- **Viruses**

A brief description of the main characteristics of these groups and sub-groups will help provide an understanding of the way diseases develop and are transmitted.

FUNGI AS PLANT PATHOGENS

Fungi are plants without chlorophyll. Therefore, they cannot use light energy to convert carbon dioxide and water into carbohydrates. They must depend on green plants to manufacture such substances (Fig. 4).

Saprophytic fungi live on dead or decaying organic matter. They are important because they eventually release the nutrients they take up from dead plants and animals and thus contribute to soil fertility. Saprophytic species can rot wood and discolor walls. They do a considerable amount of damage to buildings which are improperly constructed or ventilated. Some species,such as mushrooms and truffles, are edible and are of considerable economic importance.

About 7500 parasitic fungi are of economic importance because of the damage and yield reduction in affected crops. The four major groups are: (Table 1):

- *Phycomycetes (lower fungi similar to algae)*
- *Ascomycetes (yeasts, molds)*
- *Basidiomycetes (rusts, smuts, mushrooms, puffballs)*
- *Deuteromycetes (oak wilt, powdery mildew of peaches)*

Some fungus pathogens attack many plant species, but others may be restricted to only one. For example, one destructive oak root fungus attacks over 600 different species of plants. Conversely, the fungus that causes fusarium wilt of tomatoes is unable to attack peas or watermelons, although other forms of the same fungus can attack these and other plants.

Spores

Most fungi reproduce by means of microscopic spores. Fungus spores vary greatly both in form and action while infecting a plant. Some can survive for weeks or months without a host plant.

Fungus spores are produced in tremendous numbers (Fig. 5). Most of them do not find a host plant to attack.

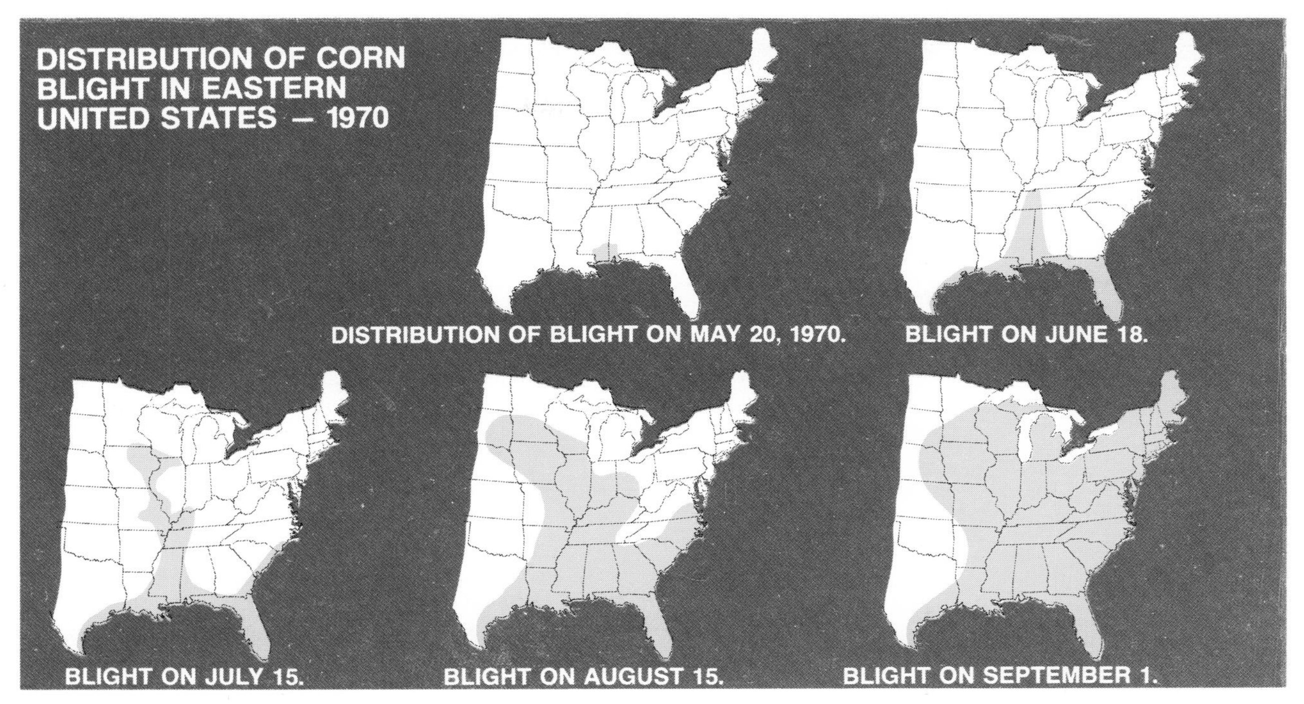

Fig. 2—Spread of Southern Corn Leaf Blight in 1970

Fig. 3—Symptoms of Common Soybean Mosaic—Normal Leaf at Bottom

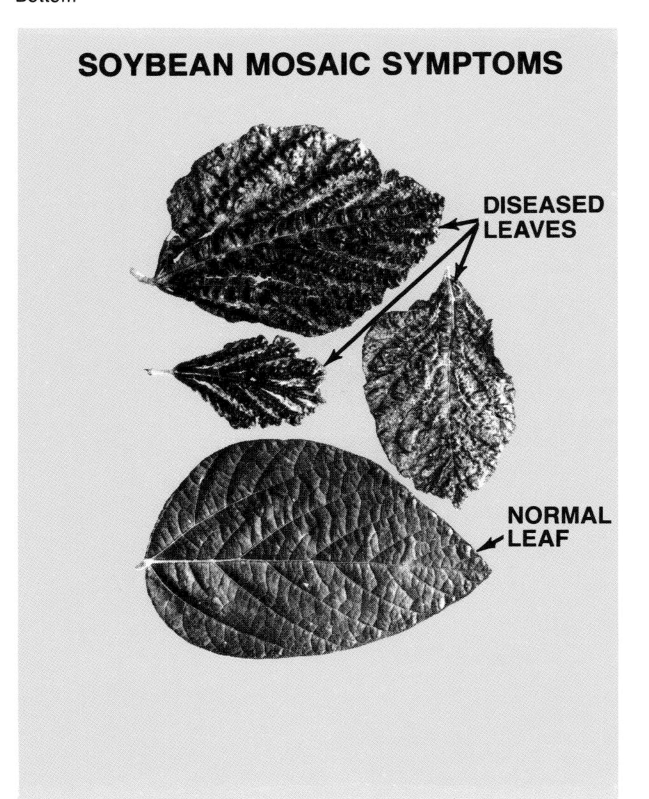

Fig. 4—Two Fungal Diseases of Soybeans

TABLE 1

DISEASES CAUSED BY FUNGAL INFECTIONS

Phycomycetes	Ascomycetes	Basidomycetes
Downey Mildews	Chestnut Blight	Black Stem Rust of Wheat
Canker Disease	Gibberella	Smut on Grasses
Late Blight of Potatoes	Powdery Mildews of Apples and Grains	Dry Rot
Late Blight of Tomatoes	Brown Rot of Stone Fruits	
	Black Rot of Sweet Potato	
	Peach Canker	

Free water or very high humidity is nearly always essential for active growth of a fungus. Often 96 to 99 percent relative humidity is needed for the spores to germinate.

BACTERIA AS PLANT PATHOGENS

In comparison to fungi, there are relatively few (about 200) major plant diseases caused by bacteria. However, a few bacterial diseases, such as fire blight of apples and pears, have been highly destructive (Fig. 6). In recent years, halo blight of beans has been an important disease. A major portion of the decay and spoilage of fresh fruits and vegetables that takes place after harvest is attributable to soft rot, which is caused by a bacterium.

Fig. 5—Brown Rot on Peaches (The Mass of Spores on the Fruit May Spread the Disease)

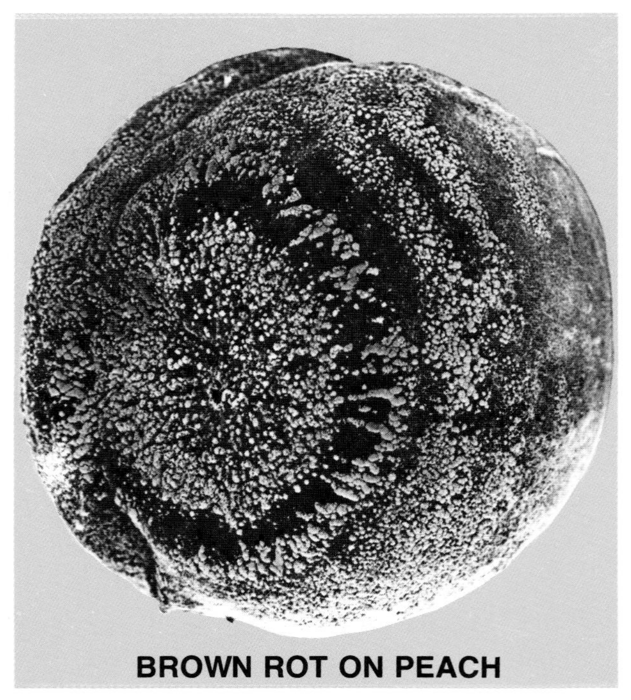

Most higher plants can be attacked by bacterial plant pathogens. Certain bacteria live in the soil and affect a wide range of plant species. The soft rot bacterium is an example. Others, such as the organism that causes bacterial pustule of soybeans (Fig. 7) appear to affect only one host species. Several bacteria cause serious losses in orchards and nurseries.

Types of Bacterial Diseases

The symptoms produced in plants by bacterial diseases are similar to those produced by other pathogens (Table 2). Each pathogen usually produces a definite pattern of symptoms in the host plant.

One major symptom is usually produced when a pathogen infects a particular plant, but there may be a combination of symptoms, known as a syndrome.

Spreading Bacterial Diseases

Plant-infecting bacteria are not spore formers and cannot remain alive very long away from a suitable host. But they can survive longer in water and are mainly dependent upon wind-driven rain for spreading from plant to plant. Other means of spreading bacterial diseases include:

- *Dust storms*
- *Irrigation water (Fig. 8)*
- *Machinery during cultivation and harvest (Fig. 9)*
- *Insects*

VIRUSES AS PLANT PATHOGENS

Viruses are sub-microscopic entities that multiply and survive only inside living cells. Few plants and animals are immune from viruses.

Fig. 6—Fire Blight of Apples

Fig. 7—Bacterial Diseases of Soybeans

TABLE 2

BACTERIAL DISEASE SYMPTOMS

Symptom	Example Disease
Yellowing	Bacterial Wilt of Alfalfa
Wilting	Southern Bacterial Wilt of Tomatoes, Potatoes and Tobacco Bacterial Wilt of cucumbers, melons, squash, and pumpkins.
Blight	Fire Blight of Apples and Pears Wildfire of Tobacco Bacterial Leaf Blight of Rice
Spot	Bacterial Spot of Stone Fruits Angular Leaf Spot of Cotton
Canker	Fire Blight Bacterial Canker of Stone Fruit
Rot	Soft Rot of Vegetables
Gall	Crown Gall of Ornamentals Leafy Gall
Scab	Potato Scab

Fig. 8—Irrigation Can Spread Bacterial Diseases

Several viruses have caused widespread damage to certain crops (Fig. 10). For example, curly top has caused extensive damage to sugar beets and still damages tomatoes, beans and other crops in the western part of the U.S.

Virus Diseases and Symptoms

Viruses can induce a wide variety of symptoms in host plants. Some, such as spotted wilt or curly top of tomatoes, can kill the plant. More commonly, the result is lowered product quality and reduced yields. There are two clearly defined groups of symptoms:

- *Mosaics cause mottling and chlorotic spotting of leaves*
- *Yellows which cause yellowing, leaf curling, dwarfing or excessive branching.*

Plants can be infected with two or more viruses simultaneously. When a plant infected by one virus is attacked by another, the symptoms may remain the

Fig. 9—Machinery May Spread a Bacterial Disease Throughout a Field

Fig. 10—Bud Blight of Soybeans Caused by the Tobacco Ringspot Virus

same, or become much more severe. The symptoms of one virus may predominate, or they may act together to produce a more severe disease than either virus produces alone. For example, when tomatoes are infected with tobacco mosaic virus or potato virus X alone, the result is mottling and some stunting. However, when both diseases attack the same plant, a marked increase in injury, characterized by the extensive killing of leaves and stems results.

Infection by more than one virus is unusual. Often, infection by one virus protects against invasion and injury from similar strains of the same or different viruses.

Virus Transmission

Any virus that spreads systemically through a host plant can be transmitted by grafting. Most viruses invade all parts of the host plant. Therefore when buds, or other vegetative parts of diseased plants are used to propagate new plants, the new plants will also be infected. This is a big problem for crops like strawberries, tree fruits, grapes, dahlias, chrysanthemums and other crops that are propagated vegetatively.

Some viruses are transmitted in infected seed, but only a small percentage of the seed from an infected plant will usually carry the virus.

The most important natural means of spreading viruses is by insect movements. Depending on the virus, the carriers may be leafhoppers, aphids, thrips, beetles, or other chewing insects (Fig. 11). The insect acquires the virus by feeding on an infected plant, then carries it to healthy plants. In general, aphids transmit mosaic-type diseases while leafhoppers carry viruses that produce "yellow"-type diseases.

NEMATODES AS PLANT PATHOGENS

More than 15000 species of nematodes have been identified, but most experts feel that presently unknown species far outnumber the known. Symptoms of nematode damage to plants are difficult to identify. The most important injury is done to the roots. The symptoms shown in aboveground parts of the plant are the same as would result from any factor which deprives the plant of a properly functioning root system. It is often difficult to recognize a nematode-damaged plant without a thorough root examination by a competent nematologist.

Nematodes are tiny, eel-like animals. They have specialized organs including both digestive and nervous systems. Nematodes move with a snake-like motion in water, even as thin as the film of moisture around soil particles or plant cells. The mouth parts of nematodes that attack plants include a spear capable of puncturing plant cells. Saliva is injected into the cell through the spear. The cell contents, after being partly digested by the saliva, are drawn back through the spear and ingested.

Nematode Damage

Nematodes can be divided into two major groups which feed on plant parts both above and below ground:

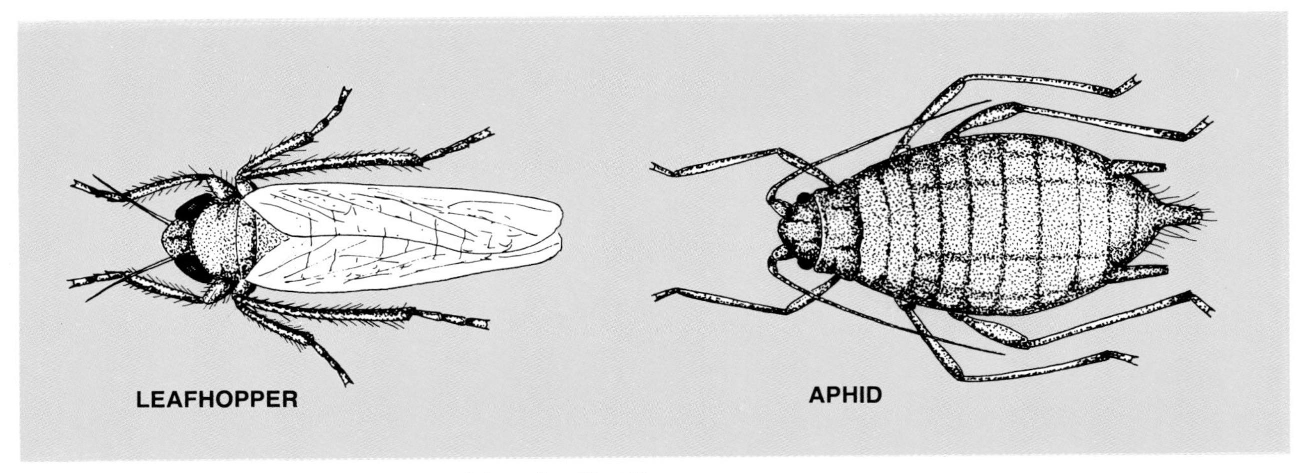

Fig. 11—Insects are the Most Common Means of Spreading Plant Viruses

- *Endoparasites–enter into and feed inside plant tissues*
- *Ectoparasites–remain outside the plant and feed on surface cells.*

Most nematodes that feed on underground plant parts attack small roots, although tubers, corms and similar organs may also be affected. Root feeding results in destruction of roots and impaired uptake of water and minerals.

Root-knot nematodes cause important plant diseases in warm climates and in greenhouses. The saliva which is injected into the plant cells causes formation of swollen and distorted roots (Fig. 12).

Cyst-forming nematodes (Figs. 13 and 14) cause infected cells to enlarge, although not as much as the root knot group. At death, the outer covering of the female cysts becomes tough and hardened. The covering, which is resistant to chemicals and other microorganisms, provides protection in the soil for several hundred eggs for as long as 10 years, increasing problem in nematode control and crop rotation.

Fig. 12—Gnarled Galls on Soybean Roots Caused by Root Knot Nematode

Other nematodes cause damage, including:

- *Root lesions*
- *Root elongation*
- *Swollen root tips*
- *Reduced or "bushy" root growth*

Fig. 13—Soybean Cyst Nematodes Can Stunt or Kill Soybean Plants

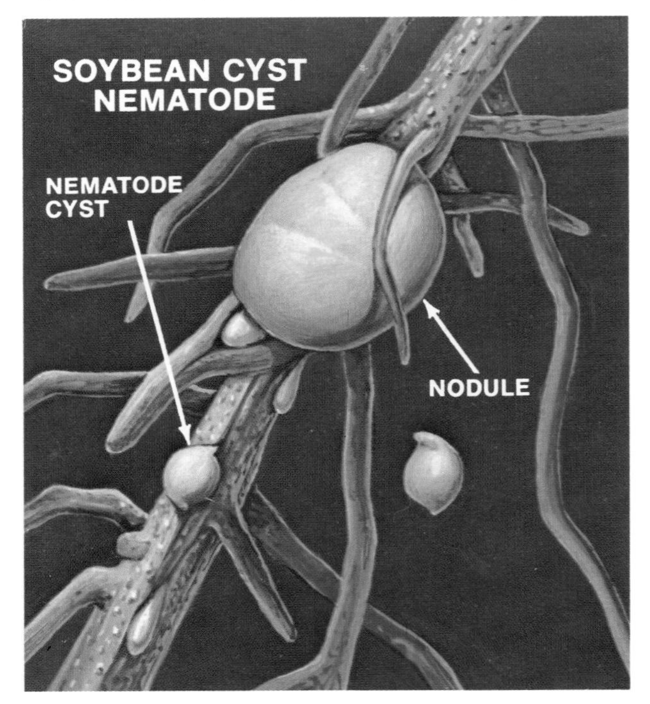

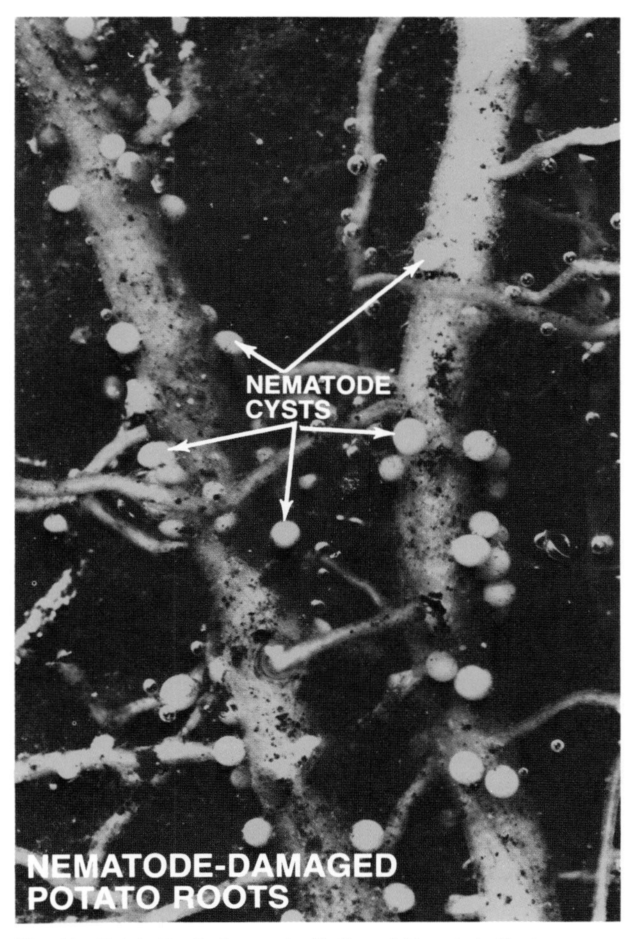

Fig. 14—Potato Roots Covered with Round Nematode Cysts

- *Crinkled and distorted stems or bulbs*
- *Leaf spots*

Certain nematodes are known to transmit viruses. Also, nematode attacks can increase the chances of a plant being invaded by fungi and bacteria that cause root diseases, such as fusarium wilt of cotton.

DIAGNOSIS OF PLANT DISEASES

Attempts to control plant diseases without sufficient information usually results in unsatisfactory control, and often in failure. For maximum effectiveness, the first step is to provide positive and correct diagnosis.

Diseased plants are recognized by comparison with healthy plants (Fig. 15). Thus knowledge of normal growth habits is necessary for recognition of a diseased condition. Under normal conditions, completely healthy plants are rare. The term "healthy" refers to plants that do not have obvious disease symptoms. Plant diseases of major economic importance usually

Fig. 15—Diseased and Healthy Plants

cause obvious damage and affect enough plants in a field (Fig. 16) to reduce the yield or lower the quality of the harvested product.

When attempting to identify a disease, weather, soil type and fertility, cultural practices and other environmental factors that affect both the pathogen and the host, should be considered important factors in determining disease severity.

GENERAL APPROACHES TO DISEASE CONTROL

Satisfactory control of most plant diseases requires the application of several control measures and usually involves an integrated program of environmental, biological and chemical factors.

As mentioned earlier, correct diagnosis is essential to plant disease control. Then, control involves the application of one or more of the following principles:

- *Avoidance–avoiding disease by planting when, and/or where, pathogens are ineffective or absent*
- *Exclusion–keeping pathogens out of a "clean" area*
- *Eradication–eliminating the pathogen source, whether an infected plant, field or region*
- *Protection–preventing an infection by using a chemical or physical barrier to keep pathogens out*
- *Resistance–using plants that tolerate, resist or are immune to the disease*
- *Therapy–reducing the severity of disease in an infected plant*

The first five of the principles are methods of disease prevention. The sixth represents the cure. At present, plant disease control measures are predominantly preventive.

Fig. 16—Infrared Photos can Detect Many Crop Diseases or Stressed Plants

From a farmer's viewpoint, the general approaches to disease control are reduced to three practical methods or systems:

- **Cultural and biological control**
- **Control through disease resistance**
- **Chemical control**

CULTURAL AND BIOLOGICAL DISEASE CONTROL

Crop management to minimize development of disease is the oldest and most generally applicable approach. A pathogen and its host must be brought together under proper environmental conditions for a disease to develop. Cultural practices are used to alter the environment, the condition of the host or the behavior of the pathogen to prevent an infection.

PATHOGEN-FREE PROPAGATION

Pathogens can frequently be carried in or on seed and other propagating materials. For many crops, it is essential to use pathogen-free planting stock. There are several techniques used to produce planting materials free of fungi, bacteria, viruses and nematodes.

Fig. 17—Furrow Irrigation Helps Reduce Spread of "Wet Weather" Diseases When Producing Seed Crops

Arid Sites

Wet weather diseases, such as anthracnose of beans and bacterial blights of legumes are characteristically seed-borne. They can be avoided if seed crops are produced in dry areas where only furrow or drip irrigation is used (Fig. 17).

Disadvantages Of Overhead Irrigation

Free moisture is necessary for the formation and release of spores from fungus pathogens such as those that cause anthracnose. It is also needed for dissemination of bacteria which cause foliage blights or fruit rots, and for the movement of nematodes. Arid conditions discourage both the development of the pathogens and seed contamination. Foliar diseases, such as black rot and blackleg of crucifers (plants such as cabbage and mustard) and bacterial blights of beans are usually of little importance when the crop is produced with furrow irrigation. However, overhead irrigation leaves free water on the foliage and these diseases have appeared with destructive intensity when sprinkler irrigation was used.

Isolation Culture

"Cultured" cuttings of chrysanthemums, carnation and geraniums must be grown in isolation if they are to be maintained for production of clean planting stock. This can be done by growing the materials in pathogen-free greenhouses (Fig. 18) or by using sites which are isolated from growing areas for these crops.

When planning for isolations, consider how far the particular pathogen may spread, how it spreads, and the distance between potential growing sites. Where alternate hosts are involved, as with cedar-apple rust of apples and junipers, isolation can be achieved by eradicating the alternate host.

Fig. 18—Greenhouses Can Be Used for Isolation Culture

Disinfection And Heat Therapy

Hot water treatments are effective in producing clean seed and planting materials: Seed and vegetative propagating materials (roots, bulbs, tubers, corms) may be treated before planting to eliminate such diseases as bacterial blight organisms in bean, pea, cabbage and similar seeds, fusarium infections in gladiolus corms and viruses in tree fruits.

Seed Aging

Some seed pathogens can be deactivated by holding the seed in storage. Seed-borne anthracnose of cotton and late blight of celery are examples. Proper storage conditions are essential to ensure that seed viability is not also lowered.

Early Harvest

Delaying the harvest of seed crops extends the possible exposure period of the seed to pathogens. The

Fig. 19—Early Harvest Reduces Disease Invasion

pathogens may build up from a local source or blow in from the outside. Mature heads of grain crops are subject to invasion by fungus pathogens, for example. Harvest as soon as possible to decrease the likelihood of exposure (Fig. 19).

CROP CULTURE

Pathogenic organisms nearly always exist in the soil or in plant debris. However, a cultivated field will not have every kind of pathogen. Each field will probably have pathogens from recent crops, but may be free from specific pathogens of other crops.

In modern greenhouse production, it is unusual to find extensive disease outbreaks where the soil is sterilized before planting and temperature and humidity are controlled. Sterilizing methods include:

- *Soil pasteurization with heat (Fig. 20)*
- *Fumigation or chemical sterilization*

Most elements of a plant is environment, including temperature, light, humidity, carbon dioxide, oxygen and nutrition, can be controlled in a greenhouse. Most pathogens cannot build up rapidly enough under these conditions to damage the crop.

There is much less opportunity to control the environment in the field. Also, grower resistance to changing cultural practices can be a major hindrance to effective disease control. The grower should be flexible and use those practices that are beneficial. Correct diagnosis of diseases present in the field will help determine cultural practices for the successful grower.

Monocultures

Continual production of the same or a closely related crop on the same piece of land can contribute to disease buildups. Maize dwarf mosiac is found where Johnsongrass is a major weed (Fig. 21). Root rot of peas is common where peas are grown on the same land year after year.

Fig. 20—Heat is Frequently Used to Sterilize Greenhouse Soil

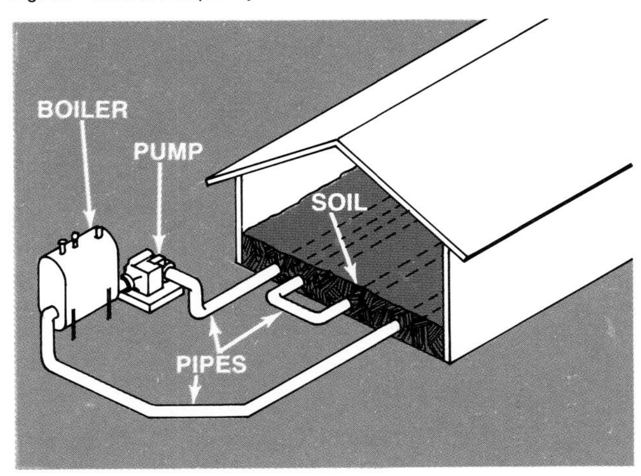

Fig. 21—Maize Dwarf Mosaic of Corn

Planting Time

Cool weather crops, such as spinach and peas, are subject to attack by certain diseases if planted too late and often emerge poorly under such conditions. Conversely, beans and melons should not be planted early if you are to avoid disease. Curly top of sugarbeets has been controlled by adjusting planting time to avoid migration of leafhoppers that transmit the curly top virus.

Plant Spacing

Close plant spacing can be an advantage or disadvantage for disease control, depending on the crop and the pathogen.

For example, tomato transplants produced in western coastal areas may include individual plants carrying the spotted wilt virus. But, when planted inland, if the stand is dense and conditions do not favor spread of the virus, a good crop can still result even though the originally-infected plants die. However, some diseases, such as botrytis blight, leaf mold and other foliar blights are spread more rapidly in dense plantings where moisture remains high for longer periods of time.

Plant spacing also affects weed infestation and nutrient uptake (Chapters 2 and 3), which, in turn, may affect various diseases.

Crop Cultivation

Wet-weather pathogens are often spread by cultivators, sprayers or other equipment moving through the field while foliage is wet. Droplets of water or pieces of plants are picked up by the machinery and transferred to nearby plants. Bacterial and fungal pathogens are spread at the same time. This process spreads tobacco mosaic, late blight of celery, bacterial blight of peas and beans and similar diseases.

Excessive Irrigation

Excessive overhead watering or heavy rains lead to water-soaked soil and suffocated roots, because oxygen is displaced. Anaerobic toxins are formed which damage the roots and allow entry of pathogens. Many phycomycetes (fungi similar to algae) are "water molds" and require overly wet soils for root infection because of spore mobility in water.

Growing plants on raised beds (Fig. 22) improves drainage and reduces infection from "water molds." Subsoiling to improve drainage has a similar effect on some soils. However, subsoiling has to be repeated periodically because the slots cut in the soil eventually seal up. The time required for sealing to occur depends on the soil characteristics.

Crop Rotation

Crop rotation usually helps prevent buildup of pathogens in the soil, particularly pathogens that survive in crop debris. Not all cropping sequences, however, are beneficial. In fact, some may actually increase disease.

Crop rotation seldom provides complete control. Usually additional methods are needed. Some examples of crop rotation that have been successful in disease control are:

- *Wheat-oats rotation against the Ophiobolus disease or take-all of wheat*
- *Soybean-potato rotation against potato scab*
- *Cotton-field pea rotation against root rot of cotton*
- *Bean-barley rotation against Fusarium, and Rhizoctonia root rots of beans.*

Fig. 22—Many Crops are Grown on Raised Beds

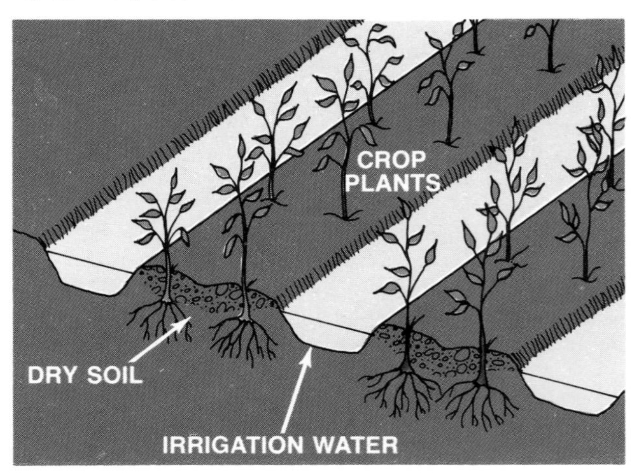

SANITATION

The grower who plants disease-free seed, and who avoids farming land infested with a damaging pathogen, should also practice crop sanitation.

Crop Residues

Infected crop residues often provide an ideal environment for winter or summer carry-over of many pathogens. In some cases, the pathogens increase greatly in the residues. There are three primary crop-residue management methods.

- *Deep plowing buries pathogen infested residues and surface soil and replaces it with soil relatively free from pathogens.*
- *Burning kills some pathogens and removes the residue that they live on (Fig. 23). This practice is no longer common in most areas.*
- *Summer or winter fallowing reduces pathogen carryover as residue decays.*

Removing Infected Plants

Systematic removal of diseased plants and plant parts has often been used to prevent disease spread, particularly virus diseases. Diseased nursery stock is discarded at the grading table. Trees are pruned to remove the source of infection before it spreads to other parts of the same tree (Fig. 24), or to adjacent trees.

Fig. 23—Burning Crop Residues

Vectors

If a pathogen is carried by an insect, or by some other means, the transmitting agent is termed a "vector". If the vector can be killed, or controlled before infection occurs, then spread of the disease can be stopped. But not all vectors can be controlled.

However, disease-free plants can sometimes be grown by timing production so that the plant is produced when vector population is low.

DISEASE RESISTANCE

Use of disease-resistant varieties (Fig. 25) is usually the most effective, long-lasting and economical means of controlling plant disease if those resistant varieties are otherwise acceptable. Resistant varieties have been one of the major factors in maintaining high levels of crop productivity in the United States. For certain crops, such as alfalfa, wheat, oats and flax, 95 to 98 percent of the acreage grown is planted with varieties that are resistant to diseases.

For many diseases of forage and field crops with relatively low values per acre, chemical controls often cannot be used because the profit margin is too low. No economical or effective control method is available for many soil-borne pathogens. For these cases, resistant varieties provide the only means of insuring continued production.

Sources of Resistance

Resistance to a specific pathogen usually occurs in plants obtained from areas where the pathogen occurs naturally. In some cases, highly resistant breeding material is found in areas where the disease does not occur, because the plants are resistant.

There are several procedures for locating resistant strains:

Fig. 24—Remove Diseased Plant Parts

- *Selection of resistant individuals from a large population subjected to heavy infection.*
- *Crossing varieties (lines) with resistance with varieties (lines) having high quality and yield, then selecting individuals with desired characteristics through a series of back crosses.*
- *Budding and grafting as a means of maintaining resistant stock.*

Fig. 25—Disease Resistant Soybeans, Right - Susceptible Variety, Left

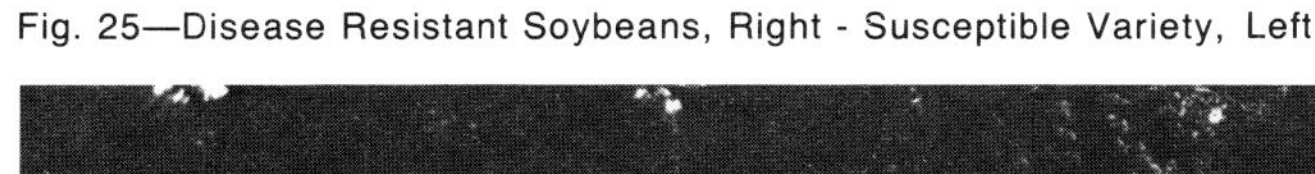

CHEMICAL CONTROL

The term **fungicide** refers to those chemicals used to control fungi and bacteria on living and non-living plants and plant parts. Fungicides may generally be classified as **protectants** or **eradicants**. Protectants are usually applied as dusts or sprays before the organism appears and kills or inhibits their growth. For protectants to be effective, they must either persist or be maintained by repeated applications.

Eradicants are less common and are applied after infection has occurred. They act on contact by killing the organism or by preventing its further growth and reproduction.

Systemic fungicides are a relatively new class of controls which can be used on living plants. Systemic chemicals are translocated in the sap stream from the site of application to other plant parts. This type may be used as both a protectant and an eradicant.

Fungicides may be grouped according to the primary ingredient:

- **Sulfur group**
- **Copper group**
- **Dithiocarbamate and Thiram derivatives**
- **Glyodin and Dodine**
- **Captan, Folpet and Captafol**
- **Dinitro Derivatives**
- **Benzimidazole derivatives**

The chemistry of these materials is very complicated and too advanced for this book. Also, their use as fungicides is dependent on the crop, the disease and the specific conditions of use. For specific recommendations, contact your county extension agent or chemical supplier.

For best results, the chemicals should be used with other control methods in a planned strategy. Efforts may be directed against a single disease or it may be designed to protect a crop from all serious diseases.

Disease control chemicals, like other pesticides, are toxic to men and animals and can be hazardous if not

Fig. 26—Oral Toxicities of Some Fungicides, Bactericides, and Nematicides

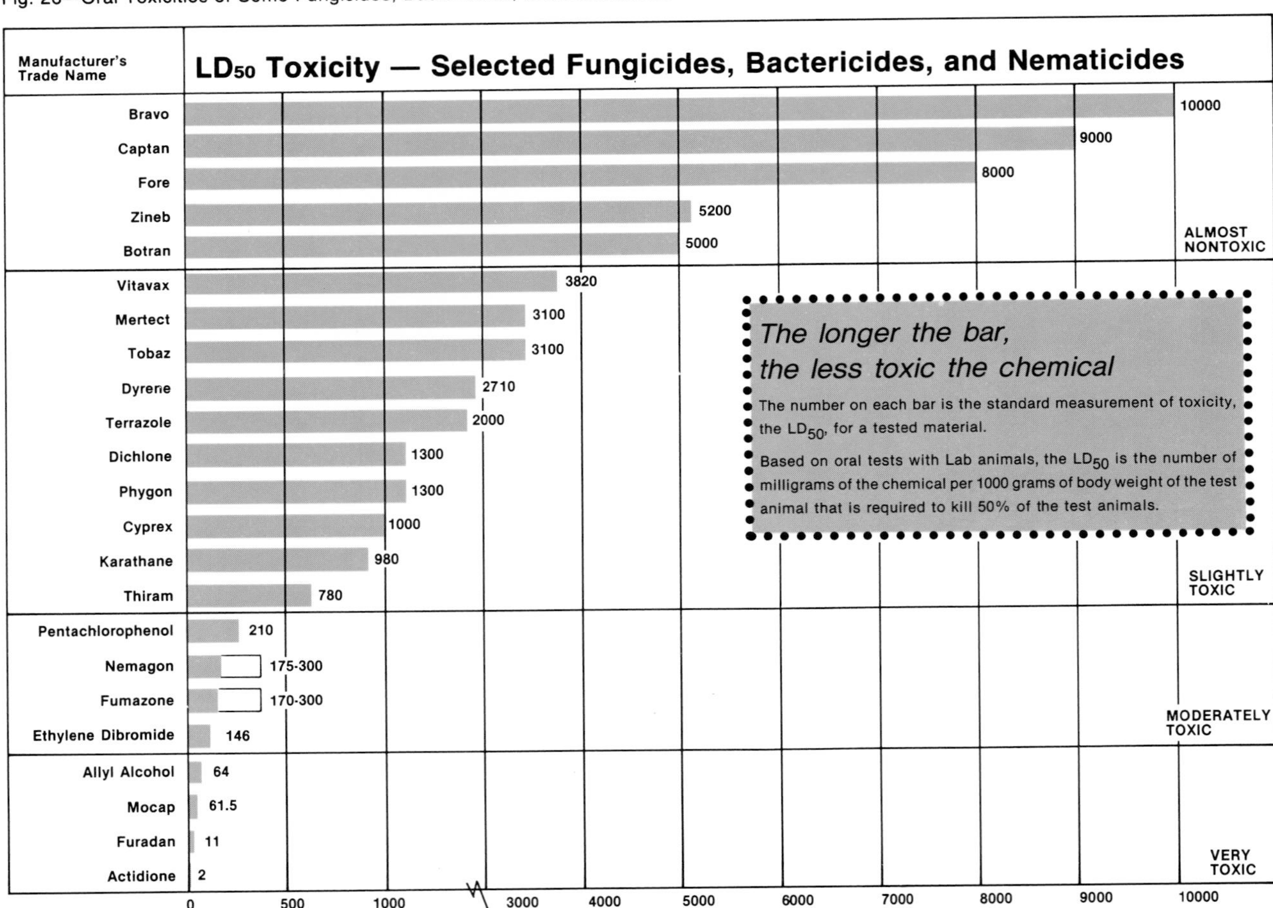

Fig. 27—Destroy or Bury Leftover Diseased Potatoes

handled properly. Safety aspects of chemical handling will be covered in Chapter 9. Figure 26 shows the toxicity ratings of a number of chemicals for disease control. In general, a long bar indicates a relatively low toxicity material.

CONTROL PLANNING FOR POTATO LATE BLIGHT

Potato late blight is a good example of a disease requiring integrated control. The disease can be avoided in part by sanitary measures such as destroying potato dumps (Fig. 27) that provide a ready source of pathogens, planting disease-free seed pieces, by growing resistant varieties, or by growing susceptible varieties in dry areas which are unfavorable to the disease.

In areas where the climate favors the late blight, resistant varieties should be grown. Regular sprayings of protectant fungicides should be applied during cool damp weather which favors the disease.

CONTROL PLANNING FOR CITRUS

Citrus trees are subject to many and varied diseases. Several control methods must be used.

Tristeza

Tristeza is a destructive virus disease that affects oranges and grapefruit. Proper rootstock and scion (grafted top) selection is important to control Tristeza. Severe symptoms develop when the virus affects a sweet orange tree or a sour orange root stock. Practical control is possible if disease-resistant rootstocks, such as rough lemon, are used. Citrus is subject to other diseases. A control strategy, combined into an overall control program for the crop, is essential.

SUMMARY

The concept of plant disease control is not as clear as control concepts for other pests. Diseases affect the ability of plants to use energy. Diseases are caused by three widespread types of pathogens:

- **Fungi**
- **Bacteria**
- **Viruses**

Pathogens can be spread by many means. Some of the most common are:

- **Wind**
- **Insects**
- **Rain**
- **Machinery**

Nematodes are tiny "eel-like" worms that attack plants. Besides the damage they do directly, nematodes may open paths for infection by many pathogens. Nematodes can attack any part of the plant, but the roots are most frequently affected.

Timely, correct diagnosis is essential to control of disease. There are five general approaches to control:

- **Avoidance**
- **Excluding pathogens**
- **Eradication**
- **Protection**
- **Resistance**
- **Therapy**

For the farmer, three primary control methods are available;

- **Cultural practices**
- **Resistant varieties**
- **Chemicals**

Many chemicals are used as fungicides. Consider all factors, then choose the best one for any specific situation.

For best results, chemicals should be used as one part of a carefully designed, integrated system of control.

CHAPTER QUIZ

1. What are nematodes?

2. Explain the term "disease".

3. ______, ______, ______ and other environment factors can all cause plant disease.

4. What is a pathogen?

5. Name the three types of biological pathogens.

6. How do fungi reproduce?

7. (True or false.) 85 percent relative humidity is needed for spore germination.

8. Name four ways that bacterial diseases can be spread.

9. The two most common symptoms of virus disease are __________ and __________.

10. Give three methods by which virus diseases spread.

11. (True or false.) The roots of a plant are most frequently affected by nematodes.

12. What are two kinds of damage done to plant roots by nematodes?

13. What are five general approaches to plant disease control?

14. Why is furrow irrigation better than overhead irrigation when producing disease-free seed?

15. How can seeds and vegetative plant parts be disinfected?

16. Give three methods of soil sterilization.

17. What is the reason for crop rotation?

18. Give three methods of residue management.

19. Give two methods of locating resistant crop varieties.

20. (True or false.) A fungicide is used to control fungi and bacteria on both living and non-living plants and plant parts.

21. What are five groups of fungicides?

5
Insect Control

Fig. 1—Effective Insect Control Is Required for High Production of Many Crops

INTRODUCTION

Insects are a part of the animal kingdom. All insects have certain anatomical characteristics in common. The most distinctive trait is that every adult insect has six legs.

Insects may be classified in groups which are similar in various ways. Three-fourths of all insects fall into four groups (Fig. 2):

- **Beetles**
- **Butterflies and Moths**
- **Wasps, Bees and Ants**
- **Flies, Gnats, and Mosquitoes**

Mites are not insects, but they are often destructive, and are combated with the same techniques and materials used to control insects. Therefore, mites are often grouped with insects by farmers. The most recognizable difference is that mites have eight legs, instead of six (Fig. 3). **For names and illustrations of many different insects, see the Appendix.**

Insects have inhabited the earth much longer than man. Insects have probably been here about 300 million years. We don't know precisely how long man has been here, but it is around 500,000 years. Insects have caused problems for man, his crops, livestock, and pets during the entire period. The Bible mentions ravages by lice, fleas, and locusts. Fleas transmit the bubonic plague organism from dead rats to humans. One outbreak of this dread disease claimed 25 million victims—one-quarter of the population of Europe at the time.

Man has also benefited from insects. A few, including grasshoppers and caterpillars, were food sources in primitive societies. For centuries, honey, made by bees, has been collected and eaten. Insects were also used in ancient religious ceremonies. The scarab beetle, carved in jade, was a symbol of eternal life to the Egyptians.

INSECTS IN AGRICULTURE

Modern man still has problems with some insects and benefits from others. Insects are essential for pollina-

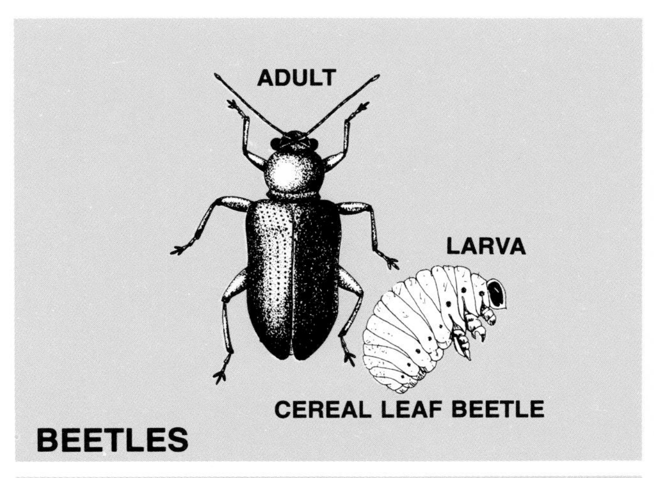

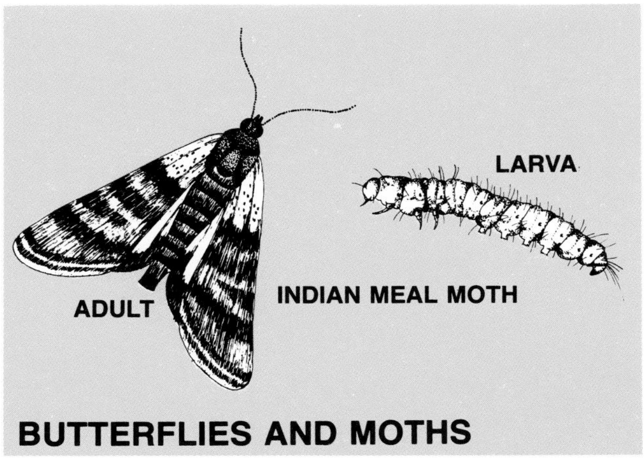

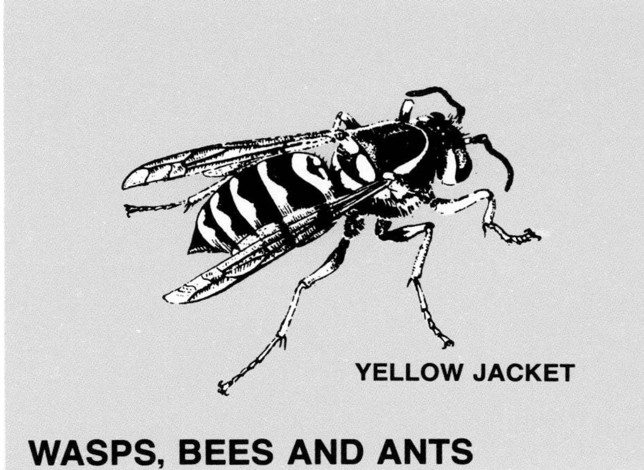

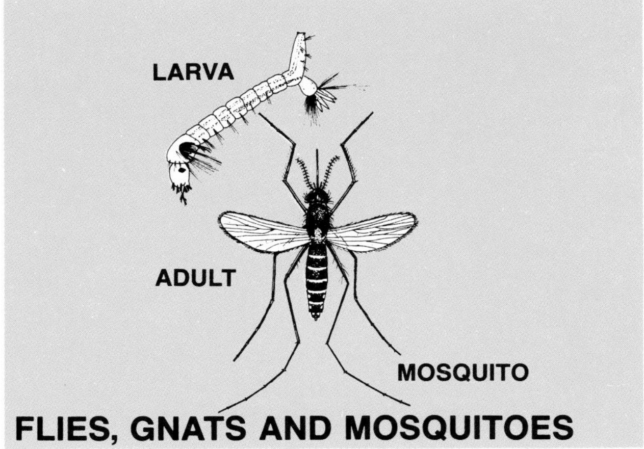

Fig. 2—Most Insects Fall into Four Groups

tion of many plants, for making honey and some kinds of wax, and as predators or parasites to control other insects. Lac is a resinous substance secreted by scale insects grown in "farms", and is the source of commercial shellac.

But, even though they recognize some beneficial insects, most farmers are more concerned with combatting insects. Insects feed on many kinds of organic matter, including crops, timber, and stored products.

Insects which damage growing crops are of two basic types:

- *Chewing insects* (Fig. 4) do direct injury to plants by eating leaves or burrowing into leaves, stems, fruit or roots. Some insects lay eggs on plants which are eaten by the larvae after they hatch.
- *Sucking insects* have mouth parts (Fig. 4) which pierce plant tissue and suck out sap. The loss of plant fluids is harmful to the plant. Viruses, fungi, and bacteria often gain entry into the plant through the puncture wounds. Diseases that result are discussed in Chapter 4.

Fig. 3—Adult of the Two-Spotted Spider Mite

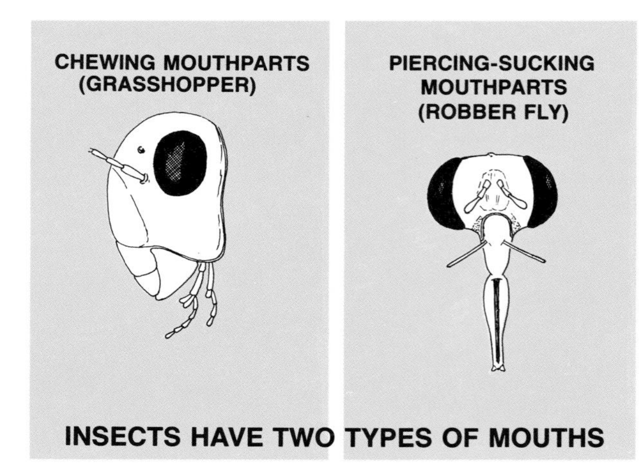

Fig. 4—Mouth Parts of Two Kinds of Insects

How serious is insect damage? A recent estimate put the annual loss in the United States from insect-caused crop damage at $5 billion. An additional $1 billion is spent annually on control efforts.

Why do insects continue to survive and attack crops in spite of efforts made to eradicate or control them? What are the special abilities and conditions that favor insects?

Many Insect Species

Insects are very diverse. No one knows how many different kinds there are. New ones are continually being identified. The total number of different species is between 600,000 and 1,500,000. There are at least three times as many species of insects as all other animal species combined.

There are over 80,000 kinds of insects in the United States. Of these, about 10,000 cause problems of some kind. Several hundred are destructive enough to warrant serious control efforts. Refer to listing in the Appendix for some typical insect pests.

Insects are highly specialized, and they have adapted to all kinds of environments. A termite was caught in a trap attached to an airplane at a height of 3½ miles (5.5 km). Some insects live far underground in caves and mines in Antarctica, in deserts, and even on the surface of the ocean, far from land. Because of the diversity and the large number of species, there are insects that can adapt to nearly any situation.

Small Size

Insects come in many sizes, from as small as a single-celled animal to as large as some birds; for instance, the Atlas moth has a wing span of 12 inches (30 cm). However, insects are small when compared to most other animals. Because of their size, insects can make a feast on a speck of grain and a drop of water which would be overlooked by other animals. Insect populations are seldom completely "starved out."

Fig. 5—Growth with No Metamorphosis

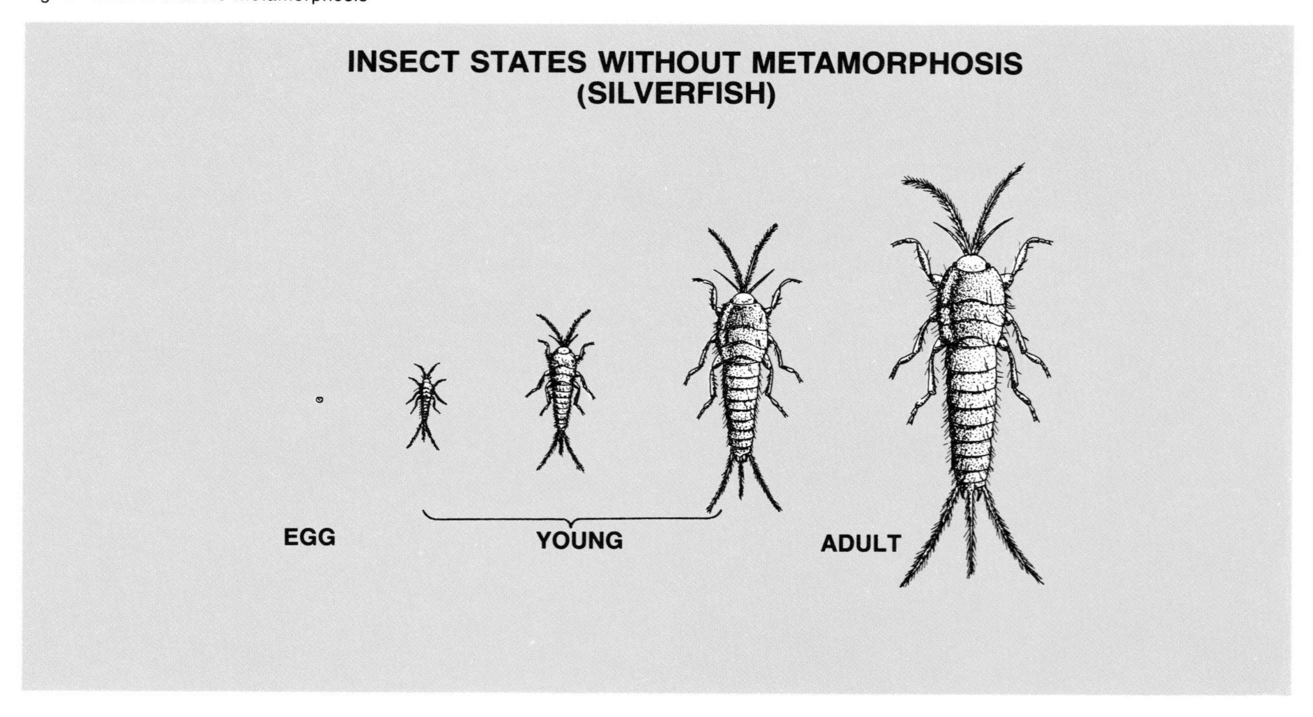

Flight

The ability to fly is an important capability of many insects. Flight allows insects to range over large areas to forage for food and escape from enemies. It also permits insects to spread into new areas. The flying capability of insect species varies widely (Table 1).

Metamorphosis

After hatching, insects grow in a series of steps. At each step, they shed the old cuticle (outside covering) and grow a new and larger one. Growth ceases after the insect reaches the adult stage. The changes which take place at each step are usually visible. This type of growth is called metamorphosis.

The changes that take place at each step vary considerably. There are three types of metamorphosis:

- *No metamorphosis* is the most primitive type of growth. The structural changes that occur during growth are almost imperceptible (Fig. 5). Silverfish are common insects in this group.

- *Simple metamorphosis* is a gradual change in external appearance as growth proceeds. The immature forms are called *nymphs* (Fig. 6) and have feeding habits similar to the adults. Wings appear at the final growth step. Grasshoppers and termites are common insects in this group.

Dragonflies and mayflies are other insects in this group. They exhibit greater changes at some steps in metamorphosis. The immature stages are aquatic and are called *naiads* (Fig. 7). Insects in this group do not damage crops.

TABLE 1

FLYING SPEED OF INSECTS

Insect	Normal Flying Speed (mph) (km/h)
Housefly	5 (8)
Butterfly	12 (19)
Wasp	12 (19)
Hornet	13 (21)
Honeybee	13 (21)
Horsefly	25 (40)
Dragonfly	25 (40)

- *Complex metamorphosis* includes four distinct growth stages: egg, larva, pupa, adult. A major change in structure and appearance occurs, during the pupal stage (Fig. 8). The larvae and adults frequently eat radically different diets and occupy different environments. Beetles, flies, moths, bees, and butterflies all undergo complex metamorphosis.

Rapid Population Growth

Insect reproductive abilities vary widely, and some species can increase their population very quickly. The population growth rate depends on:

Fig. 6—Growth with Simple (Gradual) Metamorphosis

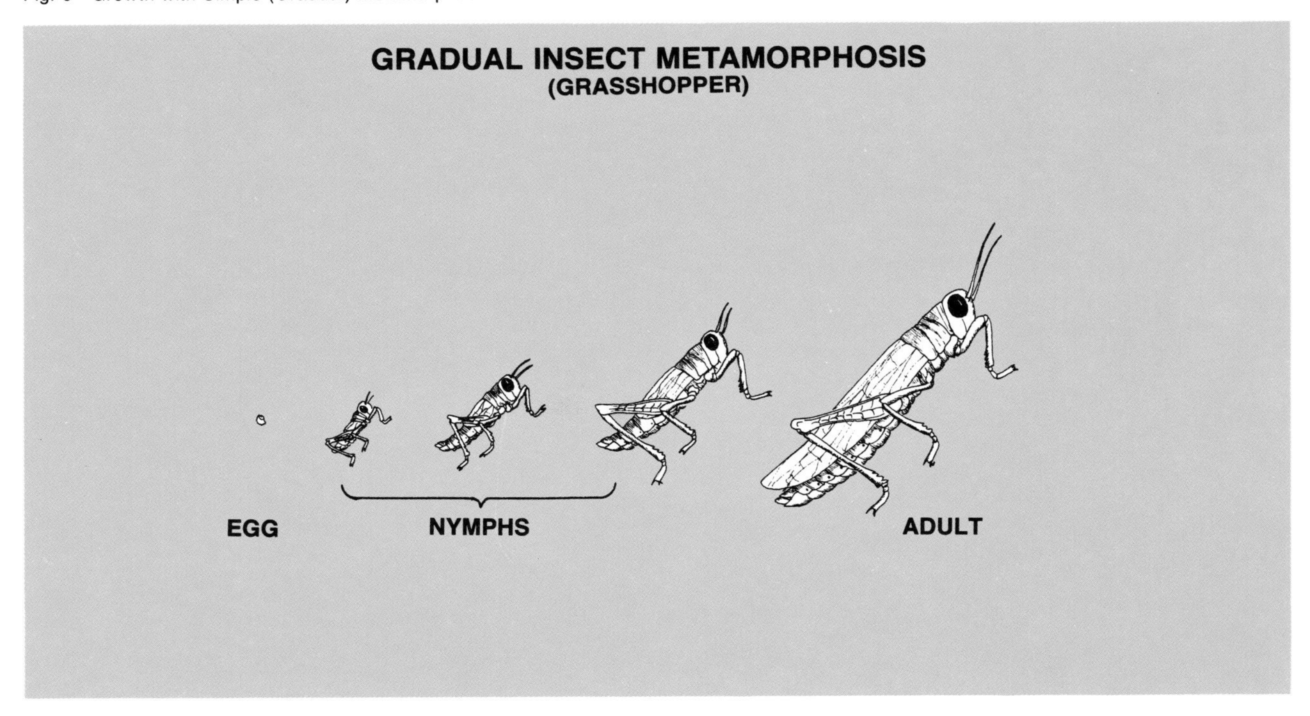

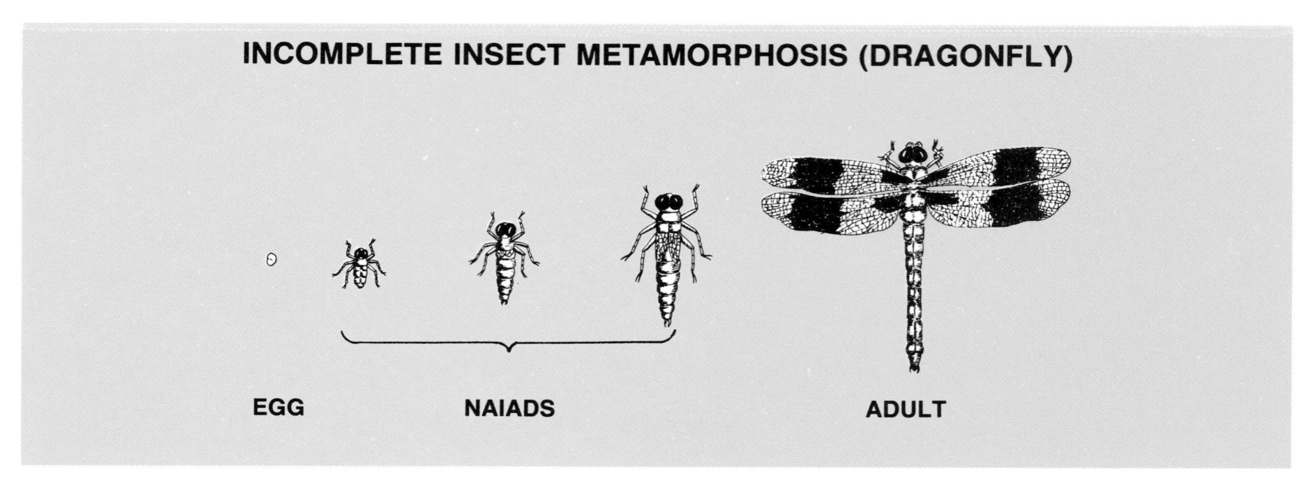

Fig. 7—Growth with Simple (Incomplete) Metamorphosis

- *Length of time to maturity*
- *Number of eggs laid and hatched*
- *The presence or absence of predators, parasites, and other control measures*
- *Supply of feed*
- *Climate*

The theoretical growth rate within a given temperature range depends on the first two factors—time to maturity and number of eggs laid and hatched (Table 2). Rapid maturity and large numbers of eggs favor rapid population expansion. A single pair of house flies in manure or garbage can increase to 1.8 million pairs with six generations or 12 weeks under ideal conditions.

Obviously, the actual insect population does not expand indefinitely. Predators, parasites, food supply, and climate all act to control the population. In a real situation, the population rises, then levels off (Fig. 9). Subsequent population increases or decreases are under the influence of the control factors.

Effect of Agriculture

In agricultural monocultures (single crop production systems), alien insect pests are often introduced along with the crop, but without the full range of their enemies. As noted previously, some insects can increase their population very quickly. If the conditions favor an insect species, a heavy infestation of that insect can develop in a short time.

Cause of Insect Outbreaks

In summary—four factors lead to insect outbreaks in agriculture:

- *Large-scale culture of a single susceptible crop in an area.*
- *Introduction of an insect pest into a favorable new area without its natural enemies.*

Fig. 8—Growth with Complex Metamorphosis

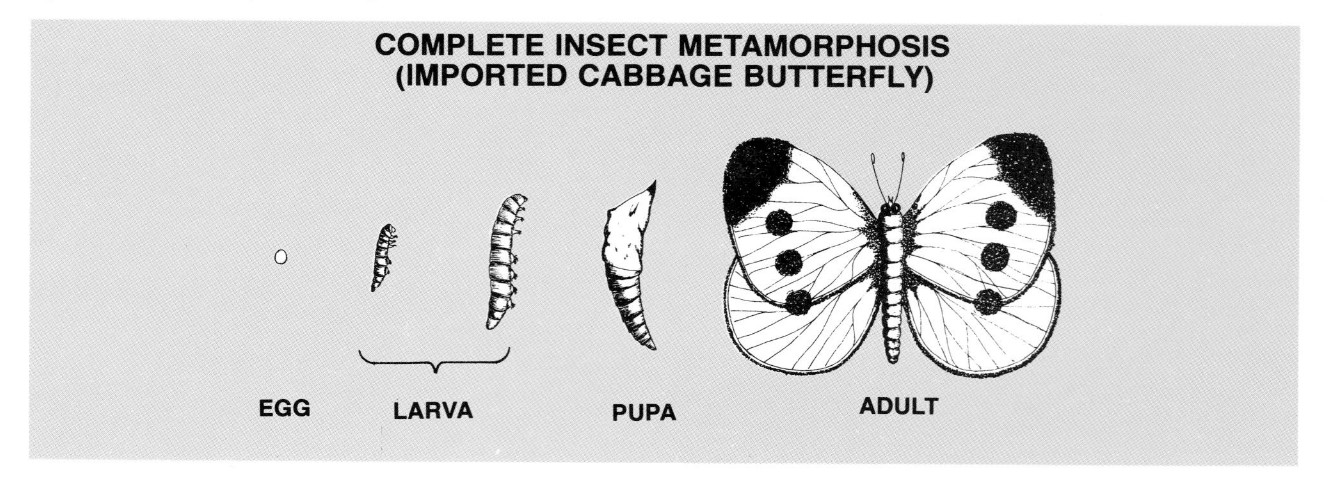

TABLE 2

THEORETICAL INSECT POPULATIONS

20 Days to Maturity*
10 Eggs Hatched Per Female

20 Days	40 Days	60 Days	120 Days
10 Insects	100 Insects	1000 Insects	1 mil. Insects

30 Days to Maturity*
10 Eggs Hatched Per Female

30 Days	60 Days	180 Days
10 Insects	100 Insects	1 mil. Insects

30 Days to Maturity*
5 Eggs Hatched Per Female

30 Days	60 Days	90 Days	120 Days
5 Insects	25 Insects	125 Insects	625 Insects

180 Days	210 Days	240 Days
15,625 Insects	78,125 Insects	390,000 Insects

*No. of females produced with adequate no. of males to reproduce again.

- *Weather conditions which favor rapid development and reproduction of an insect pest, especially if conditions are unfavorable to its natural enemies.*

- *Use of insecticides or other control measures which kill or reduce the numbers of natural enemies of the pest or exert other effects favorable to the pest.*

INSECT CONTROL

Although insects have many powerful survival abilities, there are ways to control them. The primary control is to prevent them from spreading into new areas. If an insect infestation has already been established, first decide what, if any, control measures should be used. This decision is usually dictated by economics. Controls are usually applied only if the **expected losses without control exceed the cost of the control.**

Insect problems are both complex and diverse. Large concentrations of host plants help to create and aggravate insect problems. There are no simple solutions. To cope with insect problems effectively, farmers, orchardists and ranchers should:

- *Correctly identify the insect causing the damage.*

- *Know, or become familiar with, the life cycle of the insect, particularly the times in the cycle when it is most susceptible to control.*

- *Apply effective and economical control procedures.*

- *Prevent reinfestation.*

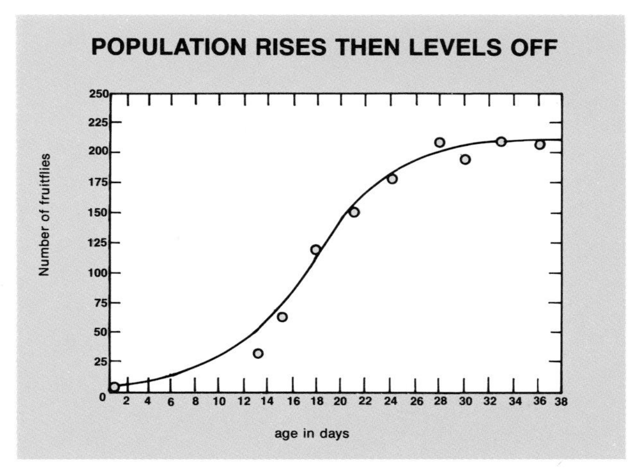

Fig. 9—Fruit Fly Population Growth

There are four methods of insect control:

- **Natural**
- **Biological**
- **Cultural**
- **Chemical**

NATURAL CONTROL

Natural control is the reduction of insect populations by the forces of nature. Natural controls include climate, topography, and natural enemies.

Climate

Weather conditions, especially temperature, affect insect activity and rate of reproduction. Climate affects insects directly and indirectly by influencing the growth and development of their host plants.

Direct killing may result if the weather acts against the insect at any stage of its life cycle. Some insects are killed by low winter temperatures. Insects with one or more stages that live in water are controlled if free water is not available because low rainfall or high temperature cause drying. Direct killing is by no means a certain process. For example, winter-killing is moderate by available protection and the inherent cold resistance of the hibernating stage. Insects seek shelter for overwintering. They burrow under bark, in plant debris, or underground. Snow also protects the insects by insulating them against extremely low temperatures (Fig. 10).

The seasonal population of plant-eating insects is closely related to growth of the host plants. Unusual weather conditions can change the normal pattern so that increased or decreased damage results. For example, if a cool spring retards the cherry fruit fly, less damage will result because fewer insects will be present to infest the fruit before it is harvested.

Fig. 10—Snow and Corn Stalks Protect European Corn Borer Larvae from Killing Winter Temperatures

Topography

Features such as mountains and large bodies of water restrict the spread of many insects. Major topographical features which affect the climate of an area also limit the distribution of insect species.

Other features of the landscape can have similar effects. Soil type is a prime factor affecting wireworms. Some species live in heavy, poorly-drained soil, others in light sandy soils. Soil type also affects plant distribution which in turn affects the insect population (Fig. 11).

Natural Enemies

Birds, reptiles, fish, and predatory insects are natural enemies which help control insect populations. Disease organisms (Fig. 12) often reduce insect populations. Of the natural enemies, predatory and parasitic insects are the most important (Fig. 13). Of all insect species more than half feed on other insects, some of which are crop eaters.

The interrelationships between pests and their enemies are complicated. In general, a population increase by pests is followed by an increase in predators. As the

Fig. 11—Topography, Climate, Soils, and Crops Affect Presence of Different Insects

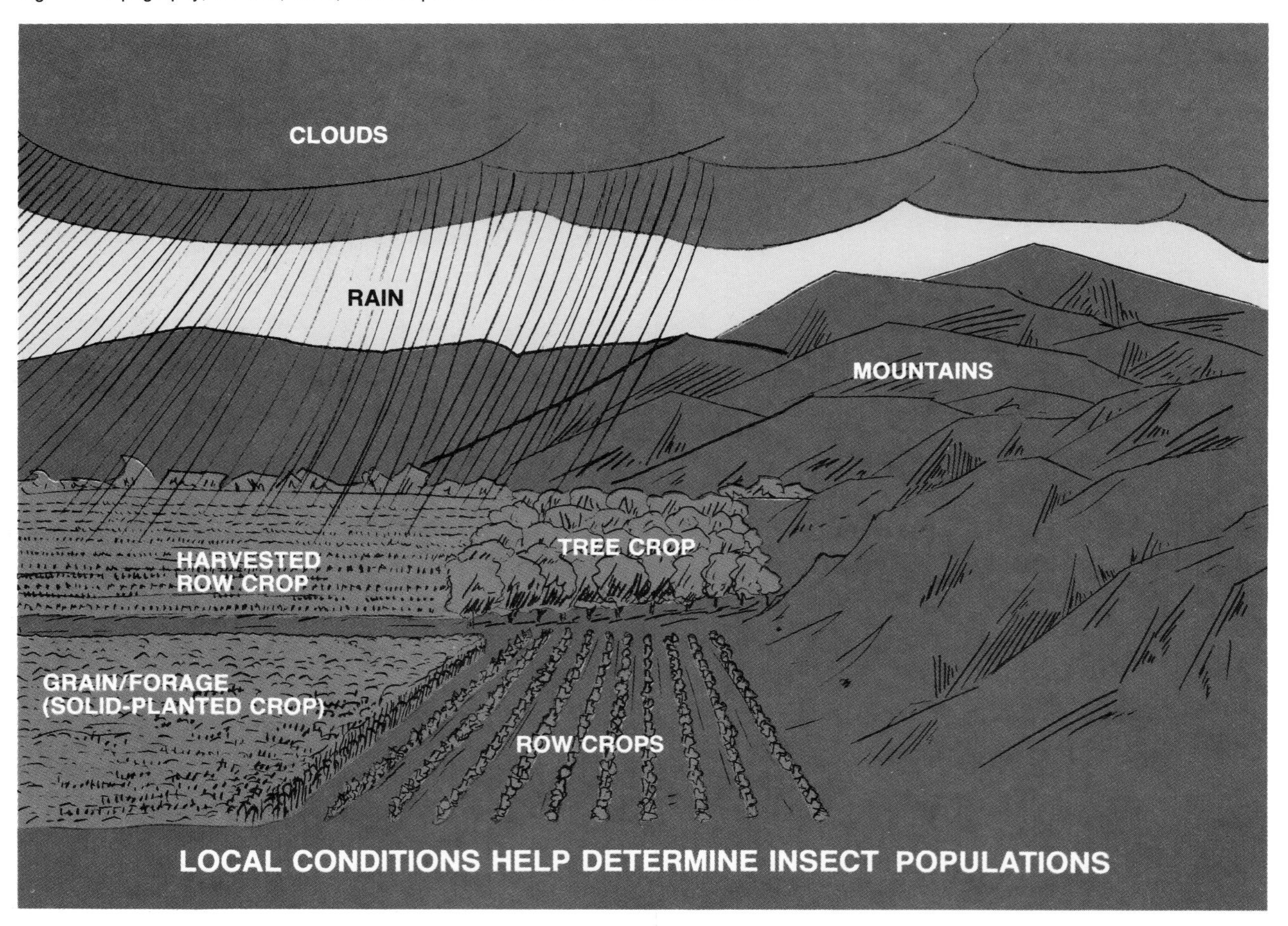

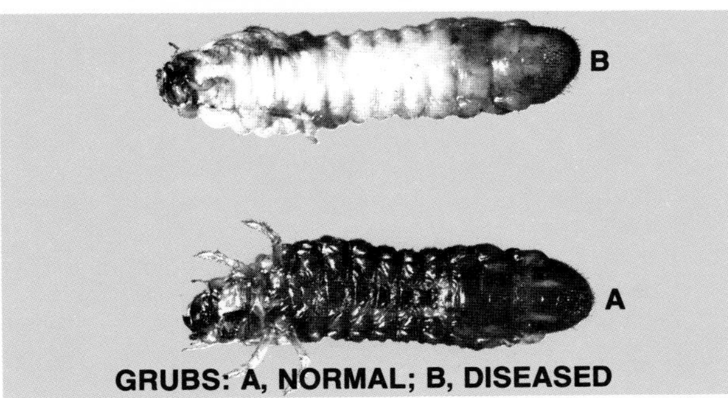

Fig. 12—Diseases Can Kill Insects

pests increase, the predators can find them more easily. As a result, the populations of predators and parasites increase. Eventually, there are so many predators that pest numbers decline. Soon the predators and parasites are themselves restricted because of a lack of feed. However, pest populations can often reach epidemic numbers before there are enough enemies to achieve worthwhile control.

A good example of a predatory insect is the ladybird beetle which feeds on aphids, scales, and other soft-bodied insects including the greenbug of wheat and sorghum and the corn borer. Adult beetles eat the aphids and lay eggs on the infested plants. When the beetle larvae hatch, the aphids are a handy food supply. Each larva can eat up to 500 aphids (300 per day or 60 per hour) before pupating (entering the next stage of metamorphosis.

BIOLOGICAL CONTROL

Biological control of pests, such as by predators, may also be achieved by insect or disease organisms that have been introduced by man.

These biological control programs involve searching for a parasite or predator for a particular insect pest. Since most major insect pests were introduced to the U.S. from foreign countries, the searches often take place in the source country. Most projects are conducted by governmental agencies.

An example of biological control is using laboratory-reared wasps to control cereal leaf beetles. The wasps lay eggs in the living beetles or in the beetle eggs. When the wasp hatches, the pest is killed (Fig. 14).

The sterile-male technique and use of baits and repellants are other approaches to biological control, as discussed in Chapter 1.

CULTURAL CONTROL

Cultural controls are agricultural practices often used by primitive as well as modern man, to control insects. Such methods may be used when other methods are not available to combat an injurious species and to supplement other controls. Cultural methods are economical and are particularly suited to pests of low-value crops.

Cultural controls include:

- **Crop rotation**
- **Trap crops**
- **Tillage**
- **Timing of operations**
- **Resistant varieties**
- **Mechanical controls**

Crop Rotations

Continuous production of a single crop on a piece of land often encourages pests to build up in the soil or plant debris. Changing crops (Fig. 15) usually aids in control of these pests. Rotations are most effective against insects which have long life cycles and infest the same crop during all stages of growth. Changing crops cuts such insects off from their primary or only food supply.

Trap Crops

Small plots of a host plant the insect favors, located near susceptible crops, are called trap crops. After the insect pests have been attracted to the "trap," they can be killed by burning or with an insecticide.

Tillage

Tillage operations can help control insects by changing the physical conditions of the soil, mechanically interfering with some stage of the insect's life cycle, removing host plants, or increasing the growth and vigor of plants to improve their resistance.

Tillage is often involved in residue-management programs. For example, shredding and plowing-under corn stalks is an important part of a control program for European corn borers (Fig. 16).

Residue Management

Shredding stalks helps eradicate plant-borne insects. Rotary-cutting and chopping action can control insects such as the European corn borer, and pink bollworm in cotton. Insects suffer greater winter-kill if the stalks

Fig. 13—Typical Predatory Insects

Fig. 14—Parasitic Wasps Kill Cereal Leaf Beetles

Fig. 15—Crop Rotations Control Some Insects

Fig. 16—Plowing-Down Corn Stalks Exposes European Corn Borers to Winter-kill

Fig. 17—Chopping Cotton Stalks with a Flail Chopper Helps Control Pink Bollworm

Fig. 18—Map Showing Approximate Wheat Planting Dates to Avoid Hessian Fly Damage

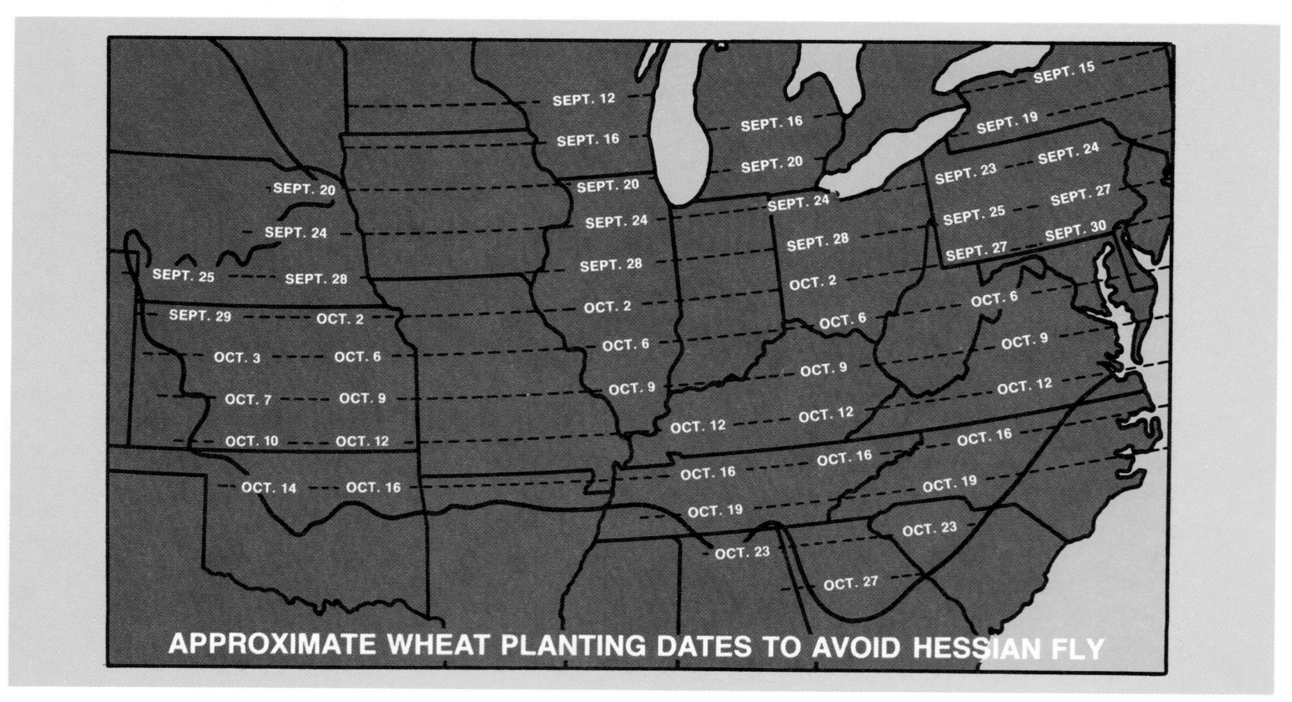

BN—21909

Fig. 19—Insect-Resistant Soybeans

are pulverized. In some states, cotton stalks must be pulverized and plowed down by a certain date. The date is set by law in a pink-bollworm control program. Plowing at least 6 inches deep immediately after shredding can reduce pink bollworm by 95 percent. A flail chopper (Fig. 17) usually kills more larvae than will a rotary cutter because of greater shredding and pounding of the debris.

Timing

Proper timing of planting or harvesting operations can be used to control insect damage if the host plant is susceptible only for a brief period or if the infesting stage of the insect's life cycle is short.

For example, prevention of Hessian fly damage in wheat is most easily achieved by delaying planting until cooler temperatures prevent further fly reproduction (Fig. 18).

At the other end of the season, crops should not be left in the field after growth is completed if they are susceptible to pest attack. For example, wireworm damage to mature potatoes causes a serious quality reduction. Since damage increases if the crop is left in the soil, harvesting should begin as soon as maturity is reached.

Resistant Varieties

Resistant strains of many crop plants are available (Fig. 19). There are three ways that insect resistance can be incorporated into a particular crop variety:

- *Including genetic changes which do not allow the insect to mature.*
- *Making the variety more vigorous so it is better able to resist insect attacks.*
- *Including characteristics that make the plant less appetizing to the insect.*

Mechanical Controls

Devices that affect insects directly (Fig. 20) or seriously alter their physcial environment are called mechanical controls. They are different from cultural controls in that the control is achieved by physical measures rather than farming practices. Screens, barriers, traps, electricity, X-rays, heaters, and refrigeration equipment are examples of physical or mechanical control devices. Both heat and cold are used to reduce insect populations in grain elevators. No insect can survive temperatures above 130 to 140°F very long (54 to 60°C).

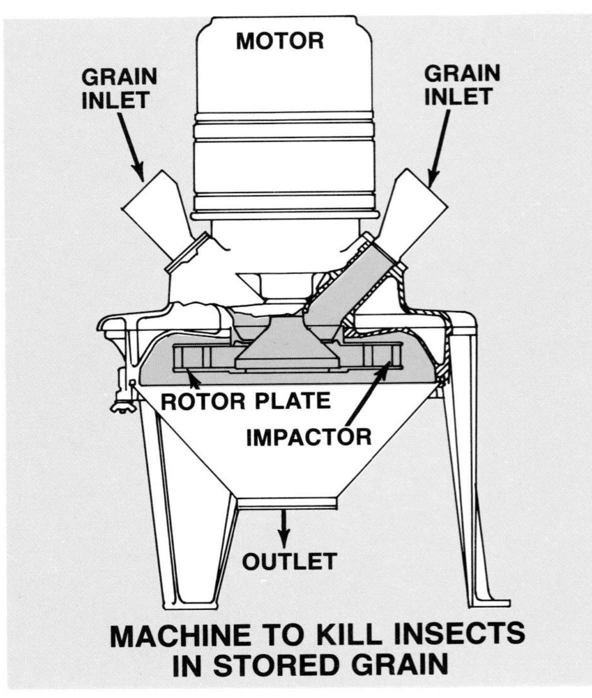

Fig. 20—Centrifugal-Force Machine Used in Flour Mills To Kill Stored-Grain Insects by Smashing Them

CHEMICAL CONTROL

Chemical control is the reduction of insect populations by the use of chemicals that:

- *Poison the insects.*
- *Repel the insects from specific areas.*
- *Attract insects to a place where they can be killed by other chemicals or some other method.*

Chemicals may be applied to seeds, plants, or soil. Effectiveness depends on the characteristics of the chemical, plus the timing and method of application. Insecticides are discussed in the following section. Application techniques are covered in Chapters 7 and 8.

INTEGRATED PEST MANAGEMENT

Integrated control of all kinds of pests, called Integrated Pest Management or simply IPM, has recently begun to receive widespread interest and support from a variety of groups and individuals. As a consequence, there are many research projects and other developmental activities underway to develop IPM techniques that can be applied on a routine basis. IPM is often thought of as a new concept, but it has actually been used informally by good managers for many years.

Fig. 21—Sampling a Cotton Field for Insects

IPM is not simply an effort to reduce pesticide chemical usage. An IPM program can use cultural, mechanical, biological, and chemical controls, usually in combination to suit a particular situation. However, insecticides and other control measures are used only if they are justified after consideration of all aspects of the situation, including:

- *Expected effectiveness of the proposed control measure*
- *Potential adverse effects of the proposed control measure*
- *Cost of the proposed control measure*
- *Timing required to apply the proposed control measure*
- *Energy costs involved in application of the proposed control method*
- *Availability and characteristics of other control measures*

It is difficult to predict at this time how much impact IPM will have on pesticide use. Current experience indicates that carefully timed pesticide use, based on reports from scouts (Fig. 22) who carefully inspect crops and soil for insects and other pests, can frequently allow one or more applications to be skipped, resulting in less expense for control. For example, in a 3-year Alabama study, pecans were sprayed an average of three times per year when applications were based on reports from the field. The conventional system, which had been in use, required an average of 9.2 applications per year. The savings amounted to $44 per year per acre ($110 per hectare) on spray materials, plus the associated equipment, fuel, labor, and other savings. However, the savings from most such efforts are more modest. The cost of control could be greater in some years when conditions are more favorable to the pests.

In summary, IPM is a management system that, in theory, considers all aspects of a system before applying any pest control method. IPM is still being developed but it is evident that certain benefits can be expected.

- *IPM programs can theoretically be designed to deal with all types of pests which adversely affect crop production.*
- *Some damage to crop plants can be economically tolerable.*
- *Pesticides will likely continue to be an integral part of most pest control programs for many years.*
- *IPM is not a panacea for pest problems and an IPM program does not mean that pesticide use will be replaced by nonchemical methods.*
- *IPM will not always reduce the cost of pest control.*
- *Regular monitoring of pests by field scouts (Fig. 22) will be essential, especially during critical periods in the life cycle of the pests and crops.*

Fig. 22—Scouts inspect crops and soil for insects.

- *More research is needed to develop better information on the economic levels of pest damage, the costs of various pest management practices, and the conditions under which the various control measures are most effective.*

Although insect control has been emphasized in some IPM programs, increased attention is being given to weed and disease control. If the theory for IPM is followed, the result should be judicious use of control measures to give effective control at an economical cost.

INSECTICIDES

Insecticides are poisons that affect specific parts or organs of an insect. For an insecticide to be useful, it must kill the insect or break its life cycle without destroying or injuring valuable plants, animals or beneficial insects.

Three systems are used for grouping. Similar insecticides may be grouped in various ways.

The "point of entry" classification has three groups:

- *Stomach poisons* are sprayed or dusted on a plant. The insect swallows the poison when it feeds on the plant. Stomach poisons are suitable for chewing insects such as codling moth, cotton bollworm, gypsy moth, and cabbage looper. Stomach poisons are "protective"—they can be applied to the plant preceding the insect infestation.

- *Contact poisons* kill when they come into contact with the insect. These insecticides kill the insect they reach during application, but usually are less effective on insects that arrive later.
- *Systemic poisons* are absorbed into the sap of plants through the roots or foliage. The material then kills insects which feed on the plant. Some systemic insecticides are also contact poisons.

An alternative classification system is based on the way the chemical affects the insect. There are four groups:

- *Physical poisons* kill by some physical action, such as by excluding air (mineral oils) or by abrasive action resulting in the loss of water from the insect's body (silica dusts).
- *Protoplasmic poisons* kill by precipitating proteins.
- *Respiratory poisons,* such as hydrogen cyanide, inactivate the breathing processes.
- *Nerve poisons* disrupt nerve operation in the insect.

These classification systems have not proven completely satisfactory because many insecticides act in more than one way. Consequently, insecticides are more frequently grouped by their chemical characteristics (Table 3). The major divisions are:

- **Inorganics**
- **Organics**
- **Synthetic Organics**

CAUTION: Use insecticides only as directed on the label for crops and purposes listed. Use protective clothing and equipment when recommended. Never apply insecticides closer to harvest than is recommended on the label.

TABLE 3

INSECTICIDE CLASSIFICATION

Inorganics	Organics	Synthetic Organics
Bordeaux Mixture	Oils	Chlorinated hydrocarbon group
Cryolite	Botanicals	Lindane
Lead Arsenate	Nicotine Sulfate	Methoxychlor
Lime-Sulfur	Pyrethrum	Toxaphene
Phosphorus	Rotenone	Aldrin*
Sodium Flouride		Chlordane*
Sodium Selenate		DDT*
Sulfur		Dieldrin*
		Endrin*
		Heptachlor*
		Organic phosphorus group
		Malathion
		Parathion*
		TEPP*
		Carbamates
		Carbaryl

*These chemicals are no longer approved for general use in the U.S.

INORGANIC INSECTICIDES

Inorganic materials are still used to some extent, even though most have been replaced by more efficient organics. *Lead arsenate* is used on trees and shrubs to control chewing insects. *Cryolite* is effective against several insects that attack truck crops. It is most suitable for plants that are sensitive to chemical damage. *Sulfur* is used as a dust to control mites. Lime-sulphur is an effective dormant spray for the control of certain scale insects, aphids, and mites.

ORGANIC INSECTICIDES

Oils are employed in various ways. They may be carriers or solvents for insecticides. Oils, by themselves, are insecticidal and are used to control aphids, scale insects, and mites on apple or citrus trees. Oil is diluted with water and applied to dormant trees as an emulsion containing 1 to 4 percent oil. Some formulations are *miscible* (can be mixed) and produce an instant "milky" emulsion when added to water. Other formulations require agitation to form and maintain the emulsion.

Botanical insecticides are derived from plants. *Nicotine* is highly toxic to many insects. It affects the central nervous system and is toxic whether eaten or absorbed through the body. *Pyrethrum* is a contact poison which acts quickly. It is quickly broken down by sunlight and does not leave a harmful residue. *Rotenone* is found naturally in many legumes. It is very safe, selective insecticide which affects the insect's respiratory system and is widely used on horticultural crops.

SYNTHETIC ORGANIC INSECTICIDES

Synthetic organics dominate the field today. New ones are constantly being developed. Three groups of synthetic organics are used most widely:

- **Chlorinated Hydrocarbons**
- **Organophosphates**
- **Carbamates**

Following are some of the characteristics of these insecticides.

Chlorinated Hydrocarbons

DDT was the most widely used chlorinated hydrocarbon insecticide for years. It is effective against a wide range of insects and acts both as a stomach and contact poison. It is a persistant material that breaks down slowly in nature. Its long life and the resulting residues have led to governmental orders for the removal of DDT from the U.S. and other markets.

Methoxychlor is a close relative of DDT. It is less toxic to mammals and is usually safer than DDT to apply to plants. **Aldrin, dieldrin, chlordane, heptachlor,** and **endrin** are all closely related. They are relatively persistent insecticides and are effective against many insects. These chemicals, like DDT, are no longer approved for general use in the U.S. (see Table 3).

Organophosphates

Parathion is highly toxic to both insects and mammals. It is used in the production of fruits, vegetables, ornamentals, and field crops. It is safe for most plants, but should not be used on McIntosh apples.

The extreme toxicity of parathion makes careful handling essential. (See Chapter 9.)

Most organophosphates are highly toxic. *Malathion* is an exception that may be safely used around animals and people if it is handled properly.

Carbamates

This group of organophosphates became popular after the development of carbaryl. The carbamates have a wide range of uses. Carbaryl has low toxicity to mammals but is toxic to bees.

SYNERGISM

Some chemicals have the ability to greatly increase the killing power of certain insecticides. If the toxicity of the combination is greater than the sum of the two chemicals used separately, it is called a **synergistic** action. Synergism is somewhat like adding 2 plus 2 and getting 6 or 9 (Fig. 23).

CHEMICAL	PERCENT OF INSECTS KILLED
A (ONLY)	30%
B (ONLY)	10%
A + B	40% (NO SYNERGISM)
A + B	90% (SYNERGISM)
SYNERGISM MULTIPLIES CHEMICAL EFFECT	

Fig. 23—Effects Of Synergism

The synergist may be inactive when used alone, or may be insecticidal. When applied together pairs of organophosphorus insecticides frequently have a synergistic effect. Synergism is important for:

- *Increase insecticide effectiveness.*
- *More economical control because less material is needed.*

Synergism can also produce a greater than anticipated hazard to people working with the chemcials. Extra caution should be exercised whenever two or more chemcials are combined.

INSECTICIDE FORMULATIONS

In the past, a grower using an insecticide could purchase the active ingredient in a more or less pure form. The spray or dust would then be prepared at the point of use.

But, the active ingredients of modern insecticides are seldom used alone as originally manufactured. Water, oil, air, or inert solids are used to dilute the insecticide so it can be handled by machinery and applied properly. The active ingredients of most insecticides cannot be readily mixed into solids or added to water. Manufacturers usually add one or more surfactants, such as solvents, wetting agents, emulsifiers, stickers, or powders. The resulting product is called a formulation. It can then be used as packaged or diluted with water or another carrier.

A single pesticide is often sold in several different formulations. The user should choose the formulation best suited to the requirements of a particular application. Things to consider when choosing a formulation (Fig. 24) include:

- *Effectiveness against the pest to be controlled.*
- *The crop to be protected.*
- *Machinery available for making application.*
- *Danger of drift or runoff.*
- *Cost of material.*

The formulation is always listed on the label. On some labels and in recommendations, the formulation may be abbreviated (Table 4).

Common insecticides are available as:

- **Dusts**
- **Powders**
- **Liquids**
- **Granules**
- **Aerosols**

TABLE 4

ABBREVIATIONS FOR INSECTICIDE FORMULATIONS

Abbreviation	Formulation
Dusts	
D	Dust
Powders	
WP	Wettable Powder
SP	Soluble Powder
Liquids	
SC	Spray Concentrate
EC	Emulsifiable Concentrate
Granules	
G	Granular

Some insecticides may be available in more than one form as explained below.

Dusts

A prepared dust is a dry, finely ground mixture of the pesticide with an inert carrier, such as clay. Dusts are ready to use when purchased. No mixing is required. Dusts are easily and quickly applied. Dusting of field crops, however, is an uncommon practice in recent years.

Dust particles, because they are finely ground, may drift long distances from the treated area to contaminate other crops. Drifting dusts are visible and may cause criticism. Never apply dusts on a windy day (above 5 to 10 miles per hour, 8 to 16 km/h).

Powders

Powders are dry formulations intended to be mixed with water or other suitable liquids and applied as a spray (Fig. 25). There are two types:

- *Wettable powders which mix with liquid and form suspensions (mixed with but undissolved).*
- *Soluble powders which mix with water and form solutions (material dissolved in the liquid).*

Powder formulations are relatively low cost and simple to transport, store, and handle. They are easy to measure and mix.

Wettable powders require constant agitation in the sprayer tank. They settle out quickly if the sprayer is turned off, and are difficult to resuspend.

Liquids

Liquid formulations are solutions of pesticide in a liquid carrier. There are three types:

- *Low-concentrate liquids, generally used as purchased.*
- *High-concentrate liquids, preparations to be diluted with water or oil before application.*
- *Ultra low volume (ULV) concentrates, nearly pure insecticide designed to be applied without dilution.*

Concentrated liquid formulations require little agitation and are well suited to low-pressure application by field sprayers and mist blowers (Fig. 26). Since they are concentrated, the quantity to be handled, applied and stored is small.

Instructions must be followed carefully to avoid improper mixing of concentrated liquid insecticides. Because of solvents in the material, some liquid concentrates may cause rapid deterioration of hoses, gaskets, and other rubber or plastic parts.

Granules

Granules are dry, ready-to-use pellets of insecticides on a carrier. Because they are relatively large, heavy, and approximately the same size, granules drift less than other formulations. They may be applied with simple equipment. Most granules are heavy enough to work their way through foliage to the ground.

Fig. 24—Considerations for Insecticide Formulations

CONSIDERATIONS FOR SELECTING INSECTICIDE FORMULATIONS

- Effectiveness
- Crop
- Application equipment
- Danger
- Cost

Fig. 25—Powders are Mixed with Water for Spraying

Fig. 26—Mist Blower can Apply Liquid Formulations

However, they will not stick to foliage, so they are not suitable for combating foliage-feeding insects.

Aerosols

Pressurized cans (known as aerosols) contain a small amount of insecticide. The insecticide is driven out of the can by a pressurized inert gas when a valve is opened. The small hand-held cans are not used in crops. However, some larger models are available that hold up to 10 pounds (4.5 kg) of material and are refillable.

INSECTICIDE HAZARDS

Anyone using insecticides must be aware of dangers in their use and the safety measures to follow for safe storage, handling, and application. Always read the label of insecticides before handling.

ACCIDENTAL POISONING

According to some reports, about 10 percent of accidental poisoning deaths in the U.S. each year can be attributed to pesticides, mostly insecticides. About 60 percent of the victims are children, most of them less than five years old. Nearly every pesticide accident results from failure to recognize the poisonous nature of these chemicals and to take proper precautions in their use.

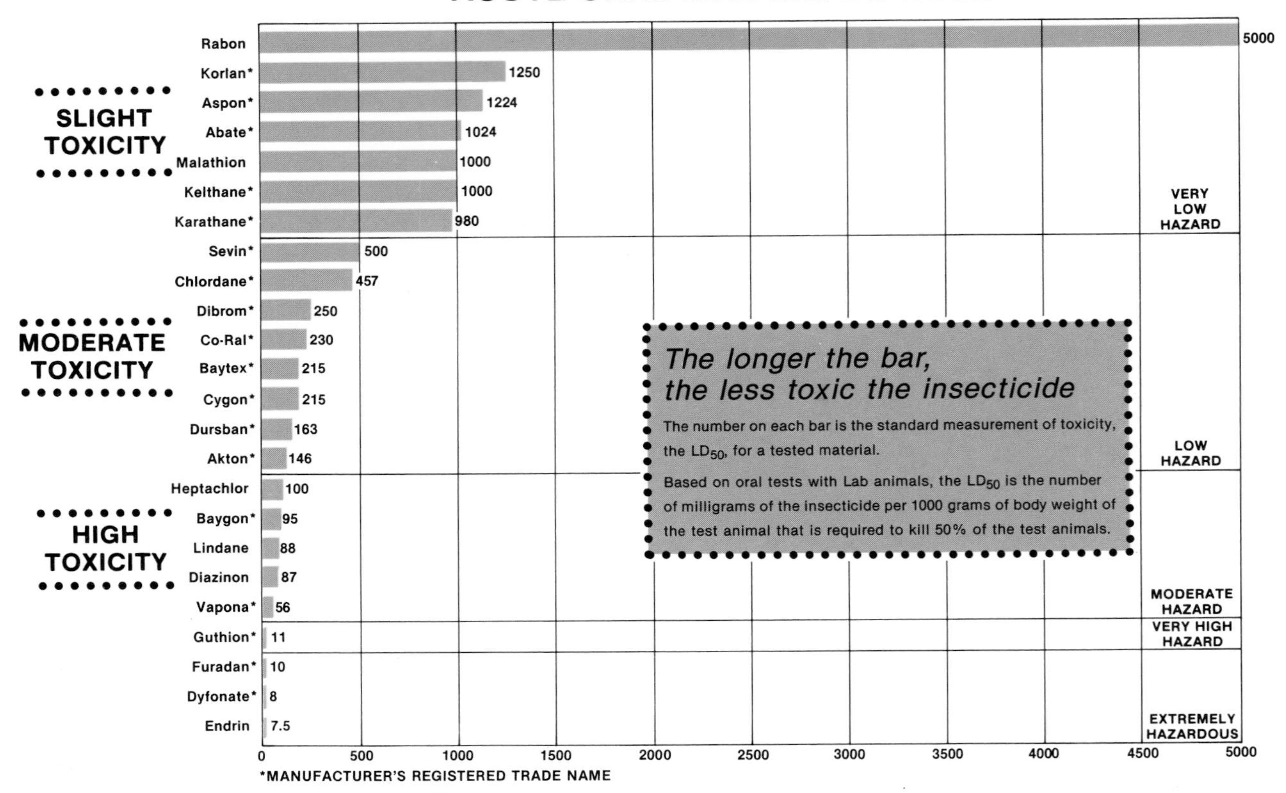

Fig. 27—Oral Toxicities of Insecticides

Pesticide poisoning can result from three kinds of exposure:

- *Oral–through the mouth and digestive tract.*
- *Dermal–through the skin.*
- *Respiratory–through the lungs and respiratory system.*

There is a wide difference in the toxicity of various insecticides (Figs. 27 and 28). For a complete discussion of insecticide hazards and safe working practices, see Chapter 7 of John Deere's FMO book AGRICULTURAL MACHINERY SAFETY.

Residues

Insecticide residues on crops are a potential cause of oral poisoning. Insecticides judged to be dangerous are subject to regulations showing a sepcified time interval between final application and harvest. When such restrictions exist, they are clearly stated on the insecticide label (Fig. 29). Always check labels for time interval between application and harvest *before* applying insecticides.

CAUTION: Insecticides may enter the body through skin or lungs, not only during application, but also from contact with treated plants after application. Definite waiting periods are specified for some chemicals before anyone can safely enter the field following chemical application.

PHYTOTOXICITY

Chemicals applied to plants to control pests sometimes are phytotoxic (poisonous to plants). The result may be temporary reduction of growth or "scorching" of leaves or fruit. Severe cases can result in plant death.

It is well known that some varieties of currants, muskmelons, and grapes can be damaged by sulfur. However, complete knowledge about phytotoxicity is not available. Always watch for phytotoxicity, particularly when new materials or formulations are applied to succulent growth. Plant injury is most likely under slow drying conditions, and in very cool or very hot weather.

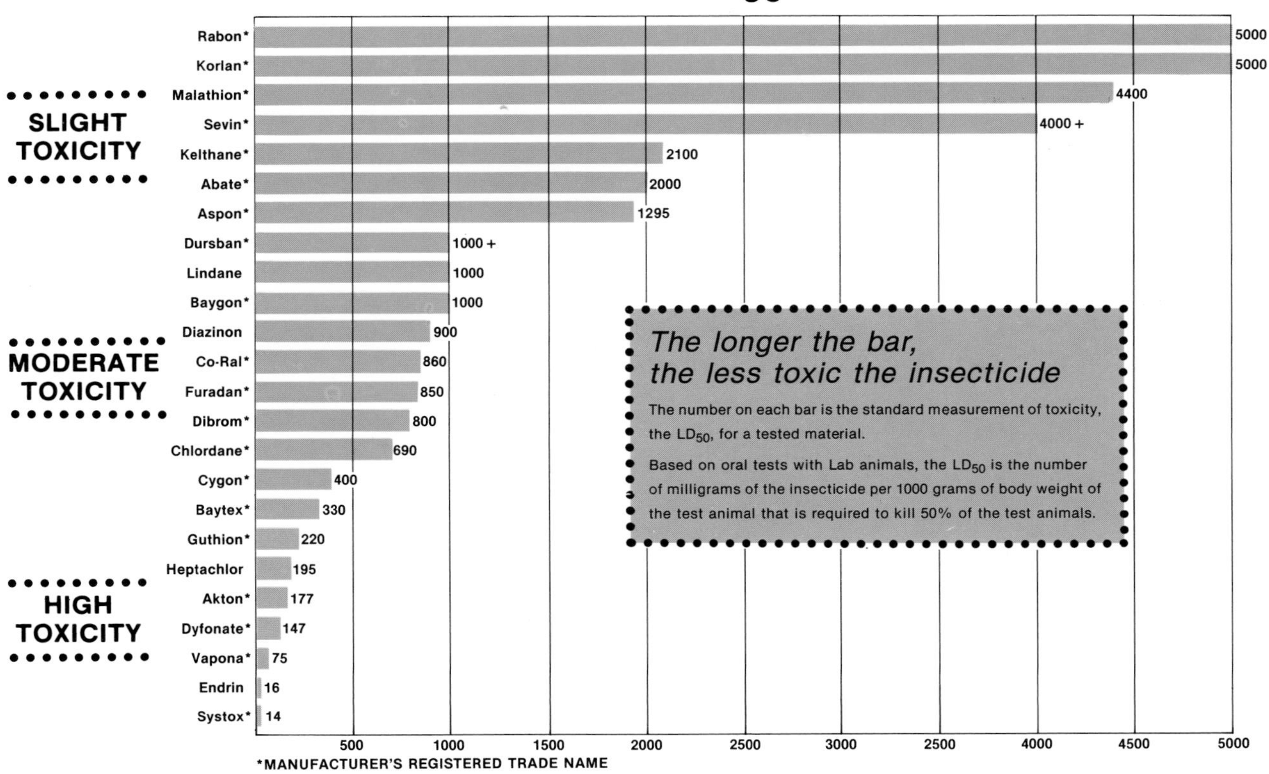

Fig. 28—Dermal Toxicities of Insecticides

DEVELOPMENT OF RESISTANCE

The insect population in a given crop is made up of individuals with varying resistance to chemicals. If repeated applications of the same insecticide are made, the most resistant individuals will be the ones that survive and reproduce. Eventually, a resistant population may develop. This process is most common in crops that receive routine insecticide applications. Thus "insurance" spraying may eventually produce resistant populations of the insect species the insecticide initially controlled.

LABELS

Pesticide labels are important to every pesticide user. The information and instructions on the label are developed at great expense during years of research by the chemical manufacturer. The label tells how to use the material correctly (see next page).

Read the label every time you use the material. Don't rely on your memory.

Label formats are similar for all pesticides and are covered in Chapter 9.

SUMMARY

Insects are a part of the animal kingdom that has been on earth much longer than man. Most insects fall into four groups:

- **Beetles**
- **Butterflies and Moths**
- **Wasps, Bees, and Ants**
- **Flies, Gnats, and Mosquitoes**

Man has some benefits from insects—primarily control of other insects. However, most concern is directed at controlling insects to reduce damage to crops and harvested products.

Sutan+ 10·G

GRANULES

A Selective Herbicide For Corn (Field, Sweet and Silage)

(Contains 10 pounds of active ingredient per 100 pounds.)

ACTIVE INGREDIENT:	
S-ethyl diisobutylthiocarbamate	10.0%
INERT INGREDIENTS:	90.0%
	100.0%

SUTAN+ — U.S. Patent No. 4,021,224

Sutan and Sutan + are Reg. in the U.S. Pat. and TM Off.

CAUTION — **Keep Out of Reach of Children**

HARMFUL IF SWALLOWED.

Avoid contact with skin and eyes. Wash skin with soap and water immediately after contact. Flush eyes with water. Avoid contamination of feed and food.

READ ALL LABEL DIRECTIONS BEFORE USING

WEEDS CONTROLLED *SUTAN+ 10·G will not control established weeds.*

ANNUAL GRASSES:

Barnyardgrass; Watergrass *(Echinochloa crusgalli)*
Crabgrass *(Digitaria spp.)*
Giant Foxtail *(Setaria faberii)*
Green Foxtail *(Setaria viridis)*
Yellow Foxtail *(Setaria glauca)*
Goosegrass *(Eleusine indica)*
Seedling Johnsongrass *(Sorghum halepense)*

PERENNIAL WEEDS:

Purple Nutgrass (Nutsedge) *(Cyperus rotundus)*
Yellow Nutgrass (Nutsedge) *(Cyperus esculentus)*

Existing stands of nutgrass must be turned under and chopped up thoroughly before treatment. Cultivation is suggested in addition to SUTAN+ 10·G treatment in fields with moderate or heavy nutgrass infestations.

GENERAL USE PRECAUTIONS

- Applied according to directions and under normal growing conditions, SUTAN+ will not harm the treated crop. During germination and early stages of growth, extended periods of unusually cold and wet or hot and dry weather, insect or plant disease attack, the use of certain soil-applied systemic insecticides, improperly placed fertilizers or soil insecticides may create abnormal conditions that weaken crop seedlings. SUTAN+ used under these abnormal conditions could result in crop injury.
- When applied according to directions and when conditions exist for normal plant growth through the season, no harmful residues of SUTAN+ should remain beyond harvest. In the **southeastern U.S.**, when SUTAN+ is used for weed control in silage corn, do not seed small seeded grains after corn harvest until September.
- Do not use on milo or sorghum. Do not use on corn seed stock.
- SUTAN+ is recommended for use on mineral soils only.
- Do not apply in combination with fertilizers, insecticides or fungicides.
- Do not store near seeds or fertilizers.
- Do not contaminate irrigation water or water used for domestic purposes.
- Keep container closed when not in use.

DIRECTIONS FOR USE OF SUTAN

SUTAN+ 10·G is a selective herbicide which is mixed (incorporated) into the soil for control of weeds listed on this label. SUTAN+ controls annual grasses as their seeds germinate by interfering with normal germination and seedling development. It does not control established weeds. All weed growth and crop stubble should be thoroughly worked into the soil before treatment.

APPLICATION DIRECTIONS

Apply only to well-worked soil that is dry enough to permit thorough mixing with incorporation equipment. Uniformly apply the recommended rate of SUTAN+ 10·G using equipment designed for application of granular herbicides. Equipment should be carefully calibrated before use and checked frequently during application to be sure it is working properly and delivering a uniform distribution pattern. Avoid overlaps that will increase SUTAN+ dosage above recommended rates as plant injury may occur.

In irrigated areas, do not apply SUTAN+ before pre-irrigation.

INCORPORATION DIRECTIONS

SUTAN+ 10·G must be incorporated into the soil immediately after application to prevent loss of the herbicide. Whenever possible, application and incorporation should be done in the same operation.

SOIL MIXING (INCORPORATION) BEFORE PLANTING: Use power-driven cultivation equipment set to cut to a depth of 2 to 3 inches; OR tandem discs set to cut to a depth of 4 to 6 inches operated at 4 to 6 mph, followed by a spiked-tooth harrow or some other leveling device which extends beyond the ends of the disc. For thorough mixing, disc in two different directions (cross disc).

PLANTING

Plant seed to a maximum depth of 2 inches within 2 weeks after treatment. To avoid removing SUTAN+ 10·G from the row (with loss of weed control in the row) do not move or shape soil after treatment, such as by planting in deep furrows.

CULTURAL PRACTICES FOLLOWING APPLICATION

SUTAN+ 10·G is not persistent in the soil and susceptible weeds germinating late in the season may not be controlled. Shallow cultivation may be necessary to control these late-germinating weeds. Do not cultivate deeper than the depth of herbicide incorporation. Pre-emergence or post-emergence herbicides may be necessary to control weeds resistant to SUTAN+.

RECOMMENDATIONS

Do not use on any type corn in Arizona or in the ten southernmost California counties (except in Kern County where field and silage corn may be treated).

BROADCAST RATES PER ACRE: Apply and incorporate 30 lbs. SUTAN+ 10·G per acre on light soils (coarse textured-sandy); 40 lbs. SUTAN+ 10·G on medium (medium textured-sandy loam) to heavy (fine textured-clay) soils. In the north central and midwest corn growing regions, 40 lbs. also is recommended on all soils with moderate to heavy infestations of nutgrass. Use 30 lbs. on blow sands in these regions. Cultivation is suggested in addition to the SUTAN+ treatment in fields with moderate or heavy infestations of nutgrass.

NOTICE—READ CAREFULLY

CONDITIONS OF SALE

Stauffer (and seller) offer(s) this product for sale subject to, and buyer and all users are deemed to have accepted, the following conditions of sale and warranty which may only be varied by written agreement of a duly authorized representative of Stauffer.

WARRANTY LIMITATION

Stauffer warrants that this product conforms to the chemical description on the label and is reasonably fit for the purposes referred to in the directions for use on the label subject to the inherent risks referred to below. Stauffer makes no other express warranties; THERE IS NO IMPLIED WARRANTY OF MERCHANTABILITY and there are no warranties which extend beyond the description on the label hereof.

INHERENT RISKS

The directions for use of this product are believed to be reliable and should be followed carefully. However, it is impossible to eliminate all risks associated with use. Buyer assumes all risks associated with use or application of this product contrary to label instructions or resulting from extraordinary weather conditions.

LIMITATION OF LIABILITY

In no case shall Stauffer be liable for special, indirect, or consequential damages resulting from the use or handling of this product and no claim of any kind shall be greater in amount than the purchase price of the product in respect of which such damages are claimed.

DISPOSAL — BURY EMPTY CONTAINER IN A SAFE PLACE AWAY FROM WATER SUPPLIES.

Made In U.S.A. By

A-1

EPA REG. No. 476-2001

STAUFFER CHEMICAL COMPANY **WESTPORT, CT 06880**

Fig. 29—Typical Insecticide Label

Insect populations can increase very quickly under suitable conditions. Modern agriculture, producing a single crop in an area, often favors certain insect pests along with the crop. Insect controls include:

- **Natural Controls**
- **Biological Controls**
- **Cultural Controls**
- **Chemical Controls**

Integrated control is a program of control specifically aimed at one insect species and uses any available, economical control technique.

Insecticides fall into three main groups:

- **Inorganics**
- **Organics**
- **Synthetic Organics**

Organics include the "natural" materials, and synthetics such as organophosphates and chlorinated hydrocarbons.

Insecticides are available in different formulations, including:

- **Dusts**
- **Powders**
- **Liquids**
- **Granules**
- **Aerosols**

Hazards from insecticides include:

- *Accidental poisoning (of humans or animals).*
- *Phytotoxicity (poisoning of plants).*
- *Development of resistance (from repeated applications of the same chemical).*

The container label is the most important source of information about an insecticide. **Read the label every time you use the material! Follow the instructions on the label!**

CHAPTER QUIZ

1. What are the four major insect groups?

2. What are mites? How are they different from insects? Why are they often grouped with insects?

3. What is metamorphosis? Why is it an important survival mechanism of insects?

4. What are five factors that affect the rate of insect population growth?

5. When should you attempt insect control?

6. Why is weather a "natural control"?

7. What is the difference between "natural controls" and "biological controls"?

8. Name five "cultural-control" and three "biological control"?

9. Name two ways to prevent Hessian fly damage?

10. Name four insect species that are natural parasites or predators.

11. What is a "resistant variety"?

12. What is integrated control?

13. What are three "point-of-entry" groups of insecticides?

14. Name three each of the following:

    (a) Chlorinated hydrocarbons

    (b) Organophosphates

    (c) Inorganics

15. What is synergism?

16. What are five factors to consider when choosing which formulation to use?

17. List five formulations.

18. Name three kinds of exposure that can result in accidental poisoning.

19. What is phytotoxicity?

20. How is resistance developed in an insect population?

21. What is the single most important source of information about a particular insecticide?

6
Fumigation

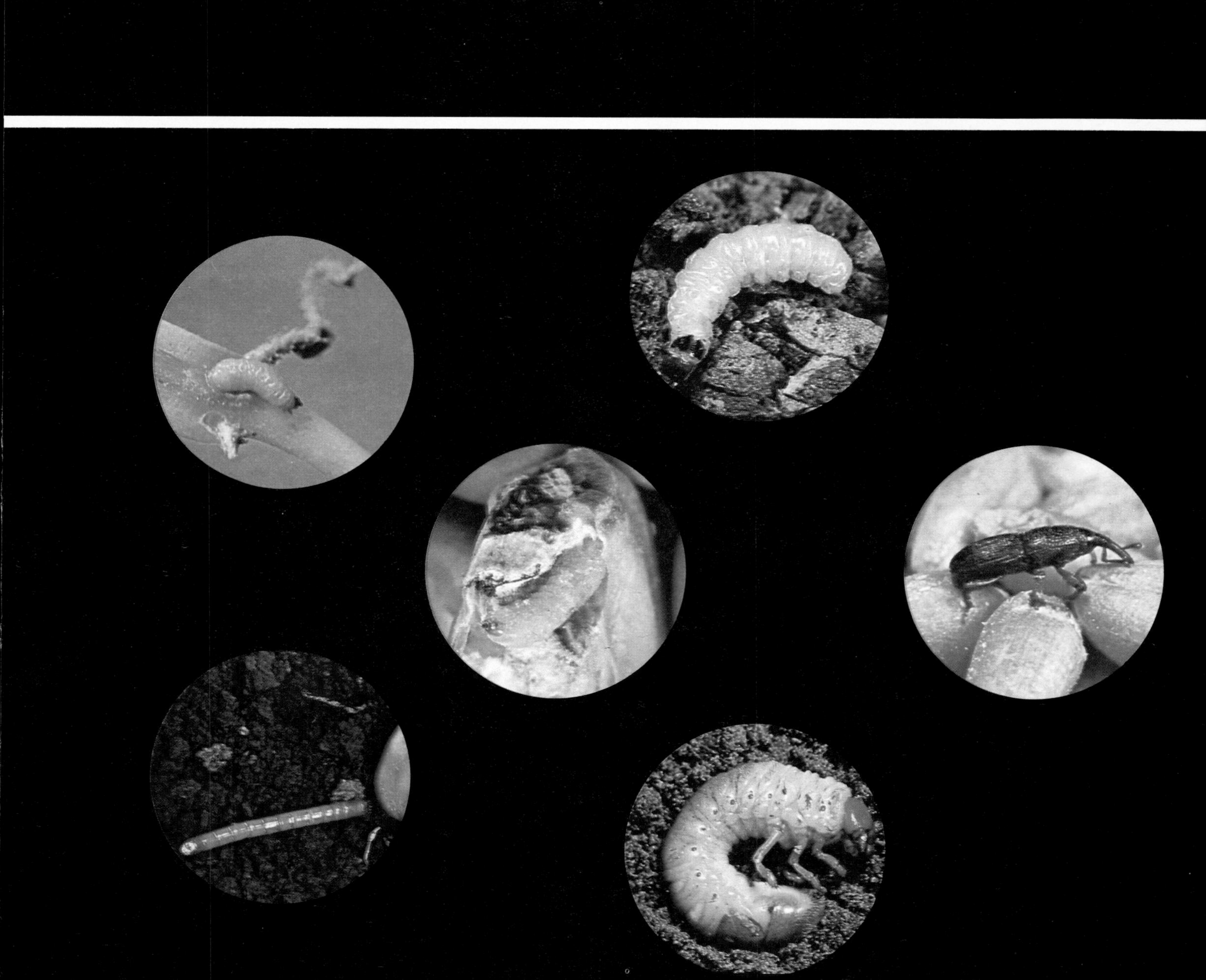

Fig. 1—Productive, Healthy Crops Require Control of Soil Pests

Fig. 2—Fumigating Harvested Products for Insect Control

INTRODUCTION

Fumigation is one of the oldest pest-control practices (Fig. 1). The method was originally applied to houses. Aromatic herbs and incense were commonly used materials.

The method has since been extended and is now used in greenhouses and for protecting stored products (Fig. 2). Although these are all significant uses, this chapter will concentrate on another application—pest control in soils.

What is Fumigation?

Fumigation is the use of a gaseous form of pesticide. It differs from control with aerosols (Chapter 7), but the uses are sometimes similar.

Fumigation requires that the application target be enclosed so that the gas cannot escape.

Where and Why Fumigants Are Used

Fumigation can kill weeds, weed seeds, insects, and all types of disease-causing organisms, including nematodes. Often, fumigation is used as a last resort against pests. Cultural practices, such as crop rotation and residue management, are usually tried first. There are two reasons:

- *Fumigation is more expensive than other methods.*
- *Most farm operators and workers are not familiar with, and are not competent to handle fumigants.*

Despite these drawbacks, the use of fumigation is increasing. But because of the cost, fumigation is still chiefly used for greenhouse soils and for high-value crops such as vegetables, strawberries, flowers and ornamentals.

In recent years, soil fumigation has become widely used to help establish grasses where a high-quality turf is desired, such as golf courses.

COMMON FUMIGANTS

Relatively few fumigants in use today can be relied on to give satisfactory results. Three of the most widely used are:

- **Methyl Bromide**
- **Chloropicrin**
- **Calcium Cyanamide**

Brief descriptions of these fumigants follow.

Methyl Bromide

Methyl bromide is a deadly gas, used for seedbed fumigation. It is fast-acting and disperses quickly. The gas is odorless and colorless, so as a safety measure, manufacturers frequently add a small quantity of another gas, usually chloropicrin, to provide an odor.

Methyl bromide is available in pressurized cans or cylinders and used with an airtight cover (Fig. 3). Mechanical equipment is available that can inject the gas and install a plastic barrier in a single operation (Fig. 4). A gas mask fitted with a back canister must be worn during application and when the cover is removed.

The gas interferes with the growth of certain crops, notably onions, carnations, and conifers. It may also retard the germination of many types of flower seeds planted in treated soil high in clay or organic matter. Other fumigants should be used when these crops are being produced.

Chloropicrin

Chloropicrin (tear gas) is an expensive soil treatment. However, many experts consider it to be the best soil fumigant, particularly in its herbicidal action and for controlling verticillium wilt.

Fig. 3—Methyl Bromide being Released under a Gas-tight Cover

The material is kept in the soil by wetting the surface with a light irrigation or sprinkling after the gas has been injected. Compacting the soil surface with a cultipacker improves the seal. Tarpaulins or gas-proof plastic sheets are also used.

Chloropicrin is relatively slow to disperse in the soil. Crops should not be planted for at least two weeks after an application. Use a chloropicrin mask and canister when working with this fumigant.

Fig. 4—Machine to Apply Fumigant and Lay a Plastic Seal in One Operation

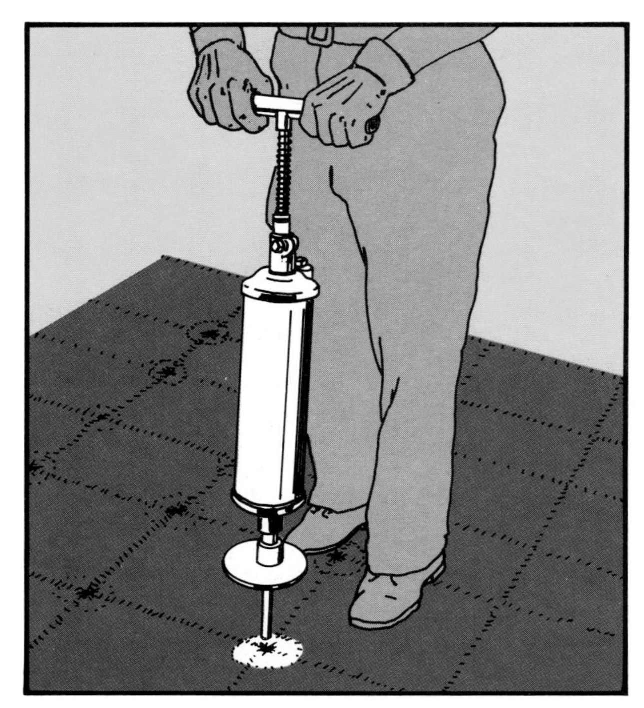

Fig. 5—Application of a Volatile Soil Fumigant with a Hand Injector

Calcium Cyanamide

Calcium cyanamide is a high-nitrogen fertilizer used in soil sterilization. It is most effective on germinating weeds while it is breaking down in the soil. Application in the spring can give good control of early-germinating weeds. It is a relatively long lasting material. Crops should not be planted in the treated soil for three or four weeks following application.

Other Fumigants

Many other fumigants are available. Check with your county extension specialist and chemical supplier for specific information.

CAUTION: Fumigants can be deadly! If you use a fumigant, follow these three steps:

- **Read the label**
- **Carefully follow all directions for use.**
- **Follow all precautions for safety.**

APPLYING FUMIGANTS

The fumigation equipment used depends on the following factors:

- **Area to be treated.**
- **Fumigant being used.**

Small areas, such as greenhouses or beds where vegetable, flowers, fruit or tobacco transplants are grown, are fumigated with hand equipment. The chemical may be injected by hand (Fig. 5) or metered under a tarpaulin (Fig. 3). In either case, some type of seal is needed to hold the gas in the soil long enough to be effective. As mentioned earlier, wet surface soil will seal tightly enough to hold certain chemicals. However, a gas-proof plastic sheet is more commonly used.

Liquid materials, such as wax, latex, or asphalt emulsions can be sprayed on the soil surface to provide a seal.

Larger areas are usually fumigated by mobile tractor-drawn equipment. Again, the specific equipment depends on the fumigant being used. However, chisels (Fig. 6) are usually used to inject the chemical into the soil. A surface sealer of either plastic film or a liquid is needed to cover the soil and retain the fumigant.

Reusing Plastic Film

The cost of polyethylene film may make recovery desirable under some circumstances. The film is difficult to recover without tearing or puncturing. It must be cleaned thoroughly, and damaged film must be repaired or replaced before it can be reused.

Reusing film is practical for small areas. Here the film can be handled in sheets. Rerolling the film for use on a mobile applicator is difficult so it is advisable to use new rolls of plastic on mobile equipment.

CLEANING EQUIPMENT

Applicator equipment should be thoroughly cleaned after each use. The main purpose of cleaning is to remove as much chemical residue as possible. The initial washing may be done with water, but it must be followed with a petroleum-solvent wash.

Water alone is not satisfactory because it may react with the remaining fumigant to form hydrochloric acid, which is highly corrosive to most metals. Therefore, use a petroleum wash as the final cleaning step.

SAFETY

CAUTION: Always read the container label and follow all directions and precautions recommended by the manufacturer. Your primary protective gear when using a fumigant is a gas mask (Fig. 7). Use only a mask approved for the chemical to be applied. Learn how to use the mask properly, and wear it whenever you work with the gas. For more information, see Chapter 9.

SUMMARY

Fumigation is the process of applying pesticide in gaseous form. Fumigation in the field requires a soil seal to prevent the gas from escaping. The seal may be formed by:

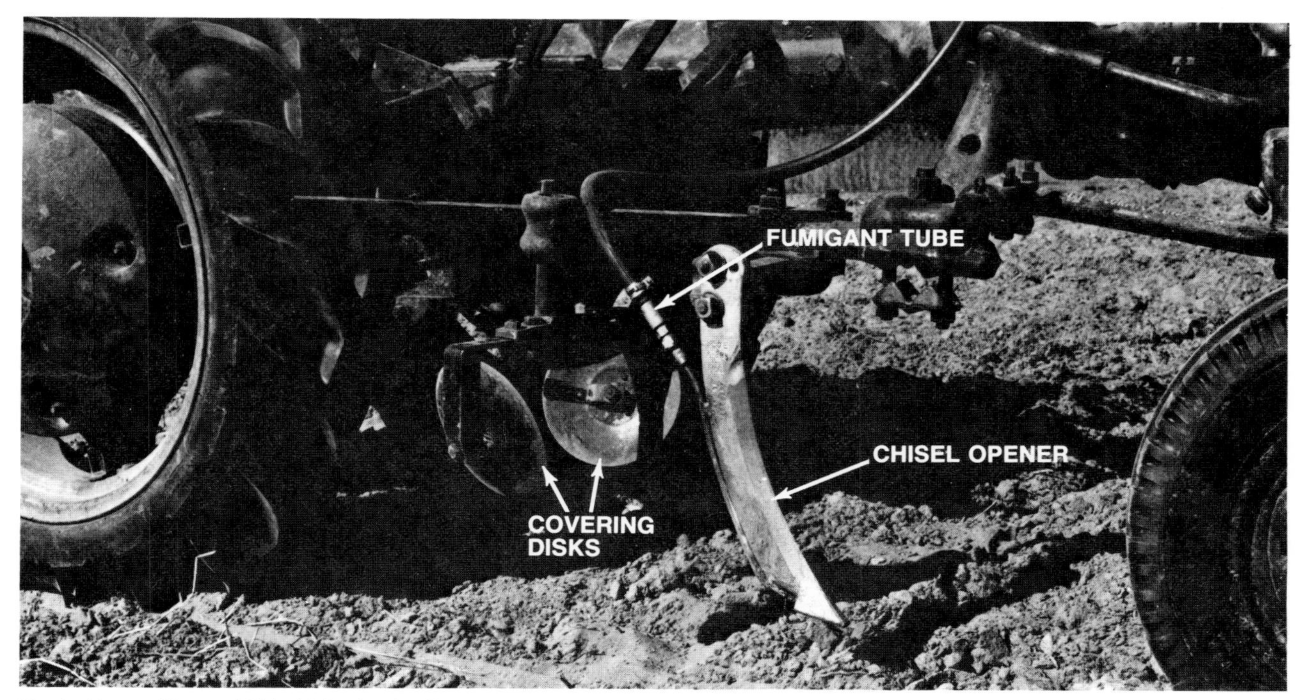

Fig. 6—Tractor-Mounted Chisel-Injection Applicator

- **Watering the soil surface.**
- **Applying a plastic film.**
- **Applying material such as liquid wax or latex.**

Any one of these methods may be used with either hand or mobile equipment. Mobile equipment is available to apply both the fumigant and the sealer in a single pass.

Fig. 7—Wear a Gas Mask when Fumigating

Three common fumigants are:

- **Methyl Bromide**
- **Chloropicrin**
- **Calcium Cyanamide**

Equipment must be thoroughly cleaned after use. Final cleaning should be done with a petroleum solvent to clean away acids which form when water is used for initial washing.

Your main piece of safety equipment is a gas mask. Get one and use it properly. Read and follow the instructions on the fumigant container label.

CHAPTER QUIZ

1. What is fumigation?
2. What are three uses of fumigation?
3. What are two reasons that fumigation is a limited practice?
4. List three of the most common fumigants.
5. What are three methods of providing a soil seal?
6. Why is it impractical to reuse plastic film from large areas?
7. Why should the final cleaning of fumigation equipment be done with a petroleum-based solvent?

7 Application of Liquid Chemicals

Fig. 1—Liquid Chemical Application

INTRODUCTION

In previous chapters, we covered the characteristics of pests and the chemicals used to control them. In this chapter and the next we will examine the characteristics and the use of equipment that applies the chemicals.

No chemical can function properly unless it is applied correctly. Good application results from observing four factors:

- *Apply the right material.*
- *Apply the proper concentration and amount.*
- *Apply it at the correct time.*
- *Apply it to the correct target.*

Good application also means that the chemical does not fall where it can poison humans or animals, kill other crops, or damage the environment.

Most pesticides can be formulated for liquid application. Equipment used to apply liquid pesticides and liquid fertilizers are similar in many ways (Fig. 1).

TYPES OF SPRAYERS

More pesticides are applied with sprayers than any other kind of equipment. There are many types and sizes of spray equipment. Pressures range from near zero to 1000 psi (6900 kPa). Application rates vary from a few ounces (milliliters) to several hundred gallons per acre (liters per hectare). In size, the equipment may be as small as an aerosol can or as large as a helicopter (Fig. 2).

Although there are many variations and combinations of types, most liquid-pesticide-application equipment falls into five categories.

- **Hand-operated sprayers**
- **Low-pressure sprayers (20-50 psi, 140-345 kPa)**
- **High-pressure sprayers (to 1,000 psi, 6900 kPa)**
- **Air carrier sprayers**
- **Foggers and Rope Wick Applicators**

Each type of equipment has distinctive features.

HAND-OPERATED SPRAYERS

Hand-operated sprayers are commonly used by home gardeners and others whose pest problems are relatively small. However, some farmers and custom operators also find hand-operated units useful for:

- *Spot treatments in a large area.*
- *Small jobs where field-sized equipment is not required.*
- *Treatment of hard-to-reach areas, inaccessible to large powered equipment.*

Compressed-Air Sprayers

Compressed-air sprayers are simple in design, easy to operate, and relatively inexpensive to buy and maintain. The capacity of the tank usually ranges from one to 5 gallons (3.75 to 19 liters). Pressure is provided by a hand-operated air pump which fits into the top of

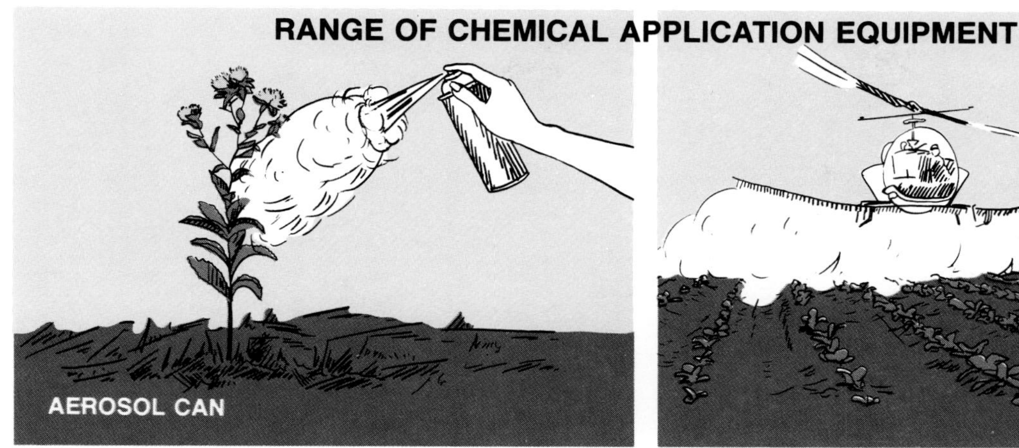

Fig. 2—Wide Range of Liquid-Application Equipment

the tank (Fig. 3). Air compressed in the tank above the spray material forces the liquid out of the tank through a tube. A valve at the end of a short length of hose controls the flow of liquid. Agitation is provided by shaking the tank. Normal operating pressure is between 30 and 80 psi (210 and 550 kPa) and is maintained by occasional pumping.

Another version of the compressed-air sprayer uses a precharged cylinder of air or carbon dioxide to provide pressure. These units include a pressure-regulating valve to maintain uniform spray pressure. Wheel mounting provides easy portability. Pesticides may be applied through a hand gun or a short boom.

Knapsack Sprayers

The *knapsack* sprayer has a small piston or diaphragm pump. The lever operated type (Fig. 4), which is powered by the operator, is widely used in all parts of the world. Some newer models (Fig. 5) are powered by a small gasoline engine. An air chamber helps smooth out pump pulsations and maintain uniform pressure. Tank capacity ranges from 2 to 6 gallons (7.5 to 23 liters) and pressures up to 180 psi (1240 kPa) can be developed.

Fig. 3—Compressed-Air Sprayer

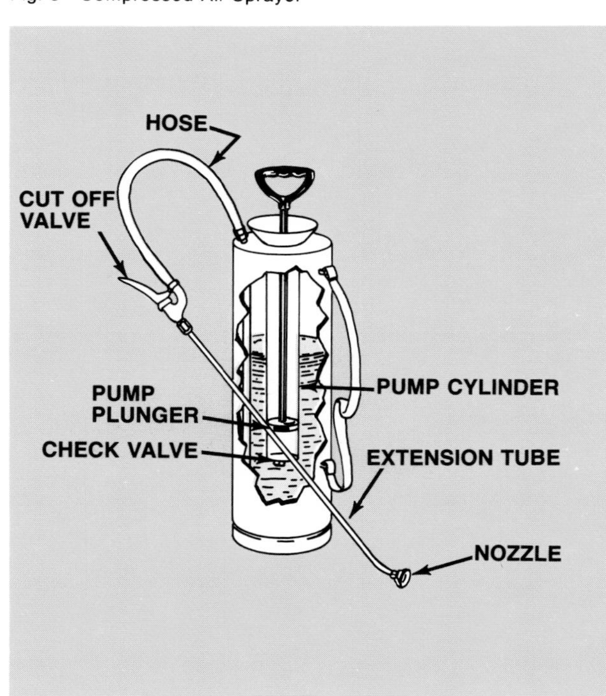

Fig. 4—Lever-operated Knapsack Sprayer

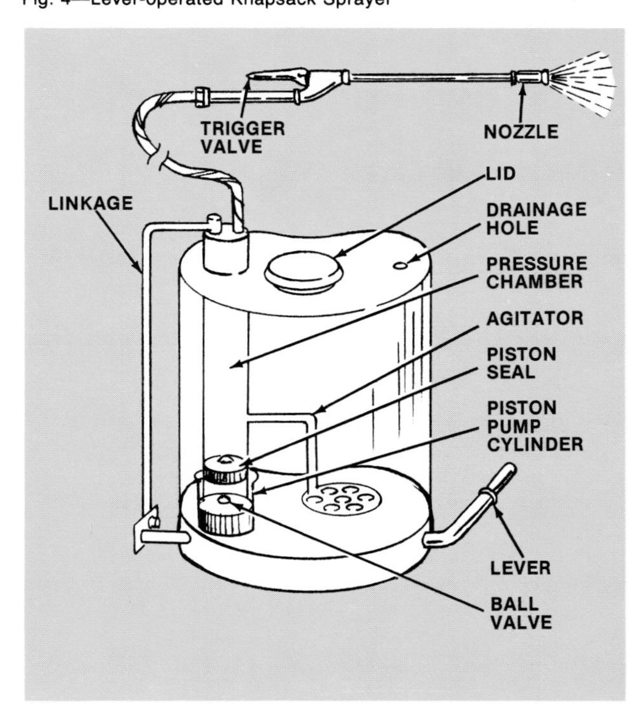

Fig. 5—Knapsack Sprayer

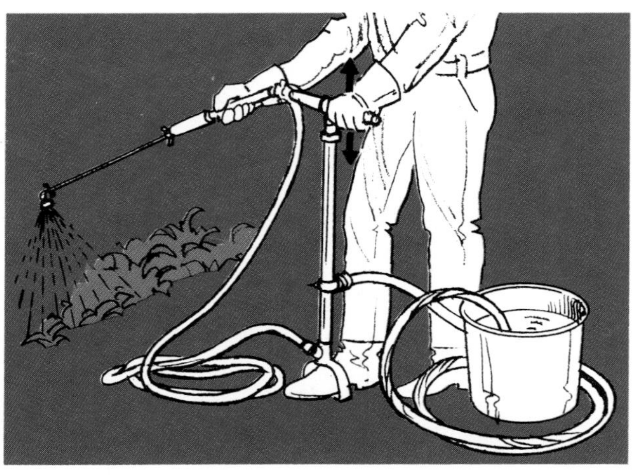

Fig. 6—Stirrup-type Pump

Agitation of spray material in the tank is provided by a mechanical agitator or by bypassing part of the pumped flow back to the tank.

As the name implies, a knapsack sprayer is carried on the operator's back. The hose and nozzle are similar to those used on compressed-air sprayers.

Double Acting Slide Pump

The double acting slide pump is a piston pump with a valve mounted at the inlet end of a cylinder and a second valve in a tube which is used as a piston. The pump is used to apply liquid chemicals from a separate container. It is operated by holding the piston handle grip still with one hand while directing the nozzle and moving the cylinder in and out regularly with the other. Because both hands are required for operation, the pump can be tiring to use. However, many thousands of them are in use.

Another version of the double acting pump is the stirrup pump. Two operators are usually involved, one each for the pump and the nozzle. Stirrup pumps (Fig. 6) are usually used with open containers and care must be taken to avoid spilling the toxic materials.

LOW-PRESSURE SPRAYERS

Low-pressure sprayers are the most widely used type of field application equipment. They are relatively inexpensive and are adapted to many uses including pre or postemergence application of chemicals to control weeds, insects, and diseases. They can also be

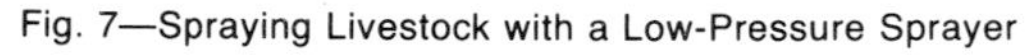

Fig. 7—Spraying Livestock with a Low-Pressure Sprayer

used for other jobs such as spraying cattle (Fig. 7). Low-pressure sprayers are available in many models and types and may be mounted on:

- **Tractors**
- **Trucks**
- **Trailers**
- **Aircraft**

Low-pressure units usually operate in the 20-50 psi (140-345 kPa) range and apply from 5 to 60 gallons per acre (gpa) (50-560 liters per hectare).

Tractor-Mounted Sprayers

Tanks on tractor-mounted sprayers usually hold from 150 to 500 gallons (570-1890 L). The tank can be mounted in one of several positions on the tractor (Fig. 8). The pump is usually attached directly to the PTO shaft (Fig. 9) but may be driven by a hydraulic motor or other means. Booms may be mounted in the front, rear, or "belly" positions (Fig. 8). Broadcast applications may also be made with a nozzle cluster (Fig. 10). Tractor-mounted units can be combined with other equipment such as planters, cultivators, or tillage implements (Fig. 11).

High-Clearance Sprayers

High-clearance sprayers have evolved from tractor-mounted sprayers. They have a frame high enough to clear corn, cotton, tobacco, and other tall crops (Fig. 12). The underslung tank fits between crop rows. The spray boom may be raised or lowered, depending on crop height and application requirements.

Trailer-Mounted Sprayers

Trailer-mounted sprayers are built on wheels and towed through the field by a tractor. Tank capacity ranges up to 1,000 gallons (3785 L). The pump is mounted on the tractor or sprayer and driven by the tractor PTO shaft or a hydraulic motor (Fig. 13). Boom lengths vary from 12 to 50 feet (3.7 to 15 meters) or more (Fig. 14). Trailer sprayers are available in high clearance models.

Fig. 9—Pump on PTO Shaft

Truck-Mounted Sprayers

Skid-mounted sprayers (Fig. 15) may be placed in a pickup or flat-bed truck (Fig. 16). Pump power is supplied by an auxiliary engine.

Fig. 8—Mounting Arrangements for Tractor-Mounted Sprayers

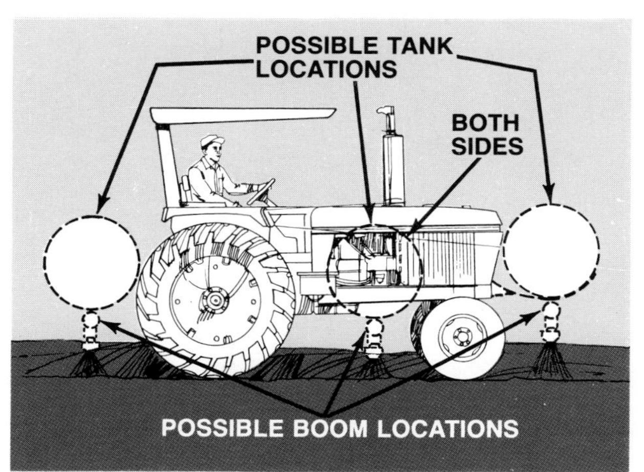

Fig. 10—Broadcast Spray Application

Fig. 11—Spraying Weeds or Insects in the Row while Cultivating

Fig. 12—Typical High-Clearance Sprayer — Tall or Short Crops

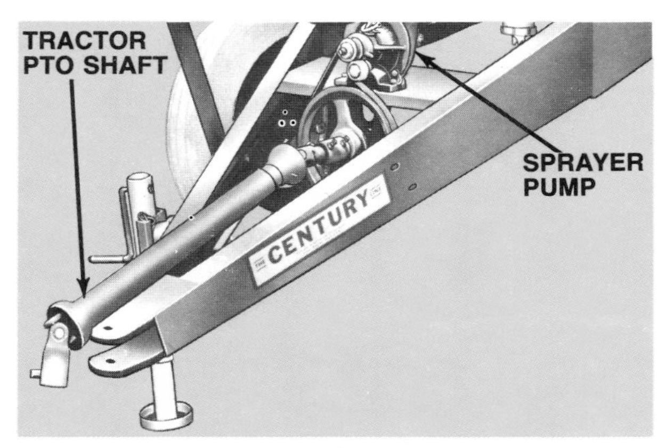

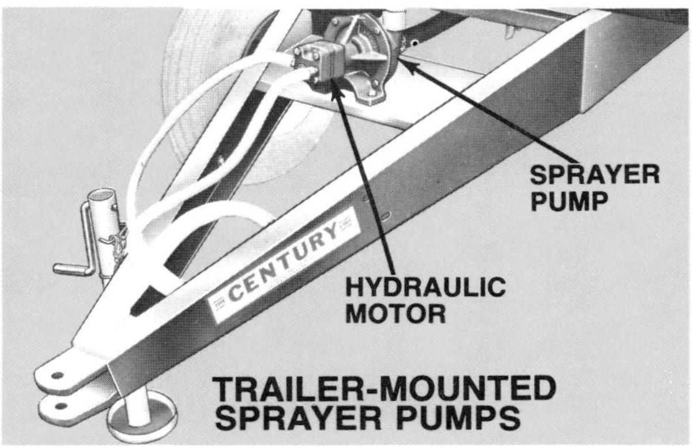

Fig. 13—Pump Mounted on Sprayer

Fig. 14—Sprayer Booms may be more than 50 Feet (15 meters) Long

Fig. 15—Skid-Mounted Sprayer

Fig. 16—Skid-Mounted Sprayer

Fig. 17—Truck-Mounted Sprayer with Flotation Tires

Fig. 18—Spraying with a Helicopter

Larger truck-mounted units (Fig. 17) can be equipped with flotation tires so they can operate under relatively wet conditions. They are best suited for large acreages or custom operation. Such units have tanks holding up to 2,500 gallons (9465 L) and booms measuring up to 60 feet (18 m).

Aircraft-Mounted Sprayers

Principal advantages of aircraft sprayers compared to ground equipment are:

- *Rapid ground coverage.*
- *Application can be made in places and at times when ground equipment cannot operate.*

Unless properly calibrated and operated aircraft sprayers are not as thorough in applying material as ground rigs, especially in tall crops. However, coverage is adequate for most needs. Both fixed-wing airplanes and helicopters (Fig. 18) are widely used.

Helicopters are more maneuverable and are not restricted to operating from a landing strip. However, they are more expensive than fixed-wing units and have a smaller payload.

Because of the limited carrying capacity of all types of aircraft, spray materials are usually applied in a more concentrated form than for ground application. Volumes of 3 to 15 gallons per acre (28 to 140 L/ha) are common. Application is made at a height of 3 to 25 feet (0.9 to 7.6 meters) above the tops of the plants, at 80 to 125 mph (130 to 220 km/h).

HIGH-PRESSURE SPRAYERS

High-pressure sprayers are similar to low-pressure

Fig. 19—High Pressure is Needed to Spray to the Tops of Tall Trees

units except they are capable of developing higher working pressures — up to 1,000 psi (6895 kPa). High pressure is used to drive spray through heavy brush or to the tops of tall trees (Fig. 19). When equipped with a boom, the high-pressure sprayer can do any work usually done by low-pressure units.

High-pressure sprayers are heavy because they are strongly built to withstand the extreme pressures. Consequently, they are more expensive than low-pressure units.

Fig. 20—Orchard Sprayer

AIR-CARRIER SPRAYERS

Air-carrier or air-blast sprayers are often called mist blowers. A high-speed air stream carries pesticide to the surface being treated. The sprayers are rated in air capacity and velocity. Capacities range from 5,000 to 60,000 cubic feet per minute (cfm) (140 to 1700 cubic meters). Air speeds range from 80 to 150 mph (130 to 240 km/h).

Because air carries the pesticide to the target, little dilution water is needed. Up to 80 percent of the water used in normal sprayers can be saved. Therefore, much less time is required for filling the sprayer, and a more efficient operation can result.

Air-carrier sprayers are used in orchards, fields, and for spraying ornamental trees. Orchard sprayers (Fig. 20) can spray to either side or both sides simultaneously. Because drift is very difficult to control, field crops are not usually sprayed with air carriers.

One type of field unit (Fig. 21) is mounted on a turntable so it can be aimed to either side. Another unit, commonly used for treating shade trees (Fig. 22), can cover foliage as much as 100 feet (30 meters) from the ground.

FOGGERS AND ROPE WICK APPLICATORS

Foggers and rope wick applicators are not really sprayers, but they are used to apply liquid herbicides.

Foggers

Foggers apply pesticides, usually insecticides, as very fine droplets called aerosols. A single aerosol droplet is too small to see, but a concentration of droplets is visible, floating in the air like smoke or a cloud. Consequently, aerosol generators are frequently called foggers.

Aerosol particles can be formed by heat, very fine nozzles, air blasts or spinning disks. Foggers may be electrical or driven by an engine. Most engine-powered foggers discharge the aerosol vapor through the engine exhaust system.

Fig. 21—Rotating-Table Blower

Fig. 22—Air-Carrier Sprayer for Tall Trees

Electrically-powered foggers may operate on standard 110-120 volt household current, or from the electrical system of a car, truck or tractor (Fig. 23). On some units, oil-based material is pumped into a heated barrel where it is vaporized and discharged as a vapor cloud. Air currents then distribute the vapor.

Fig. 23—Foggers Vaporize Pesticide for Complete Coverage of Plants, Animals or the Inside of Buildings

Fig. 24—Fogger Used in Greenhouse

Foggers are used to completely fill or blanket an area with insecticide. They are most often used in enclosed spaces such as greenhouses (Fig. 24), but are sometimes used outdoors for insect (mostly mosquito) control. Weather conditions must be suitable for a fogger to be used outdoors. The fine particles are easily moved by air currents. Rising air currents or a strong wind can easily carry the fine aerosol particles away before they contact all target insects.

Rope Wick Applicators

Rope wick applicators apply liquid material by wiping it onto the plants. The pesticide liquid in a pipe soaks out through rope segments (Fig. 25). When the tractor carries the applicator across the field, plants which reach above the pipe are contacted and the pesticide is wiped off the rope onto the plants. Rope wick applicators are well suited for the situation where scattered tall weeds such as johnsongrass, shattercane or volunteer corn are growing in a field of a shorter crop such as soybeans. The rope wick applicator can be used to treat such weeds with herbicide so that they die back and cause fewer problems at harvest time.

Many farmers construct rope wick applicators to suit their own needs and desires. More recently, small manufacturers have begun to market the equipment.

LIQUID FERTILIZER APPLICATORS

Liquid fertilizers are applied in three ways:

- **Surface applicators**
- **Subsurface applicators**
- **Irrigation water injectors**

The characteristics of the fertilizer material being applied determines the proper choice of surface or subsurface applicators. There are three types (Table 1), differentiated by the vapor pressure (non-pressure, low or high pressure) developed.

Vapor Pressure

Ammonia is the primary material source for nitrogen in liquid fertilizers. A tank containing ammonia is never completely full of liquid. The top portion contains ammonia gas under pressure. The vapor pressure depends on the percentage of ammonia in the liquid and the temperature (Fig. 26). Liquids with lower ammonia concentrations develop lower vapor pressure.

Fig. 25—Rope Wick Applicator

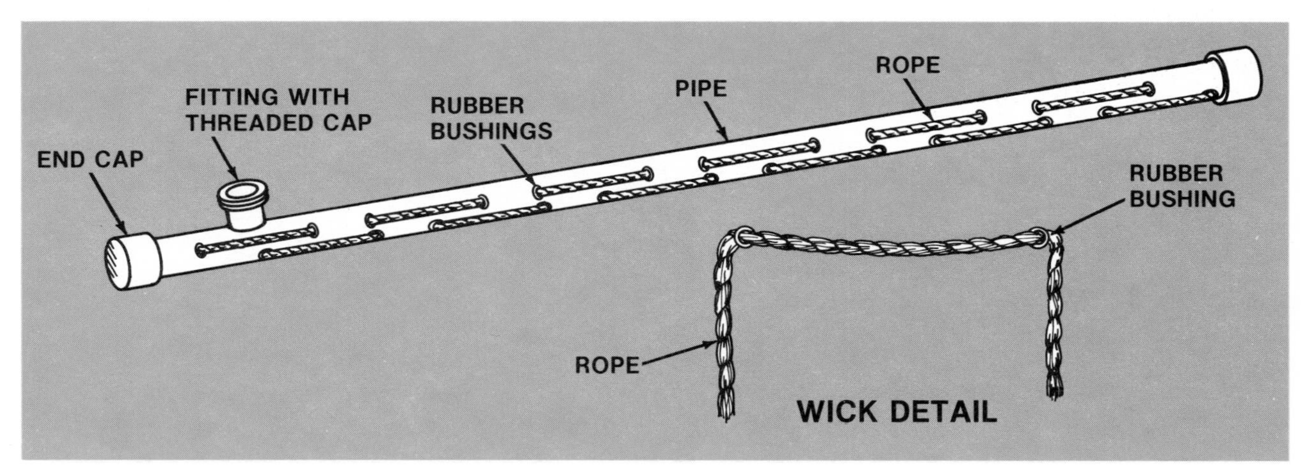

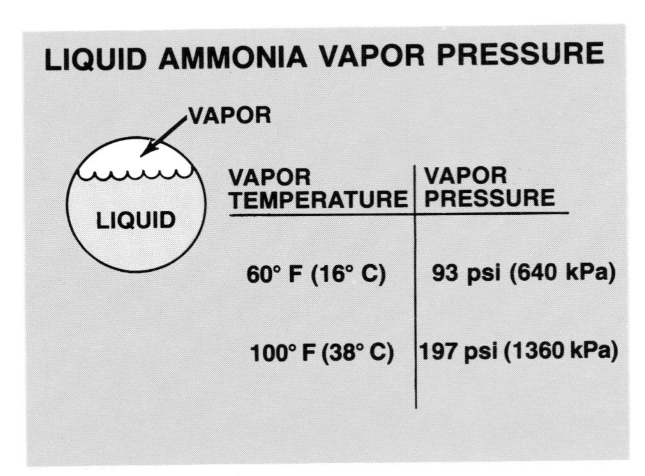

Fig. 26—Vapor Pressure in a Liquid Ammonia Tank

SURFACE APPLICATORS

Surface liquid-fertilizer applicators, which are similar to sprayers, are used to apply unpressurized liquid fertilizer formulations to the soil surface. They have higher application rates and lower power requirements than subsurface applicators. Three general types of surface applicators exist:

- **Boom-spray applicators**
- **Boom-stream or dribble-tube applicators**
- **Boomless spray nozzles**

TABLE 1

LIQUID FERTILIZER APPLICATION METHODS

Fertilizer Type	Example Solution	Application Method
Non-pressure	Urea in water	Surface
Low-pressure	Aqua ammonia	Subsurface (2 in. or 5 cm)
High-pressure	Anhydrous ammonia	Subsurface (4 in. or 10 cm)

The first two are the most widely used. A boom-spray applicator may be converted into a dribble-tube applicator by removing the nozzles. Boom applicators may be mounted on trucks (Fig. 27), tractors, or trailers.

Boom lengths up to 60 feet (18 meters) are available. Aircraft application can be used when the soil conditions prevent ground applications.

Liquid-fertilizer applicators are often used as attachments on other equipment (Fig. 28).

SUBSURFACE APPLICATION

Low- and high-pressure solutions are applied below the soil surface to prevent loss of ammonia. Sub-

Fig. 27—Boom-Type Liquid-Fertilizer Applicator

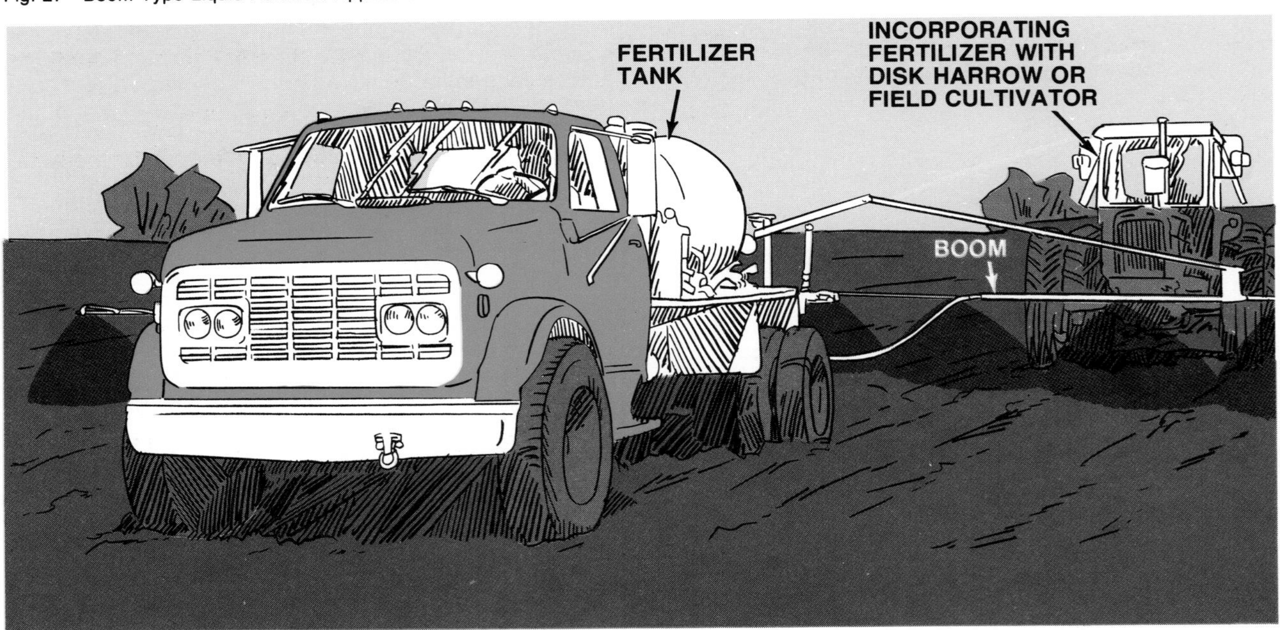

Fig. 28—Liquid-Fertilizer Attachment on a Conventional Corn Planter

Fig. 29—Types of Knife or Chisel Openers

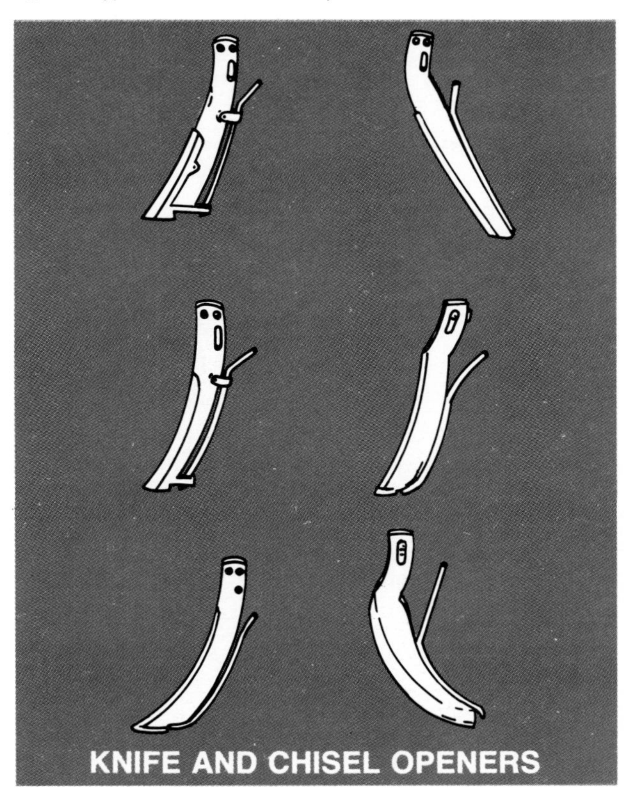

surface applicators have knives or chisels to open furrows in the soil. Liquid fertilizer is then metered into the furrows and soil is pressed over them to seal the surface.

Subsurface application can be made with the following kinds of equipment:

- *Injector blades or knives with injector tubes attached (Figs. 29 and 31).*
- *Disk openers with injector tubes (Fig. 30).*
- *Cultivator shovels, chisels, or moldboard plows equipped with injector tubes.*

Fig. 30—Disk Openers

Fig. 31—Subsurface Anhydrous-Ammonia Application

ELEVATION FURNISHES PRESSURE

Fig. 33—Elevating the Nurse Tank Furnishes Pressure to Load the Applicator Tank.

Fig. 32—Direct Application Methods

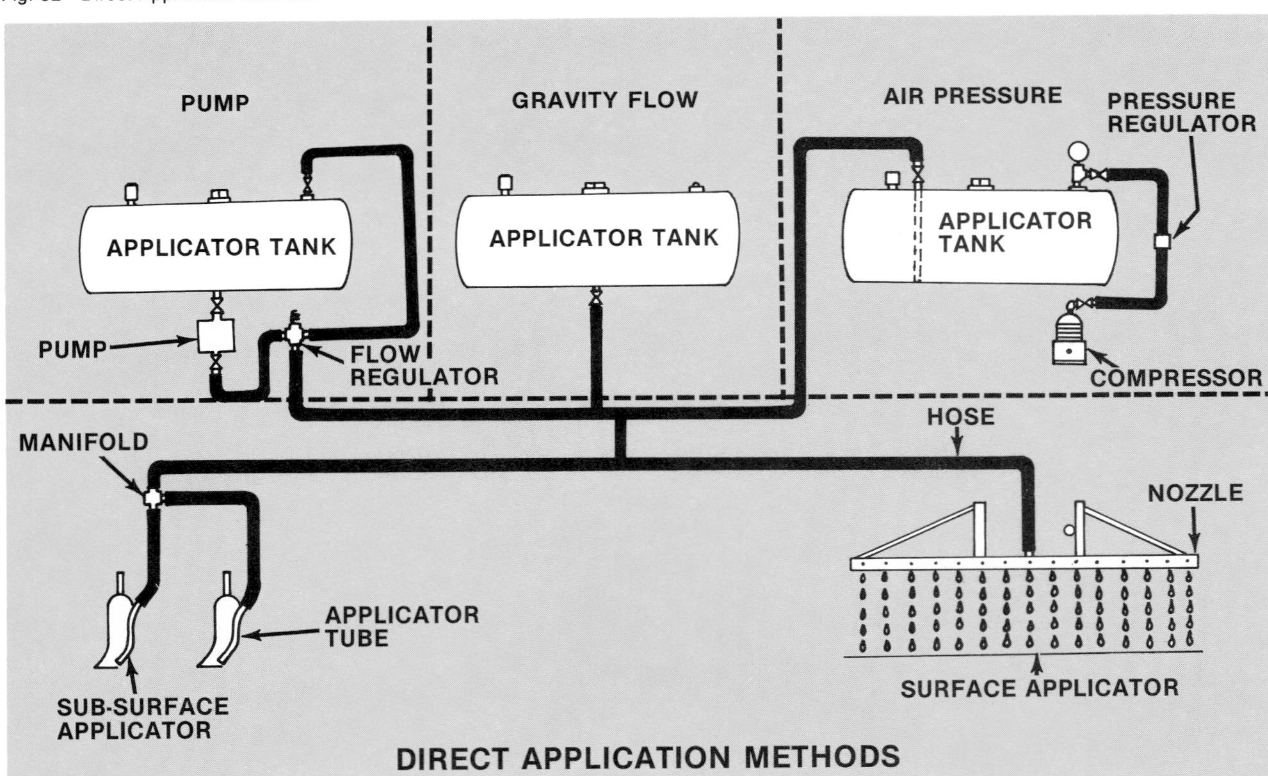

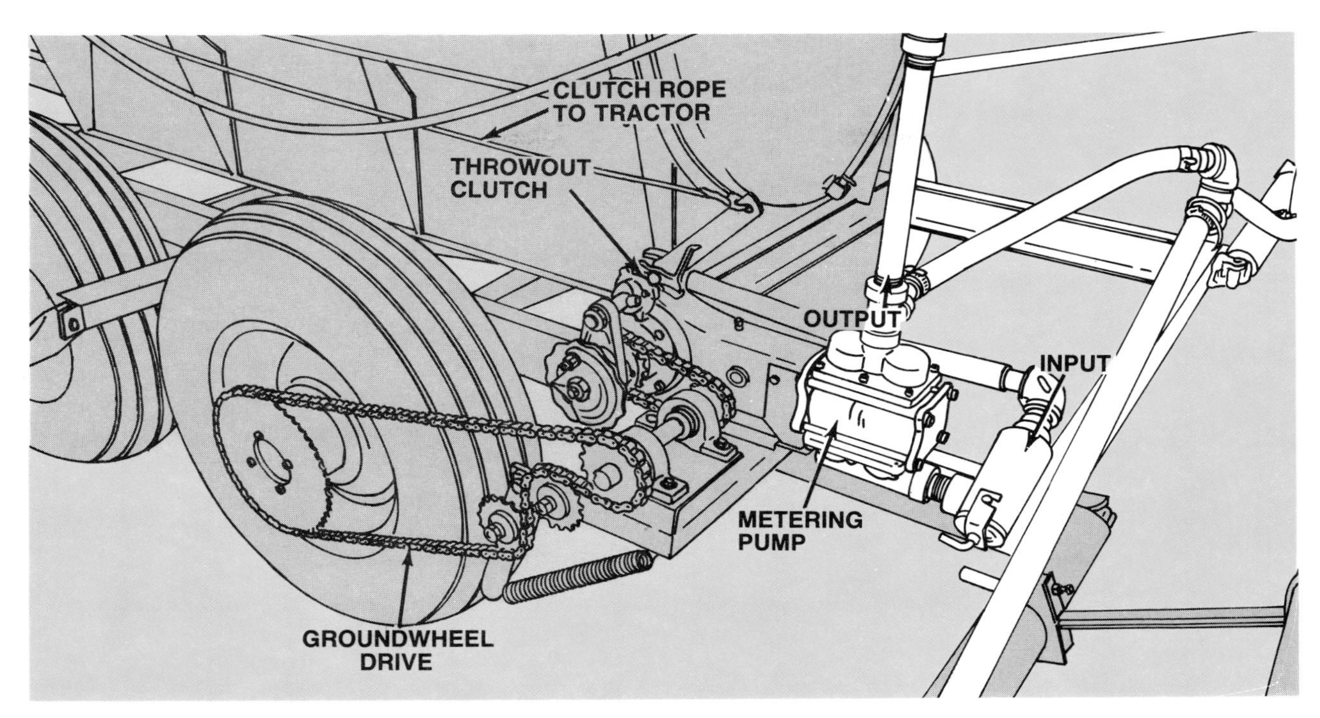

Fig. 34—Metering Pump with Ground Drive on a Trailer-Type Surface Applicator

LIQUID-FLOW SYSTEMS

There are three flow systems for moving liquid fertilizer from the tank to the distribution system for each of the preceding applicator types (Fig. 32):

- **Gravity**
- **Pump**
- **Compressed Gas**

Descriptions of these systems follow.

Gravity

Liquid flows from the tank due to weight of the liquid. Gravity-flow applicators are the simplest and lowest in cost for surface or subsurface use. However, they are inherently inaccurate because of pressure change as the tank empties.

Gravity flow is sometimes used to transfer the solution from a nurse tank to the applicator storage tank. Elevating the tank increases the pressure and reduces the time needed to complete the transfer (Fig. 33).

Metering orifices are used to control the rate of fluid flow for both surface and subsurface applicators. The metering orifice mounts in a standard nozzle body (discussed later). Interchangeable plates with different orifice sizes are available. These metering devices give accurate flow control as long as the pressure is held relatively constant. They can be used in many different locations, but are usually mounted near the final outlet point.

Pump

Pumps are used on many applicators to supply liquid to the nozzles. Roller or diaphragm pumps are driven by the tractor PTO or by an auxiliary engine.

When a pump is used, application rate depends on pressure and ground speed, as with sprayers.

A metering pump, driven by a ground wheel, assures a uniform application rate at any speed (Fig. 34).

Fig. 35—Anhydrous Ammonia Tanks are Built to Withstand High Pressure

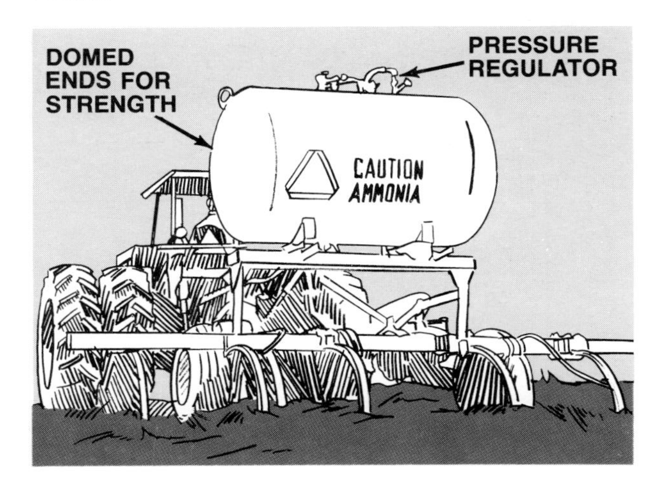

Pressurized Gas

Anhydrous ammonia is pushed out of the tank by the vapor pressure of the gas that forms in the tank (Fig. 35). A similar system, using compressed air or propane, can be used to pressurize any liquid. A regulator maintains a constant pressure. The tanks used must be strong enough to withstand the pressure.

IRRIGATION-WATER INJECTORS

Applying fertilizer materials through irrigation equipment is frequently called "fertigation". There are advantages and disadvantages to the method.

Advantages include:

- *Double use of equipment.*
- *Labor saving because the system does not need to be continuously observed.*
- *Water helps prevent "burning" of plants.*
- *Less fertilizer may be needed during the season.*

Disadvantages include:

- *Fertilizer is applied only as uniformly as the water is spread.*
- *Application is only possible when irrigation is needed.*
- *Only certain fertilizer materials can be used.*
- *Adding the fertilizer may slow down the irrigation.*
- *Only broadcast applications can be made.*

Three methods are used to inject fertilizer into the stream of irrigation water:

- **Suction-Intake**
- **Aspirator**
- **Auxiliary Pump**

Explanations of these methods follow:

Fig. 36—Typical Suction-Intake System

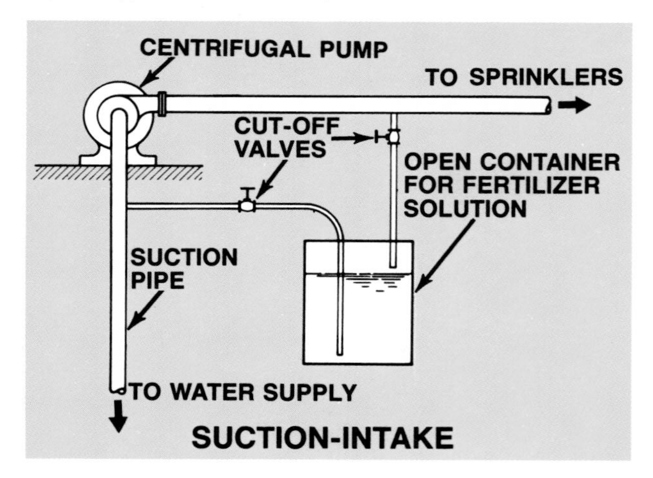

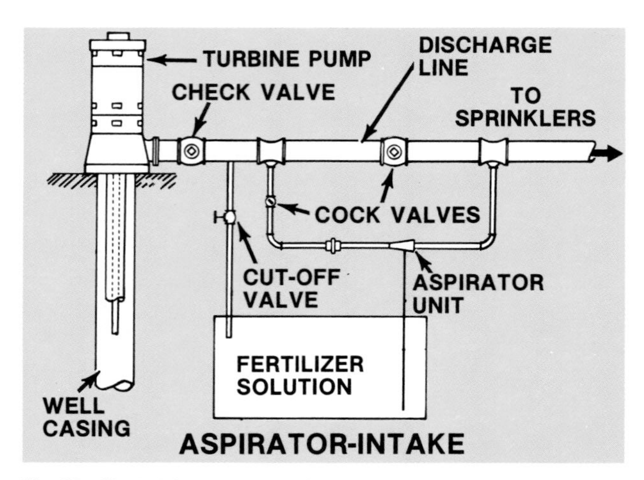

Fig. 37—Typical Arrangement of Aspirator-Intake System

Suction-Intake

The suction-intake method is applicable only for shallow-well systems. An inlet, regulated by a valve, is installed on the suction side of the irrigation pump. The fertilizer solution is then drawn into the water (Fig. 36). Application rates are often difficult to control with this method.

Aspirator

Aspirator injectors may be used with any pumping system. A portion of the water from the outlet side of the pump is bypassed through an aspirator (contriction) (Fig. 37). The resulting suction draws fertilizer solution into the line.

Auxiliary Pump

Auxiliary pumps are frequently used to inject fertilizer into irrigation water. The fertilizer pump has to deliver a higher pressure than the irrigation pump (Fig. 38). The auxiliary pump can be driven electrically, by a gasoline or natural gas engine, by a water-driven motor, or by the main irrigation pump motor.

Fig. 38—Typical Arrangement of Auxiliary Pump-Injection System

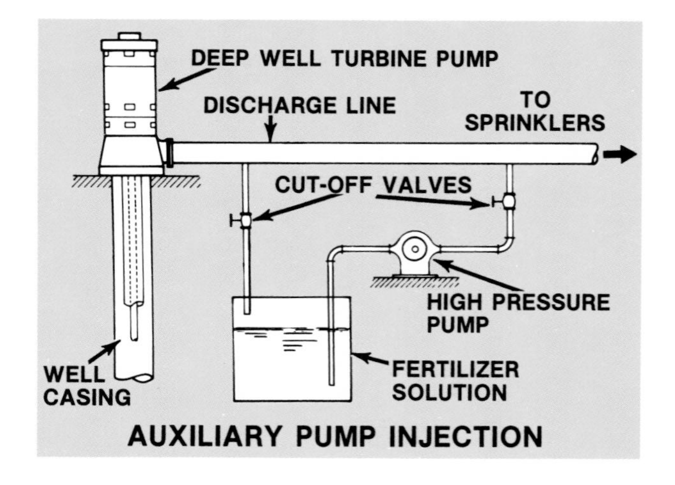

SELECTING A SPRAYER

Sprayers and other application equipment have three basic functions to perform:

- *Distribution of the material accurately into the desired pattern.*
- *Metering the quantity of material being applied.*
- *Storage of chemicals prior to application in the field.*

Sprayers are composed of numerous components. Many different arrangements and combinations may be used. For a particular situation, the best combination depends on:

- **Chemical Being Applied**
- **Application Rate**
- **Crop Being Treated**
- **Required Accuracy**

Sprayers may have the following components (Fig. 39):

- **Tank**
- **Agitation System**
- **Pump**
- **Pressure Regulator**
- **Pressure Gauge**
- **Strainers and Screens**
- **Pipes and Hoses**
- **Frame**
- **Control Valves**
- **Nozzles**

TANKS

Sprayer tanks should meet the following requirements;

- *Sufficient capacity*
- *Easy to fill and clean*
- *Corrosion resistant*
- *Shape suitable for easy mounting and effective agitation*

Galvanized Steel Tanks

Galvanized steel tanks are inexpensive and can be made in almost any shape or size. They are easy to repair or modify. **The biggest drawback is corrosion.** Even with protective coatings, chemicals cause rapid rusting. Rust flakes off, plugs nozzles, clogs strainers, and damages pumps. Galvanized tanks and recycled barrels are suitable for most pesticides. But they should not be used with the more corrosive liquid fertilizers and insecticides and nematicides.

Recycled barrels must be thoroughly cleaned before being used for spraying and should be replaced immediately if rusting or corrosion begins.

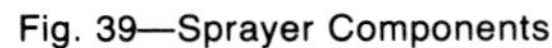

Fig. 39—Sprayer Components

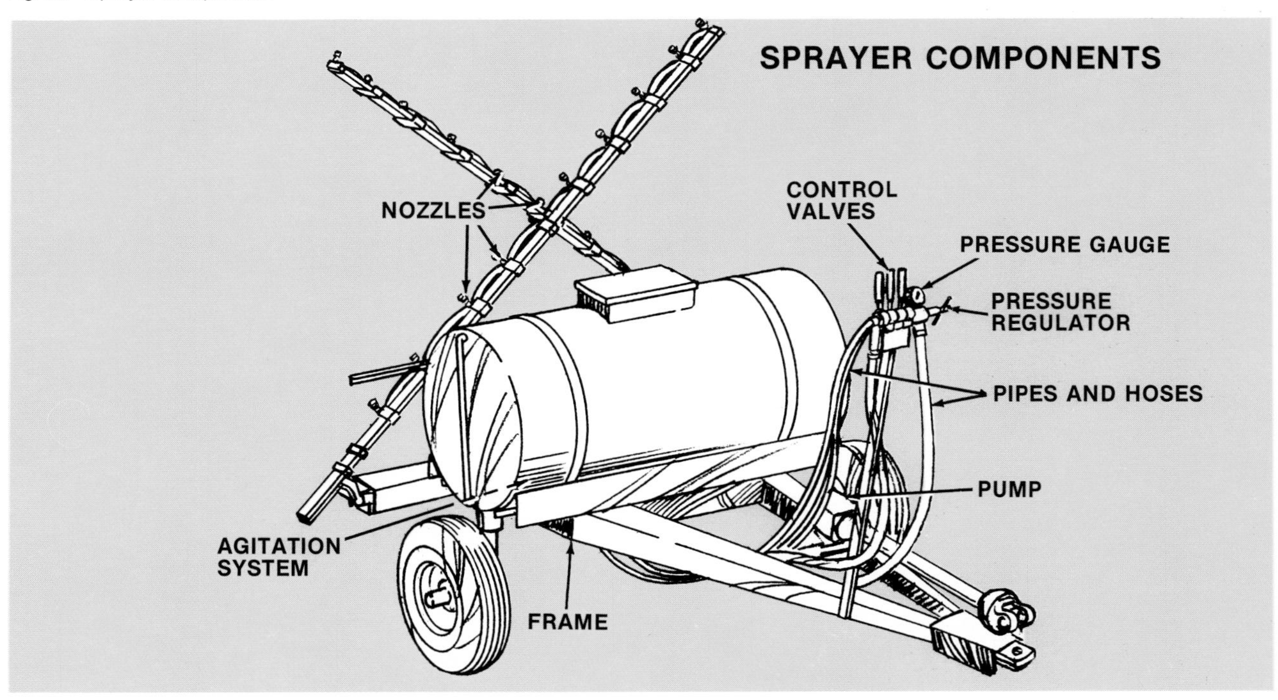

Fig. 40—Polyethylene Tanks are Tough

Polyethylene Tanks

Polyethylene tanks are relatively inexpensive and can be made in many sizes and shapes. They are non-corrosive and can be used with liquid fertilizers except ammonium phosphate solutions or complete-analysis liquid fertilizers. Stainless-steel tanks are needed for these materials. Polyethylene tanks are tough and durable (Fig. 40). However, if one is cracked or broken, it must be replaced. There is no effective way to repair it. Polyethylene breaks down under ultra-violet light so these tanks should be kept inside out of the sun when not in use.

Aluminum Tanks

Aluminum tanks are medium in cost, resist corrosion, and are suitable for many chemicals. However, they should not be used for liquid-nitrogen solutions with a phosphoric-acid base.

Fig. 41—Typical 3-point Hitch Sprayer with Fiberglass Tank

Fig. 42—Stainless Steel Tanks can be used with any Crop Chemical

Fiberglass Tanks

Fiberglass tanks are widely used on all types of sprayers and applicators and as nurse tanks (Fig. 41). Fiberglass is a strong, durable material. However, it will break or crack under impact. One advantage of fiberglass over polyethylene is that there are repair kits for "on farm" use, or a dealer can make repairs. Fiberglass tanks are about equal to aluminum in cost and can be used with most kinds of chemicals. They may, however, be affected by some kinds of solvents.

Stainless Steel

Stainless steel is the highest quality material for pesticide and fertilizer applicator tanks. It is strong, durable, and resistant to corrosion by any crop chemical (Fig. 42). It is the most expensive material commonly used for tanks. As a result, only equipment with high annual use is equipped with stainless-steel tanks.

Fig. 43—Polyethylene and Fiberglass Tanks must be Saddle-Mounted

Fig. 44—Mechanical Agitators

Tank Mounting

When barrels or small metal tanks are used, tank mounting is not critical. However, polyethylene and fiberglass tanks must be properly mounted and supported on a "saddle" (Fig. 43). The saddle supports the tank over a large area. Without it, weight of the liquid may break the tank as the sprayer bounces over obstructions or rough terrain.

AGITATION SYSTEM

Most sprayers are equipped with some kind of agitator. Agitation is important to maintain a uniform mixture when emulsions or wettable powders are being used. Three agitation systems are used and will provide adequate mixing if properly designed and operated.

- **Mechanical**
- **Hydraulic**
- **Air Sparging (Bubbling)**

Following are descriptions of these systems.

Mechanical Agitation

Mechanical agitators are propellers or paddles mounted on a shaft near the bottom of the tank (Fig. 44). The shaft commonly rotates at 100 to 200 rpm. Excessive agitator speeds can cause foaming in some spray mixtures. If this happens, slow the agitator by changing the drive ratio.

Hydraulic Agitation

Hydraulic agitation returns a portion of the pump output to the tank and discharges it through a series of orifices in a boom along the bottom of the tank (Fig. 45) or through a volume-booster nozzle. Volume boosters increase the movement of fluid. Liquid flowing through the nozzle draws additional fluid from the tank into the moving stream through openings in the side of the nozzle (Fig. 46). Up to twice as much volume can be drawn into the stream from the tank as is pumped into the volume booster nozzle.

Fig. 45—Agitation by Fluid Recirculation

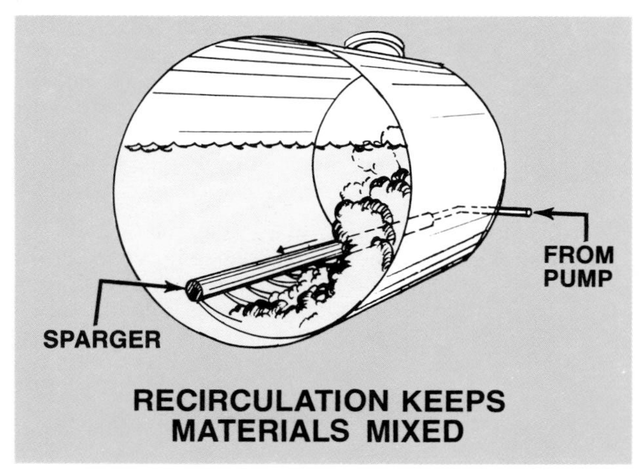

Fig. 46—Volume Booster

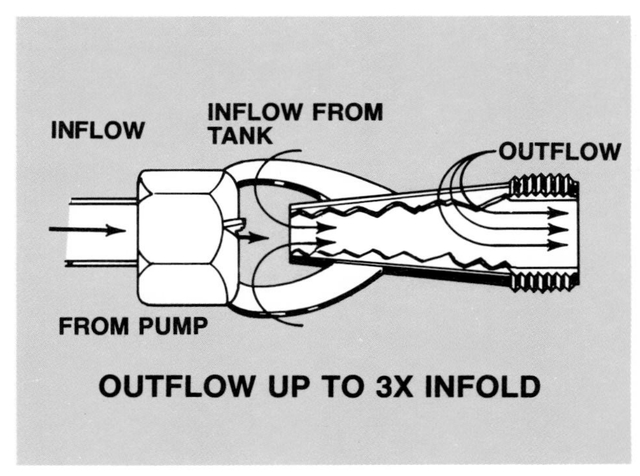

Air Sparging

Air sparging is agitation by bubbling air through the liquid (Fig. 47). A compressor supplies air which is discharged from a sparger tube at the bottom of the tank. As bubbles of air rise to the surface, they create turbulence which keeps the fluid well-mixed.

PUMPS

Six kinds of pumps are used on sprayers. Each kind has certain capabilities and limitations that determine when it should be used (Table 2). The kinds of pumps are (Fig. 48):

- **Gear**
- **Diaphragm**
- **Flexible Impeller**
- **Roller**
- **Centrifugal**
- **Piston**

Of these, the last three listed are the most widely used on agricultural spray equipment.

Gear Pumps

Many early weed sprayers used gear pumps. However, they are seldom used today because of the high wear rate incurred when pumping abrasive fluids. The pump cannot be reconditioned and must be discarded after it is worn.

Diaphragm Pumps

The pumping action in a diaphragm pump is produced by the movement of a flexible diaphragm. Liquid is drawn into one chamber on the downstroke and forced out of another on the upstroke. The diaphragm is resistant to wear by abrasives but may be attacked by certain chemicals.

Fig. 47—Air Sparging

TABLE 2**

PUMP CHARACTERISTICS

Pump	Capacity (gpm)	Speed (rpm)	Max. Press. (psi)	Material
Gear	5-20	500-1800	100	Non-abrasive
Diaphragm	1-10	200-1200	100	Abrasive
Flexible Impeller	0-30	500-1500	50	Mildly Abrasive
Roller	0- 35	600-1800	300	Non-abrasive
Centrifugal	0-150	600-4000	50*	Abrasive
Piston	0- 60	500-1000	800	Abrasive

*Multi-stage units develop higher pressures

**For equivalent table with metric units see page 202, Table 7.

Flexible-Impeller Pumps

Flexible-impeller pumps have a series of rubber "paddles" attached to a rotating hub. The pump housing squeezes the paddles, as the rotor turns causing the pumping action. This pump has an automatic pressure relief. The paddles will not return to the radial position if the pressure is too high.

Roller Pump

The "rollers" of a roller pump fit into slots of a rotating hub. The slots allow the rollers to follow the eccentric shape of the housing. As the rollers pass the inlet port, the space between rollers and the housing becomes larger and draws fluid into the pump. The fluid remains between two rollers as it moves to the outlet port. As the rollers near the outlet port, the spaces become smaller and the fluid is expelled from the pump.

The output from a roller pump decreases substantially as the operating pressure increases because the rollers tend to let more fluid "leak back" between rollers (Fig. 49).

Centrifugal Pumps

Centrifugal pumps have become increasingly popular in recent years. They handle abrasive materials well, and high capacity provides plenty of hydraulic agitation. Centrifugal pumps must be driven at high speed to develop pressure. Belt, gear, or hydraulic drives are used to provide the high speed.

Pump output falls off rapidly at 30 to 40 psi (200 to 275 kPa), (Fig. 49). The steep performance curve is an advantage as it permits controlling pump output without a relief valve. However, it also produces uneven pump output under some conditions.

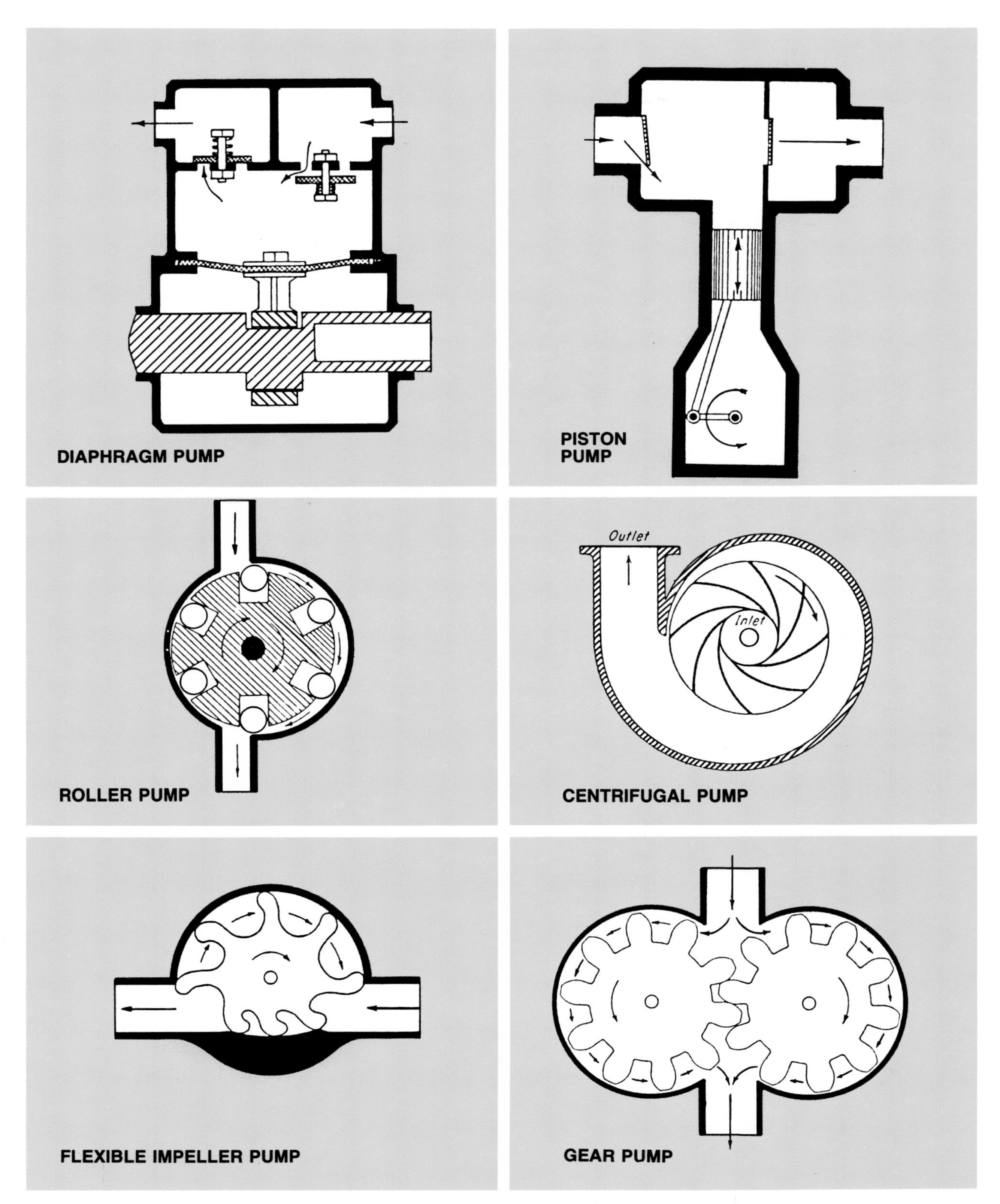

Fig. 48—Six Kinds of Pumps used on Sprayers

Piston Pumps

The piston pump output is virtually unaffected by pressure (Fig. 49). Pump output is usually low and may not be sufficient for hydraulic agitation. It is a good pump for wettable powders and other abrasive mixtures.

Pump Selection

Different pump types have specific performance characteristics which suit them to particular types of spraying (Fig. 49).

Pump selection depends on five factors:

- *Required flow rate for boom and agitation.*
- *Characteristics of materials to be pumped.*
- *Speed of drive shaft*
- *Drive direction and coupling.*
- *Horsepower to drive pump.*

Example

You want to apply a wettable powder at 80 psi (550 kPa). Boom flow rate is 4 gpm (15 Lpm) and the agitation system requires 12 gpm (45 Lpm). Which would be better, a roller, centrifugal, or piston pump?

Answer

Select a piston pump. Many wettable powders are abrasive, so a roller pump should not be used. The operating pressure exceeds the normal capability of centrifugal units.

Pump Power

Pump power is given by the formulas:

$$Hp = \frac{(gpm)\ (psi)}{1714\ (Eff)}$$

Efficiency includes all losses, including factors such as drive friction and pump leakage.

When estimating pump horsepower needed for an application, use an efficiency of 50 to 60 percent.

Metering Pumps

Metering pumps are driven by a ground wheel. When ground speed changes, the rate of pumping increases or decreases proportionately. Thus, the application rate is held constant.

Two types of metering pumps are normally used on liquid-fertilizer applicators:

- **Variable-Stroke Piston**
- **Hose Pump**

The variable stroke piston pump is designed to change the length of the piston stroke to adjust the application rate. Increasing piston stroke increases flow; reducing stroke cuts flow.

The hose pump (Fig. 50) consists of a series of flexible plastic hoses stretched over a roller reel. A ground wheel drives the reel. A fixed amount of solu-

Fig. 49—Typical Pump-Performance Curves

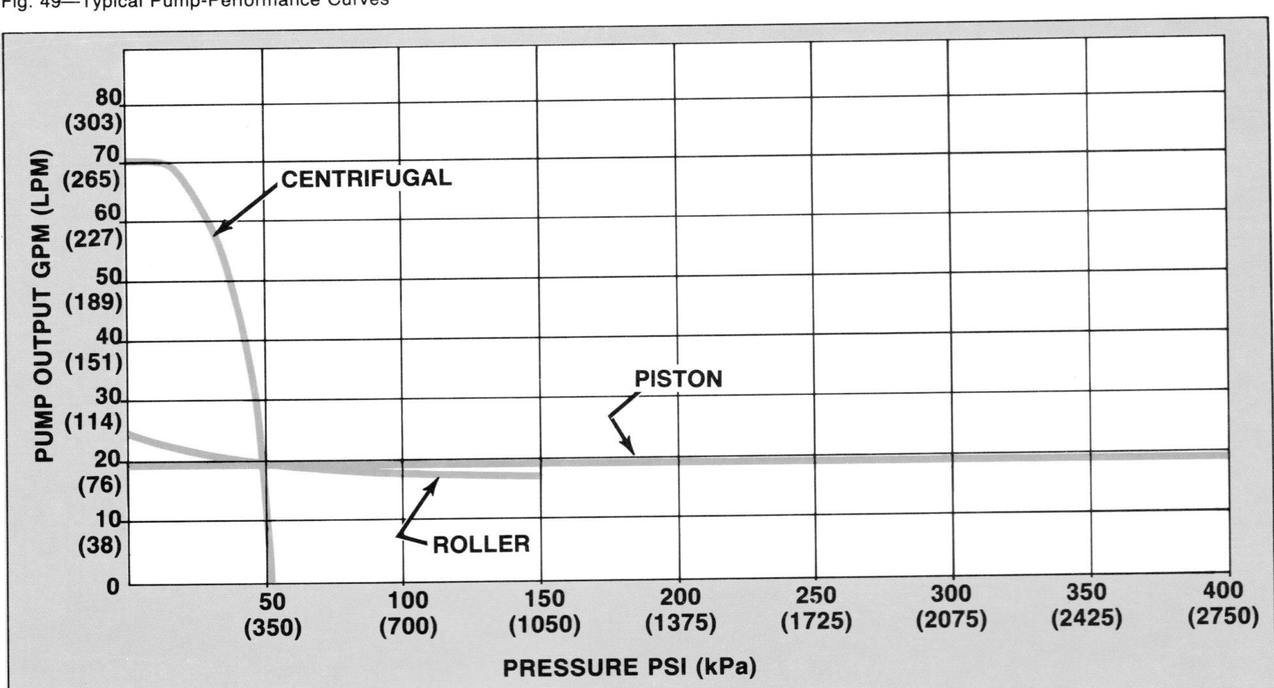

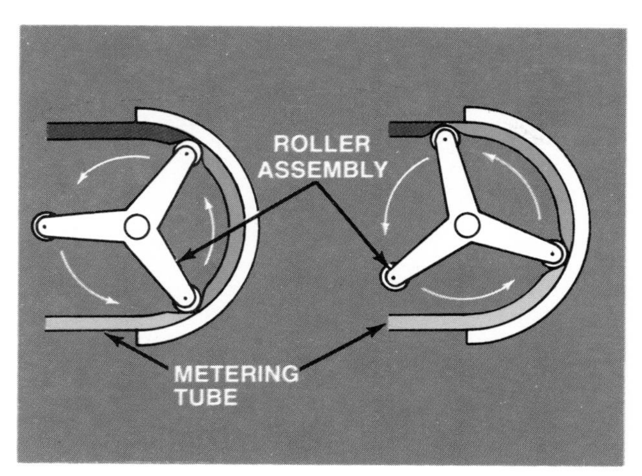

Fig. 50—Hose Squeeze Pump

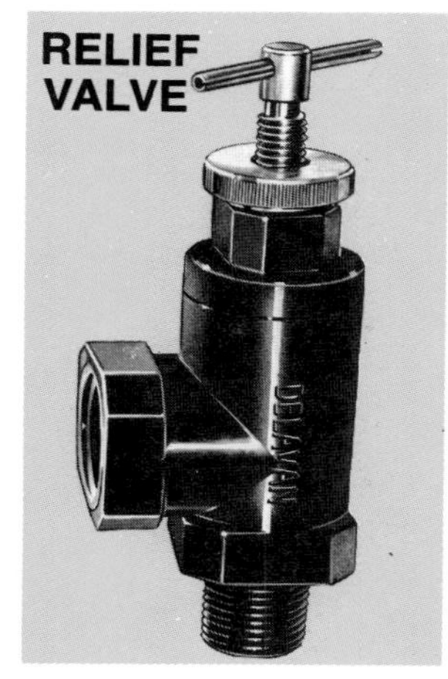

Fig. 51—Pressure Regulators

tion is squeezed out with each turn of the reel. Corrosion is not usually a problem because fertilizer does not come into contact with moving parts of the pump.

PRESSURE REGULATORS

A relief valve is a safety device that releases liquid when the pressure exceeds a safe level. Relief valves can be used to regulate sprayer pressure by adjusting them to open as the desired setting (Fig. 51). When used this way, the valve is always partly open while the sprayer is operating. The excess flow is bypassed back to the tank.

Unloader Valves

Spraying systems operated at pressures over 200 psi (1375 kPa) may use an unloader valve in place of a relief valve to unload the pump when the distribution system is turned off. An unloader valve (Fig. 51) opens and enables the pump output to flow back to the tank at low pressures. This saves fuel and reduces wear.

PRESSURE GAUGE

The importance of a pressure gauge is often underestimated. You cannot regulate pressure if you cannot measure it. The pressure gauge is also a valuable tool for diagnosing sprayer problems.

Pressure gauges (Fig. 52) should have a total of at least twice the expected operation pressure. A pulsation damper is needed to smooth pressure surges from piston pumps.

STRAINERS AND SCREENS

Strainers are commonly used in three places on sprayers. They are:

- **Tank-Filler Screens**
- **Line Strainers**
- **Nozzle Strainers**

Tank screens are coarse screens that remove twigs, leaves, and other large foreign material when the tank is filled.

Line strainers (Fig. 53) are necessary to prevent rust, scale, sand, or other small foreign material from entering and damaging the pump. Location of the strainer depends on the kind of pump. Normally, a 40 or 50-mesh strainer with ports at least as large as the hose size, is adequate.

Fig. 52—Pressure Gauge

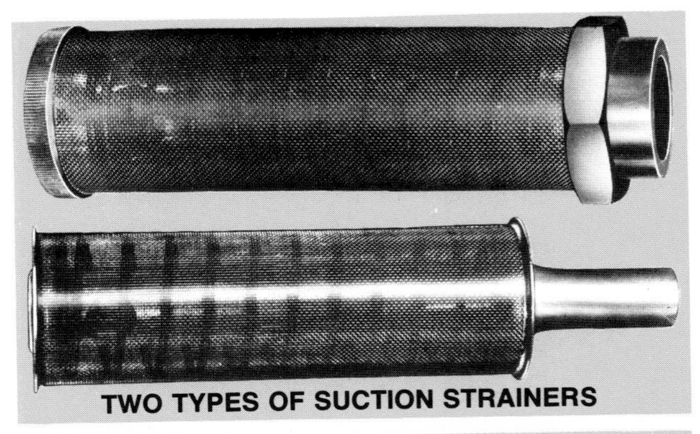

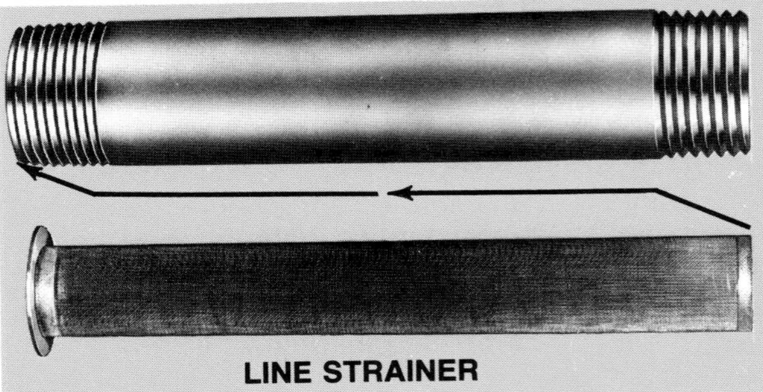

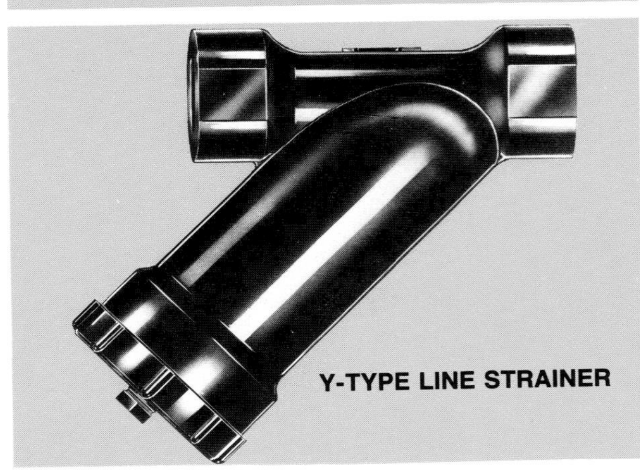

Fig. 53—Typical Strainers

Nozzle Strainers remove fine particles that could clog nozzles. They are discussed in the section on nozzles in this chapter.

PIPES AND HOSES

Pipes and hoses convey the liquid through the sprayer. Liquid pressure varies at different points on the sprayer. Hoses and pipes must be strong enough to prevent bursting. Different types of hose construction can be used (Fig. 54). The rated working pressure of a hose decreases as the diameter increases.

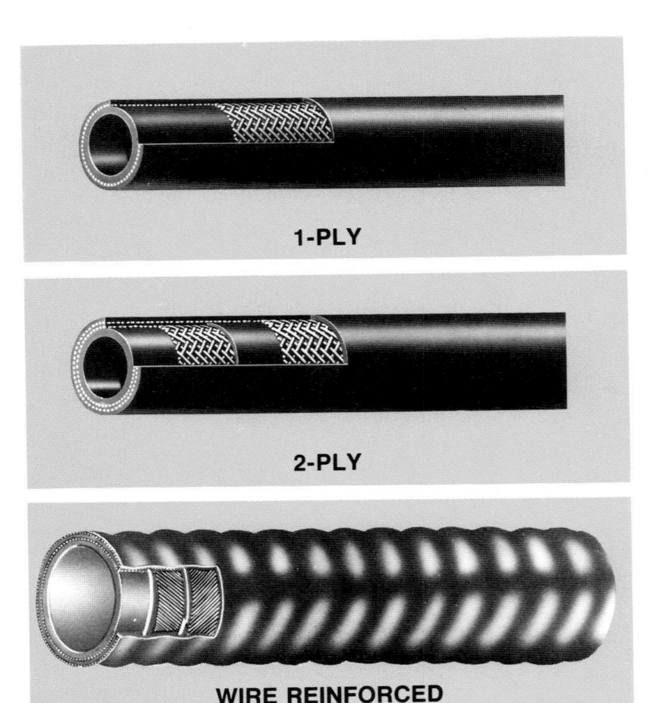

Fig. 54—Sprayer Hoses

Be sure hoses are rated for higher capacity than the expected operating pressure to provide a margin of safety and avoid bursting from pressure surges.

Suction hoses are not pressurized and will not burst, but they can collapse, if the inlet becomes plugged. Suction-hose diameter should be at least as large as the pump inlet port. The hose must be chosen carefully and should be of the noncollapsing, wire-reinforced type (Fig. 54). A collapsed suction hose can restrict flow of liquid and "starve" a pump, causing decreased outflow and greatly accelerated wear.

The inner and outer layers of all hoses should be resistant to the chemicals to be used. Check with both the chemical supplier and the hose supplier if there is any doubt. A hose weakened by chemical attack can leak or burst unexpectedly.

Hose size is important because pressure losses affect flow rates. Pressure loss depends on hose diameter, length and flow rate (Fig. 55). Although pressure losses may not seem significant in hoses shorter than 10 feet (3 meters), it's wise always to use hose of the recommended size to minimize pressure and power losses.

Galvanized pipes resist corrosion. Pipes and fittings used on liquid-fertilizer equipment should not be brass or copper. These materials are weakened by ammonia and can present a safety hazard.

TABLE 3		
SUGGESTED HOSE SIZES		
Pump Output	Hose Size — inch (cm)	
(Lpm)	Suction	Pressure
Up to 12 (45)	¾ (1.9)	⅝ (1.6)
12-25 (45-95)	1 (2.5)	¾ (1.9)
26-50 (98-190)	1¼ (3.2)	1 (2.5)

Fig. 55—Typical Pressure Drop in Hose

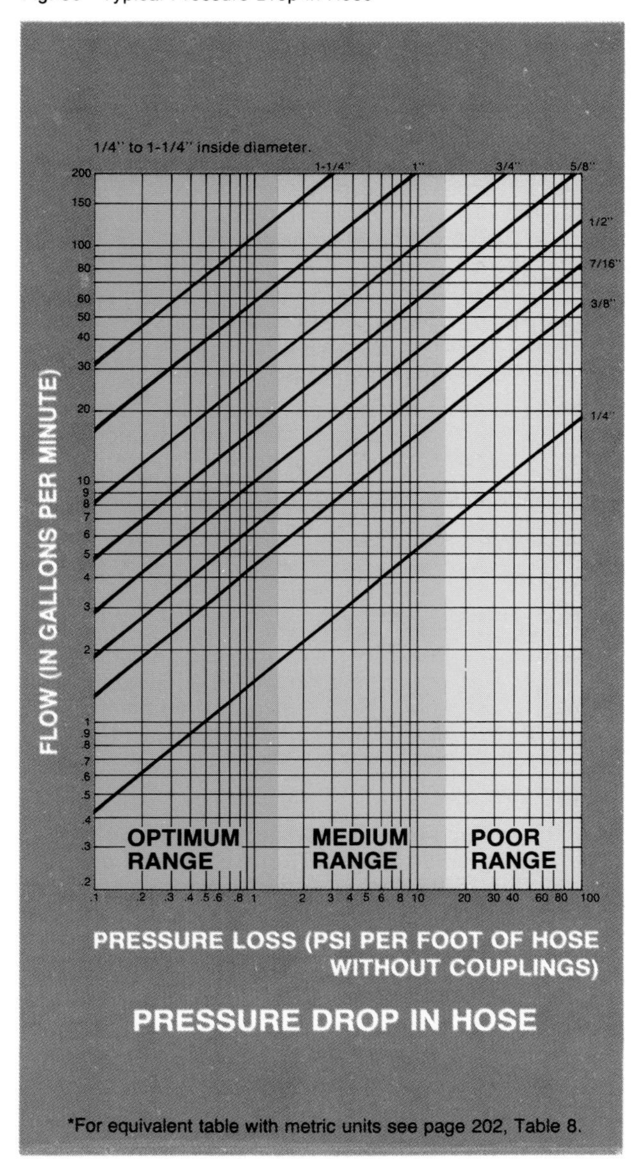

Fig. 56—Sprayer Bouncing on Test Track

FRAME

Sprayer frames must be strong and durable (Fig. 56), and provide convenient points for attaching the boom and mounting other components. With proper design the load is balanced and easy to control.

CONTROL VALVES

Control valves used to regulate the flow of liquid are those valves other than the pressure regulator. One of the most common is the 3-section boom control. This type of control valve permits eight different spraying patterns (Fig. 57).

The type of control valves on a particular sprayer depends on operating requirements and the preferences of the manufacturer.

Fig. 57—Typical Boom-Control Valve

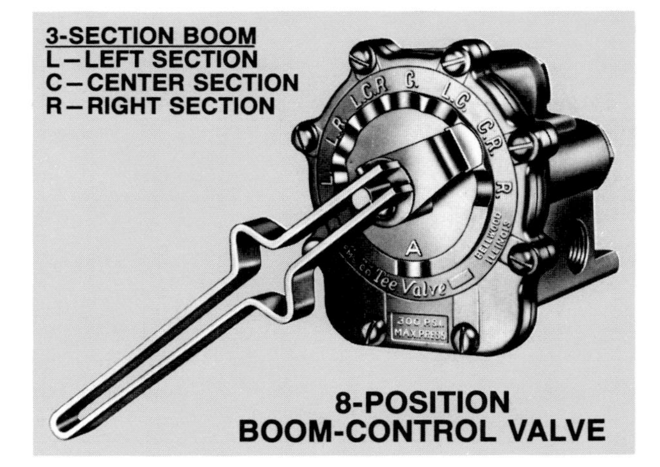

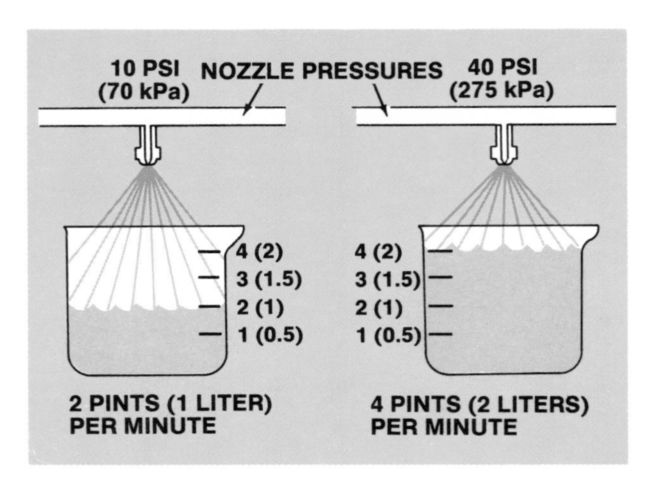

Fig. 58—Flow Rate Is in Proportion to Pressure

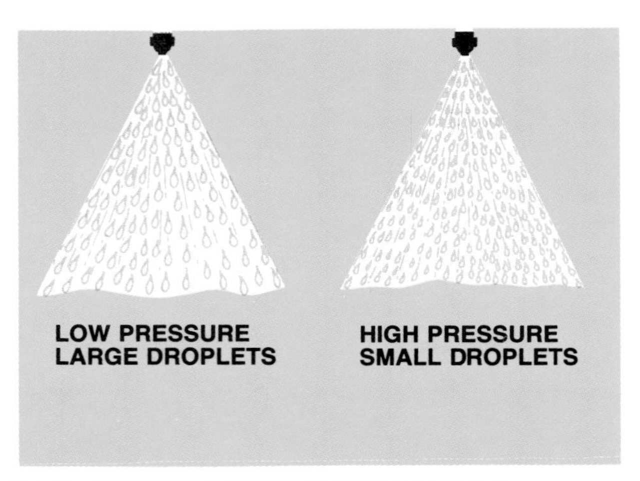

Fig. 59—Droplet Size Decreases as Pressure Increases

NOZZLES

The nozzles on a sprayer serve three functions:

- *Meter liquids*
- *Atomize the liquid stream into droplets*
- *Disperse droplets in a specific pattern*

Flow Rate

For a liquid, nozzle flow rate depends on the effective size of the orifice and the pressure. With most nozzles, flow rate increases as pressure increases. However, it is not strictly proportional. Doubling the pressure does not double the flow rate. In fact, doubling the pressure increases the flow rate to 141 percent of the original. Pressure must be increased by a factor of 4 to double the flow rate (Fig. 58).

TABLE 4

TYPICAL DROPLET SIZES

Spray	Spray Droplet Size (Microns)
Fine aerosols	Up to 50
Coarse aerosols	50-100
Fine sprays	100-250
Medium sprays	250-400
Coarse sprays	400-500
Minimum drift jet stream nozzles	600-900
Low turbulence nozzles	800-1000

Tables of nozzle capacity at various pressures are supplied by manufacturers. These tables are developed by measuring the flow rate of *water* through the nozzle. When other liquids are used, the flow rate will be different. Most dense or viscous liquids have lower flow rates. This is one reason why sprayers must be calibrated for each type of liquid applied.

Atomization

Atomization is the liquid breakup caused by the tearing action of the air. Liquid exits from the nozzle in an unstable sheet which breaks up into droplets.

A nozzle produces a range of droplet sizes from very small to large. Droplet size is measured in microns—one micron is one millionth of a meter (Table 4).

The average size of droplets is affected by:

- **Nozzle Type and Size**
- **Pressure**
- **Liquid Characteristics**

As pressure increases, average droplet size decreases (Fig. 59) until a limit is reached where additional pressure has no effect on particle size. If droplets are too small, drift may be excessive; if droplets are too large they can roll off plants without sticking.

Nozzle Patterns

Spraying nozzles are described according to the shape of the application pattern. There are five common patterns (Fig. 60). Each nozzle type is available in various sizes and spray angles. Each is adapted to a particular type of operation.

Hollow Cone and Solid Cone nozzles are popular for row-crop spraying of herbicides, fungicides, and insecticides (Fig. 61). They are used on spray booms

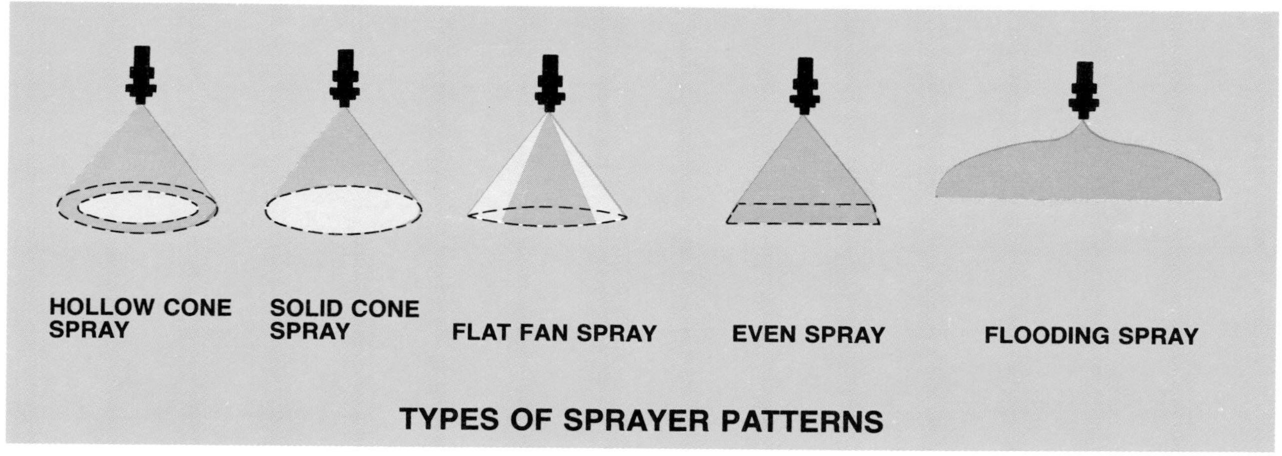

Fig. 60—Types of Spray Patterns

Fig. 61—Row Crop Spraying with Cone-Type Nozzles

Fig. 63—Band Spraying with Even Spray Nozzles

Fig. 62—Even Distribution with Flat Fan Nozzles

Fig. 64—Broadcast Application with Flooding Nozzles

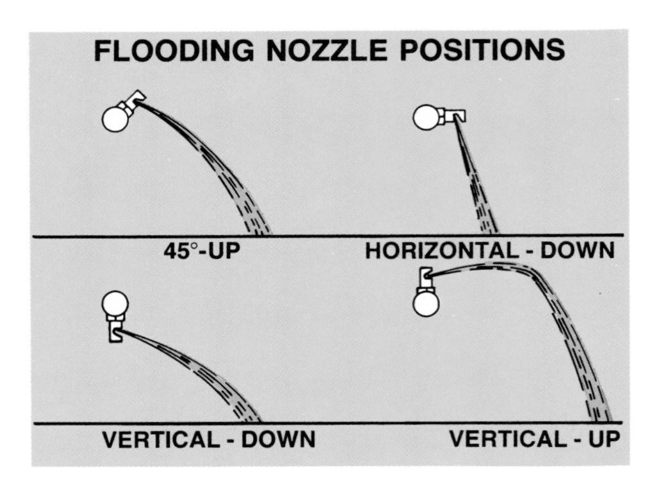

Fig. 65—Mounting Positions for Flooding Nozzles

and handguns. The spray angle may be from 30 to 120 degrees. Hollow cone nozzles generally produce a finer, more uniform spray than the solid cone type.

Flat spray nozzles are used for broadcast application of fertilizers, insecticides, and herbicides (Fig. 62). The pattern is fanshaped with gradually tapered edges. Relatively uniform coverage across the sprayer width is achieved by installing nozzles on the boom with tapered edges of adjacent patterns overlapping.

Even spray nozzles produce a narrow elliptical pattern with a relatively sharp cutoff at the edges of the area sprayed (Fig. 63). They are used for band application of chemicals and are frequently used with planting equipment.

Flooding nozzles deliver a wide flat spray with large droplets. They may be mounted in various positions on a boom to produce different patterns (Fig. 65). The most uniform application results when the nozzle is mounted about 45 degrees above horizontal. Flooding nozzles are most often used for broadcast application of fertilizers and defoliants. They are sometimes used alone for boomless broadcast spraying (Fig. 64).

Parts of a Nozzle

Nozzles have four basic parts (Fig. 66):

- **Body**
- **Cap**
- **Tip**
- **Strainer**

Nozzles Bodies and Caps are made of brass, aluminum, stainless steel, zinc-placed steel, ceramic or nylon. The body and cap make a unit that holds a strainer and a tip (Figs. 66 and 67). The nozzle body may be mounted on either a wet boom (material flows through the boom) or a dry boom (material flows through hose attached to the boom) (Fig. 68). Swivel adapters may be attached to drop pipes to permit pointing nozzles in any desired direction.

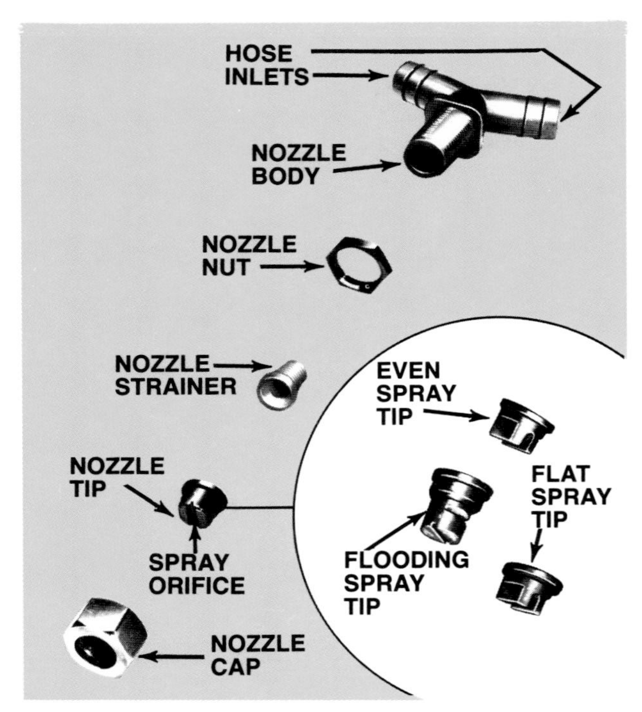

Fig. 66—Typical Spraying Nozzle Assembly

Nozzle Tips form liquid spray into a distinctive pattern. Most agricultural sprayers use interchangeable tips (Fig. 66). Tips are available with a wide range of capacities and distribution patterns for specific spraying requirements. Tip orifice size and sprayer pressure determine the rate of flow.

Fig. 67—Other Spraying Nozzle Parts

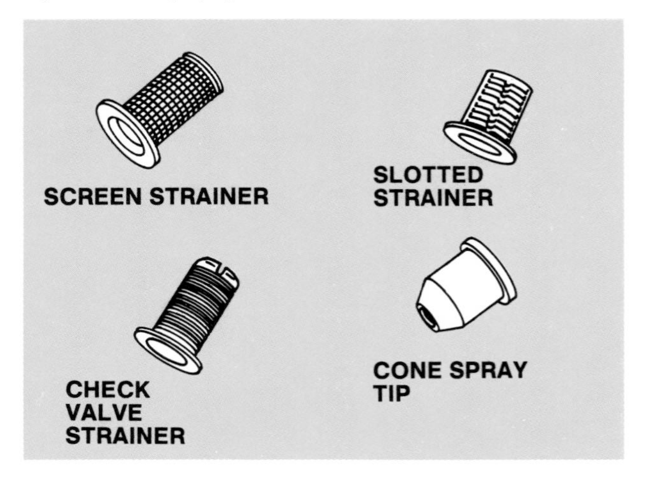

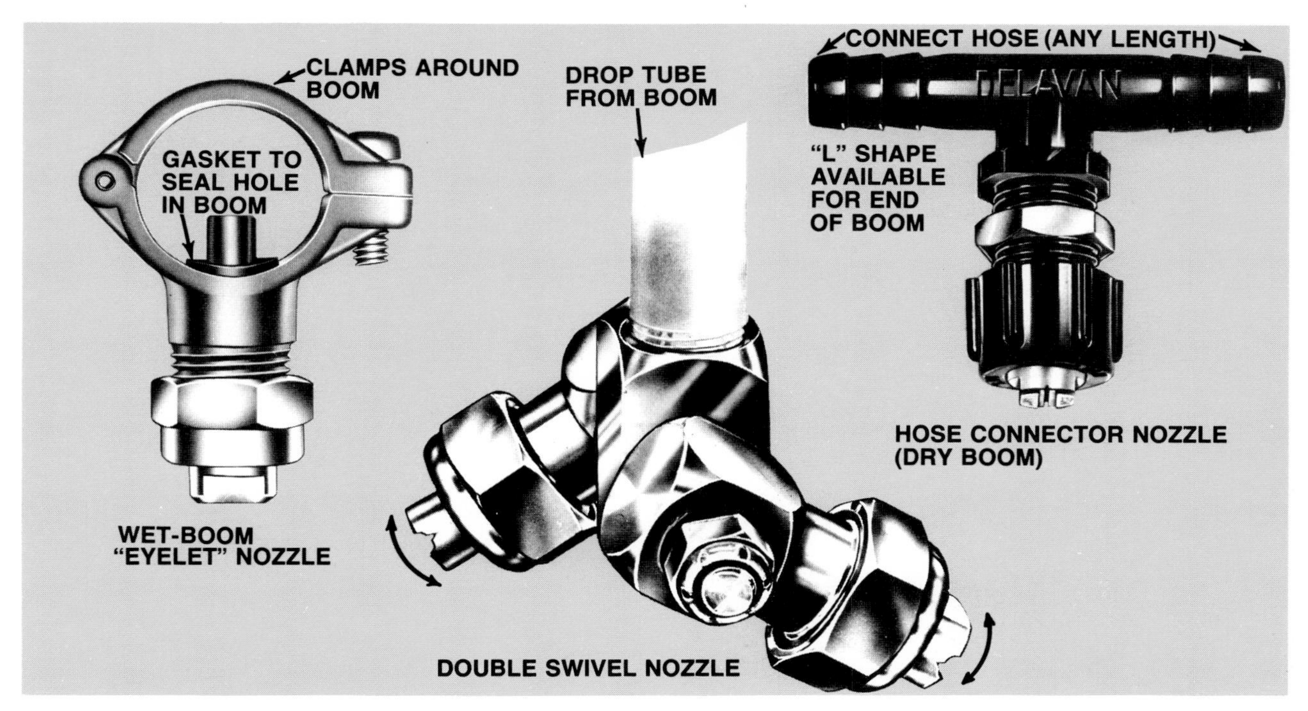

Fig. 68—Typical Nozzle Bodies and Mountings

Strainers are placed ahead of the nozzle tips and provide final screening of the liquid to protect tips from plugging (Figs. 66 and 67).

Most nozzle screens are cylindrical in shape and fit inside the nozzle body. Screens are usually made of stainless steel or brass and have from 50- to 200-mesh openings (Fig. 53). For spraying abrasive materials, a slotted strainer with openings equivalent to 16- to 50-mesh screens may be used.

When selecting a nozzle screen, choose one with a mesh opening smaller than the nozzle orifice. Of course, finer screens clog more readily. Screens must be kept clean for efficient and accurate distribution of spray material. Clean and check them often. Use a soft brush or compressed air to clean screens.

Caution: Do Not Try To Clean Screens With Your Breath. You Could Get Harmful Pesticide into Your Mouth, Nose or Eyes.

In some spraying operations, it is desirable to have a quick shut-off at each nozzle to prevent dripping while turning at ends of rows, filling the tank, or transporting the sprayer. Special strainers equipped with check valves are available for this purpose (Fig. 69).

ROTARY NOZZLES

Rotary nozzles use centrifugal force rather than hydraulic energy to break the material into droplets. Spray material is fed to the bottom of a flat spinning disk or the bottom of a small spinning cup. It is forced to

Fig. 69—Plunger-type (Nozzle) Check Valve

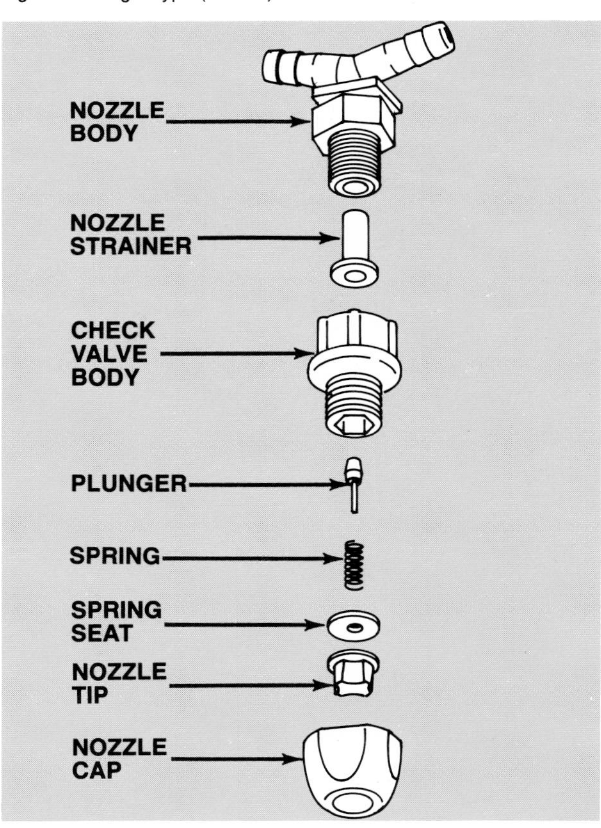

Fig. 70—Rotary Spray Nozzles

the edge of the disk or cup by centrifugal force (Fig. 70). When it reaches the edge, it breaks up and is thrown out in a hollow cone pattern.

Fig. 71—Battery-powered Rotary Nozzle

Rotary nozzles produce more uniform droplets than hydraulic nozzles. Most units produce a primary and a smaller secondary particle size. Most of the spray material is in droplets of the primary size. A lesser amount is in particles of the secondary size. Relatively little is discharged in particles of other sizes. Droplet size can be changed by altering the rotational speed. For example, one two-speed unit puts out mostly 250 micron particles at a low speed setting and 75 - 100 micron particles at a higher speed.

Rotary nozzles can be battery powered and mounted on the end of a long handle for hand use (Fig. 71) or attached to conventional sprayers (Fig. 72). They may be used in applications where careful droplet size control is needed to obtain maximum effectiveness, drift control, or conservation control.

To date, there has been relatively little experience on farms with rotary nozzles. If the rate of adoption increases, the best uses for the nozzles and the best management of the equipment may be sorted out.

SPRAYER CIRCUITS

A sprayer circuit is the path followed by the liquid as it pumps through the sprayer and out the nozzles. All circuits are similar and involve the same basic components. However, a few important differences exist.

Fig. 72—Rotary Nozzles Attached to Conventional Sprayer

Circuit variations depend primarily on the kind of pump (Figs. 73, 74, and 75). Each new sprayer has a specific hydraulic circuit. If the type of pump is changed, always change the hydraulic circuit to match the new pump.

SPRAYER OPERATION

Sprayer operation includes all activities—before, during, and after field use—that affect the quality of field application. Eight separate activities are:

- **Planning**
- **Preliminary Adjustment and Setting**
- **Calibration**
- **Loading**
- **Transport**
- **Field Operation**
- **Cleaning**
- **Storage**

These activities are discussed in detail.

PLANNING

Preliminary planning begins with the decision to apply a pesticide and to choose the particular chemical to be used. Consult bulletins from your cooperative extension service, chemical advertisements, and the farm press for advice.

While selecting the chemical, decide how it is to be applied (sprayer, duster, granules, aircraft, ground application). The chemical label will help. You then have four more decisions to make:

- *Application rate — gpa (L/ha)*
- *Number and placement of nozzles*
- *Proper operating speed*
- *Pressure required for proper application*
- *Nozzle tips needed for desired application rate, pressure, and speed*

Application Rate

The application rate should be within the range recommended on the chemical label. Low-concentration mixtures (lots of water) usually drift less because the spray particles are larger. On the other hand, such heavy applications require more water, more frequent tank filling, and more unproductive time.

Nozzle Arrangement

The arrangement of nozzles will depend on how the material is to be applied and the size of the target. Material for foliar applications is usually broadcast with nozzles uniformly spaced on the boom. Directed sprays for row crops may require one or more nozzles

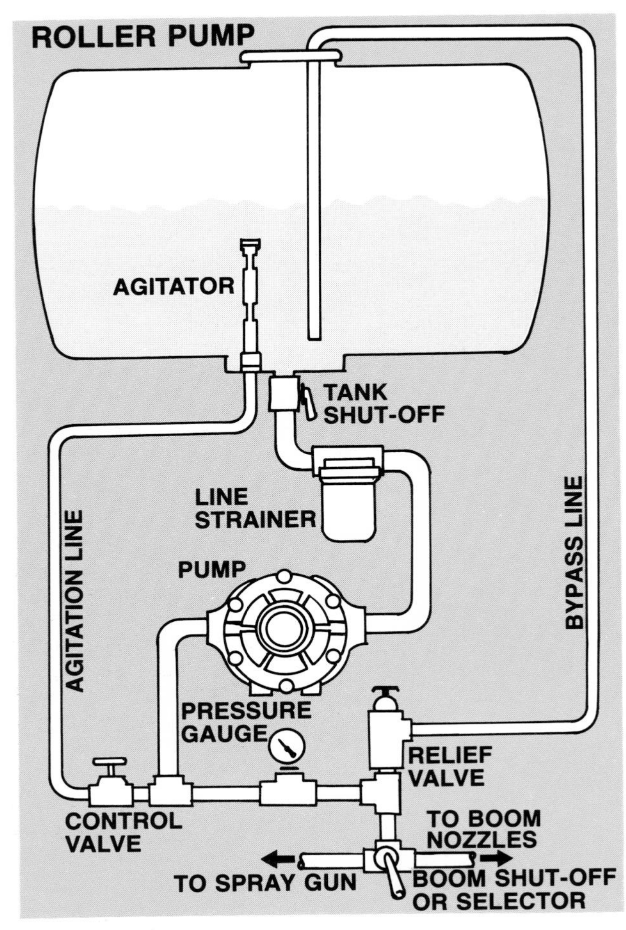

Fig. 73—Sprayer Circuit with Roller Pump

on each side of the row, depending on material to be applied and height of crop (Fig. 76).

Operating Speed

Operating speeds range usually from 1 to 10 mph (2 to 16 km/h), depending on the equipment and terrain. Higher speeds allow quicker, more timely application of the chemical. But, to maintain the same application rate, higher pump capacity and more power is needed.

Selecting a Tip

Nozzle tip selection is a key planning step. The selection should always be checked by careful calibration (discussed later). Both tip size and material

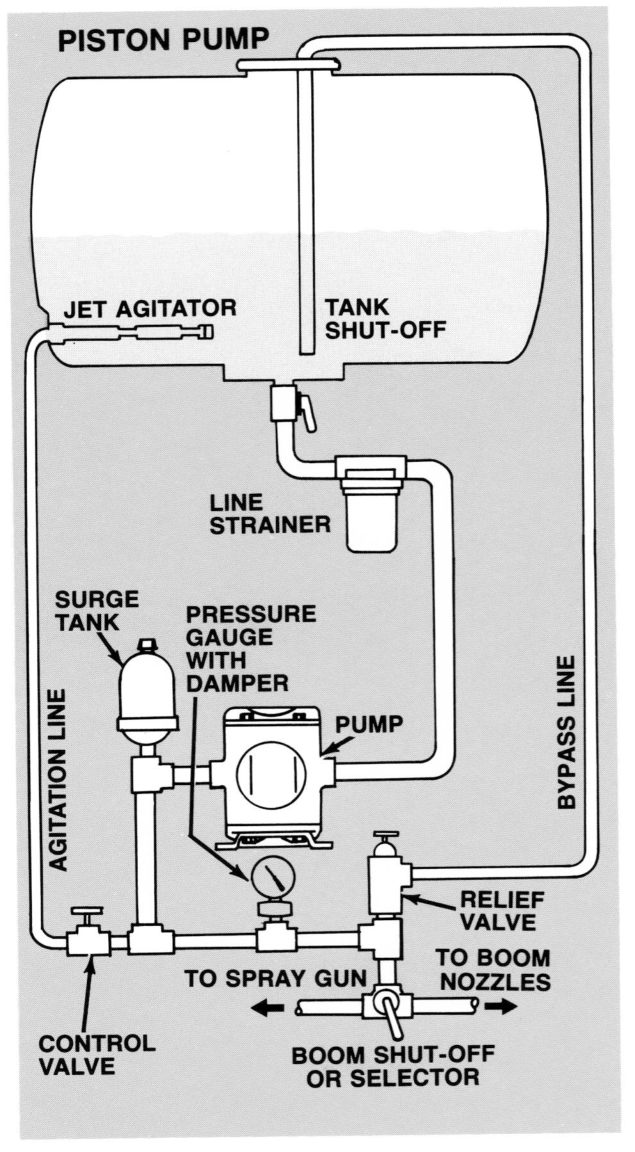

Fig. 74—Sprayer Circuit with Piston Pump

should be specified. Tip size depends on the application rate, row spacing or width of spray pattern and ground speed. Some simple calculating will be required. The basic relationship is:

$$\text{Flow Rate (gph)} = \frac{\text{Application Rate (gpa)} \times \text{speed (mph)} \times \text{width (ft.)}}{\text{8.25 (a constant)}}$$

$$\text{Flow Rate (Lph)} = \frac{\text{Application Rate (L/ha)} \times \text{speed (km/h)} \times \text{width (m)}}{\text{10 (a constant)}}$$

The following example shows how to use the relationship.

Example:

Apply 12 gpa, using 2 nozzles per 36-inch row (effective spraying is 1.5 ft/nozzle), ground speed of 6 mph and pressure of 40 psi or less.

Solution:

Substitute the appropriate numerical values into the formula.

$$\text{Flow rate (gph)} = \frac{(12\ \text{gpa}) \times (6\ \text{mph}) \times (1.5\ \text{ft})}{8.25}$$

$$\text{Flow rate (gph)} = 13.09\ \text{gph}$$

$$\text{Flow rate (gpm)} = \frac{13.09}{60} = 0.218\ \text{gpm}$$

Fig. 75—Sprayer Circuit with Centrifugal Pump

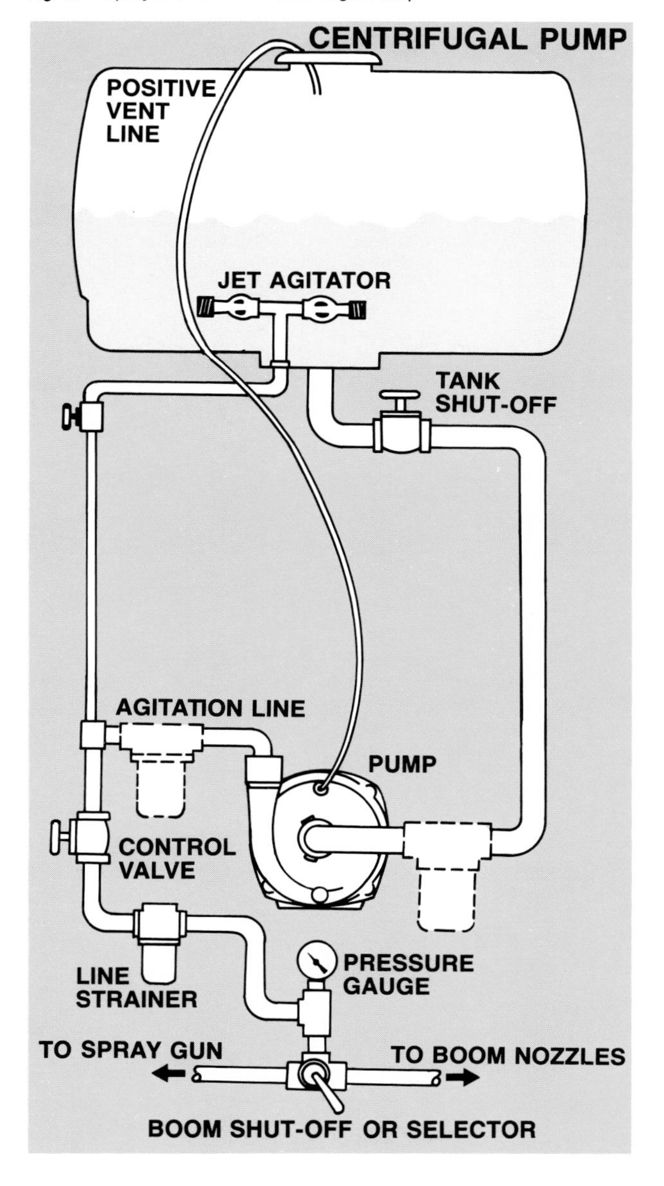

Example (metric):

Apply 110 gpa, using 2 nozzles per 1-meter row (effective spraying is 0.5 m/nozzle), ground speed of (10 km/h) and pressure of 275 kPa.

Solution:

Substitute the appropriate numerical values into the formula.

$$\text{Flow rate (gph)} = \frac{(110\ \text{L/ha}) \times (10\ \text{km/h}) \times (0.5\ \text{m})}{10}$$

$$\text{Flow rate (gph)} = 55\ \text{L/h}$$

$$\text{Flow rate (gpm)} = \frac{55}{60} = 0.92\ \text{L/min}$$

Conclusion: Select nozzles which apply at least 0.218 gpm (0.92 L/min) at a pressure of 40 psi (275 kPa) or less.

If you find the mathematical relationship confusing or difficult, you can approximate the same process by using the graphs in Fig. 77. Comparable graphs with metric measurements appear in the Appendix at top of page 203. Simply select the set which most closely approximates the situation. The previous example is diagrammed on chart C. If you are selecting a tip for a speed, width, or application rate which is not specifically covered by one of the graphs, simply estimate the result. You may then want to work through the equation to double check the result.

Important — the graphical result gives the required nozzle capacity in gallons per hour. Specifications for most nozzles are in gallons per minute. If that is the case, convert by dividing by 60.

The type of spray material used affects tip life. Abrasive materials will wear nozzles faster than oils or non-abrasive materials. When selecting nozzle tips, keep in mind the spray material to be used and relative life of different tip materials (Table 5).

PRELIMINARY ADJUSTMENTS AND SETTINGS

Preliminary adjustments and settings include all of the adjustments that are made when the machine is being prepared for use.

Before starting to spray, check wheel bearings and tire inflation, and lubricate moving parts as recommended in the operator's manual. Tighten any loose bolts or nuts.

Install tips, screens, check valves, and any other equipment that has been selected. Be sure fan nozzles are aligned so patterns overlap slightly but do not interfere with each other (Fig. 78). If multiple nozzles are being used on each row, aim them to obtain maximum coverage and so that the patterns do not interfere with each other (Fig. 79).

Boom height depends on the spray angle of the tips selected (Table 6). Set the boom at the required height and level it from side to side (Fig. 80). Proper

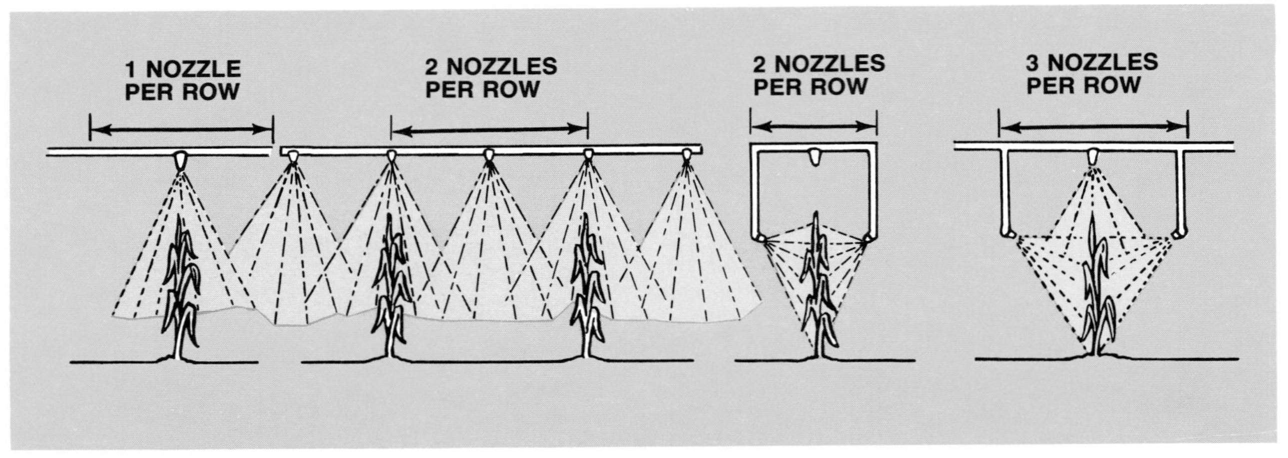

Fig. 76—Nozzles per Row

Fig. 77—Application Rate Charts—Nozzle Spacing in Green Type (see page 203 for comparable charts in metric)

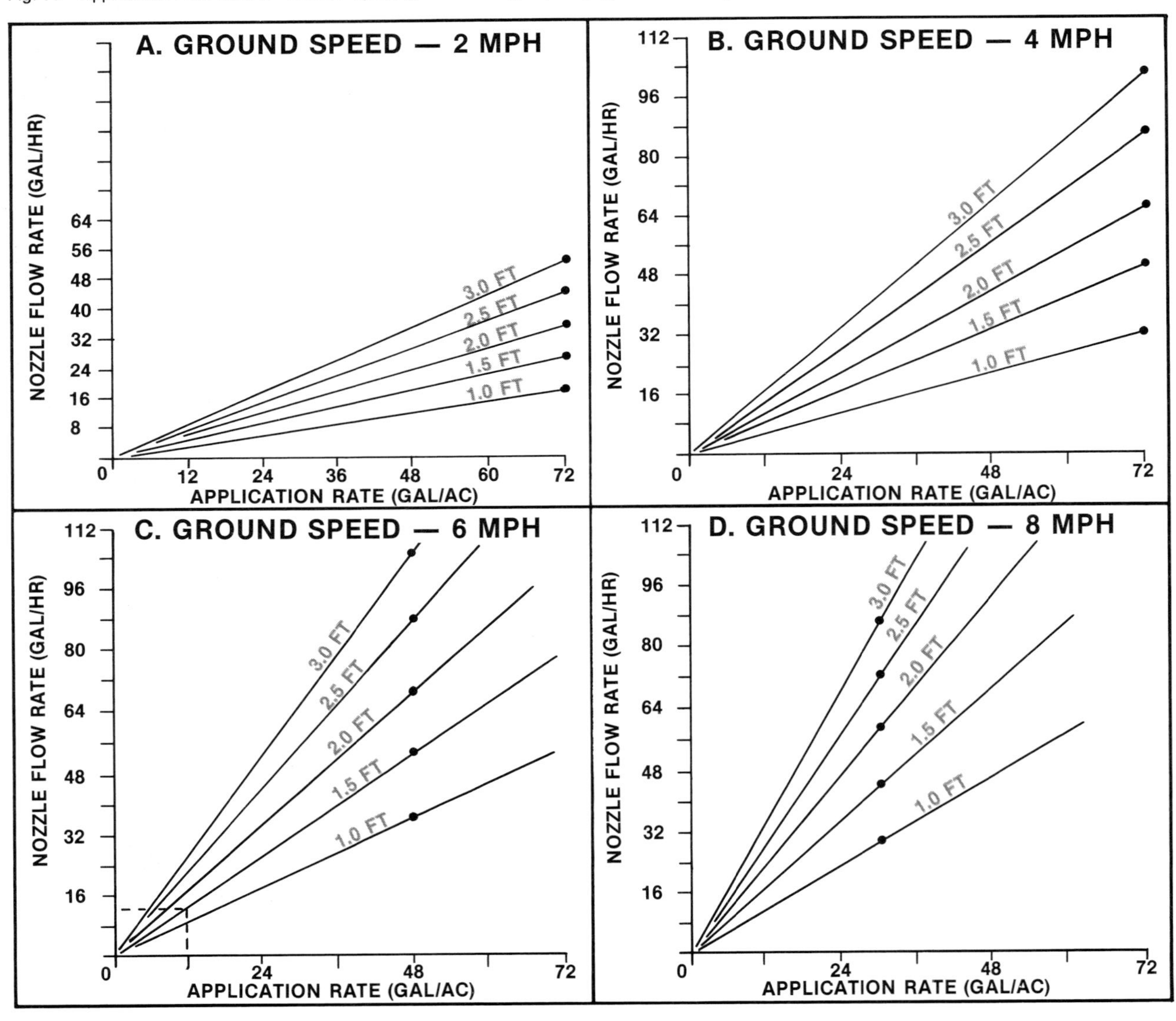

TABLE 5

RELATIVE NOZZLE TIP WEAR LIFE*

Material	Life (year)
Brass-Aluminum	1
Stainless steel	2 to 3
Hardened stainless steel	10 to 15
Ceramic	Lifetime
Carbides (Tungsten, Chrome)	Lifetime

* Life number is that in relation to the wear expectancy of Brass and Aluminum.

Fig. 78—Let Flat Fan Spray Patterns Overlap Slightly

boom height is high enough to clear plant tops when spraying. When material is applied directly to the soil, set the boom to the height recommended by the nozzle manufacturer. Improper height causes nonuniform application.

CALIBRATION

Calibration is the determination of sprayer output or application rate. Sprayers must be calibrated periodically for two reasons:

- Most application rate charts are based on the flow characteristics of water. For accuracy the sprayer must be calibrated to check flow rates after chemicals are added.
- Many studies have shown that the volume of flow and the spray pattern both change as nozzles wear.

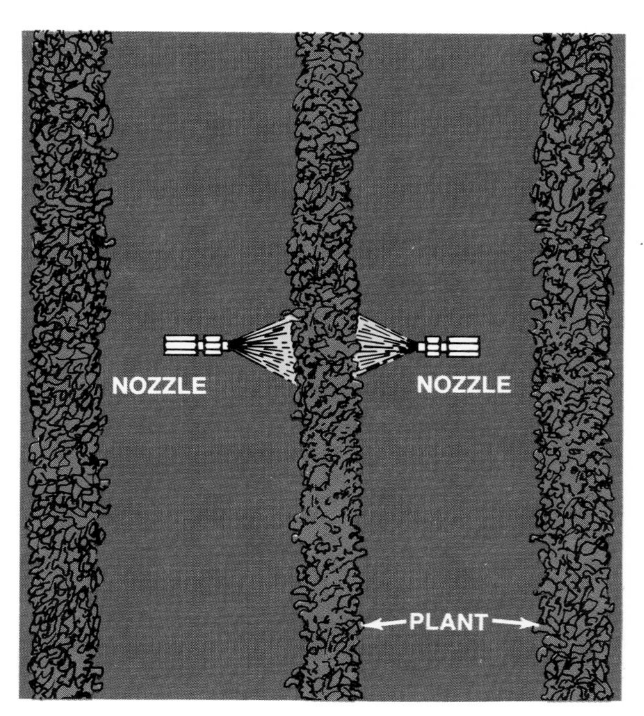

Fig. 79—Aim Nozzles for Row-Crop Application

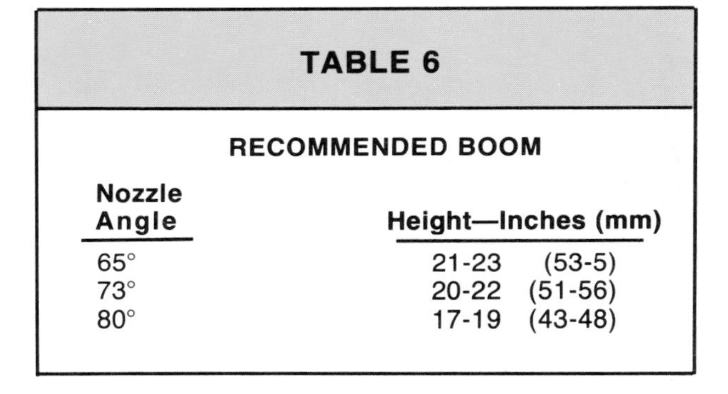

TABLE 6

RECOMMENDED BOOM

Nozzle Angle	Height—Inches (mm)
65°	21-23 (53-5)
73°	20-22 (51-56)
80°	17-19 (43-48)

Fig. 80—Level the Boom for Uniform Application

Fig. 81—Collecting Spray from Each Nozzle

The wear rate is highest when wettable powders are being applied. Therefore, the sprayer should be calibrated frequently when applying such materials.

Check Nozzle-Tip Flow Rates

Before calibrating the entire sprayer, check each nozzle for equal flow rates. Install the selected tips and partly fill the tank with water. Run the sprayer a few minutes to make sure the lines are full and all air has been expelled. During this time, adjust the pressure regulator or the flow-control valves so the desired operating pressure is developed, and observe the pattern of the spray on the ground.

With the sprayer stationary and operating at the proper pressure, collect spray from each nozzle (Fig. 81) for an equal amount of time (30 to 120 seconds). Weigh or measure the volume of each sample. Record each sample and calculate the average.

Clean or replace any tip having an uneven spray pattern or an output 10 percent above or below the average (Table 7).

After obtaining uniform nozzle output, calibrate the sprayer. Any of several methods can be used. Any calibration procedure rests on the relationships between the following four measures of sprayer operation:

- **Application rate (gal/ac or L/ha)**
- **Ground speed (ft/min or m/min)**
- **Boom width (ft or m)**
- **Pumping rate (gal/min or L/min)**

If any three of these are measured, the fourth can be easily calculated. The calibration procedures that will be discussed determine the application rate. Procedures available from other sources may have the goal of specifying one of the other three variables. For best results, you should understand what you are doing

TABLE 7

CHECKING TIP FLOW

Nozzle Number	OUTPUT ounces	(mL)	Comments
1	27.6	(815)	OK
2	25.4	(750)	OK
3	28.5	(840)	OK
4	24.2	(715)	Low-clean and re-check or replace
5	29.3	(865)	OK
6	26.6	(775)	OK
7	27.6	(815)	OK
8	30.1	(890)	High-replace
9	26.3	(777)	OK
10	27.4	(810)	OK
Total	273.0	(8073)	
Average	27.3	(807.3)	
10% of Average	2.7	(80.7)	

Acceptable from 24.6 to 30.0 (725 to 885)

TABLE 8

SHORT COURSE CALIBRATION PROCEDURE

1. Measure 163 1/3 feet in the field to be treated.
2. Adjust tractor speed, pressure, and orifice size according to manufacturer's directions.
3. Spray over the measured course, catching the discharge from one nozzle.
4. Measure discharge with a standard measuring cup.
5. $\frac{\text{Number of cups} \times 200}{\text{Nozzle spacing in inches}} = \text{gallons/A}$
6. Make up spray solution with the correct amount of chemical in the amount of water that will be applied to each acre.
7. This procedure may also be used in calibrating for band treatments. For band application, substitute band width for nozzle spacing.

rather than just follow a "cookbook" procedure. However, a simple procedure is provided for those who wish to use that option (Tables 8 and 9).

Calibration Procedure

The following procedure requires that you make a few simple measurements and do some simple calculations to determine the following three essential pieces of information.

Ground Speed (ft/min or m/min) can be determined by measuring the distance traveled in one minute. Repeat the test several times and average the results. Remember to use the same throttle setting (tachometer) and transmission gear each time. Run the tests in the field to be sprayed and have the sprayer tank half full. Soil surface and load can affect ground speed and a half-full tank represents the average load.

TABLE 9

SHORT COURSE CALIBRATION PROCEDURE (Metric)

1. Measure 50 meters in the field to be treated.
2. Adjust tractor speed, pressure, and orifice size according to manufacturer's directions.
3. Spray over the measured course, catching the discharge from one nozzle.
4. Measure discharge with a standard milliliter measuring cup.
5. $\frac{\text{Number of mL} \times 20}{\text{Nozzle spacing in centimeters}} = \text{liters/hectare}$
6. Make up spray solution with the correct amount of chemical in the amount of water that will be applied to each hectare.
7. This procedure may also be used in calibrating for band treatments. For band application, substitute band width for nozzle spacing.

Effective Boom Width (ft or m) is found by multiplying the number of nozzles or drop pipes by the spacing. Remember, the application area may be wider than the boom or the distance between outer nozzles.

Pumping Rate (gal/min or L/min) can be found by collecting the output from one nozzle for one minute and multiplying by the number of nozzles. Again, to be more accurate, repeat the test several times (collecting output from different nozzles) and average the results. Make sure the operating pressure is held constant.

After collecting the needed information, make the necessary calculations (Table 10 or 11). The result of the calibration procedure is the rate of total liquid application per acre (hectare) — it does not tell the rate of

TABLE 10

CALIBRATION CALCULATIONS

Average distance travelled in one minute = 512 ft.

Spray boom width = 20 ft.

Pumping rate = 2.5 gal./min.

Figure area sprayed per minute:

(512 ft./min.) x 20 ft. = 10,240 square feet per minute

Figure minutes to spray one acre:

$\frac{43{,}560 \text{ sq.ft./acre}}{10{,}240 \text{ sq.ft./min.}} = 4.25 \text{ min. per acre}$

Figure spray applied per acre:

(2.5 gal./min.) x (4.25 min./acre) = 10.625 gal./acre

TABLE 11

CALIBRATION CALCULATIONS — METRIC

Average distance travelled in one minute = 156 m

Spray boom width = 6 m

Pumping rate = 9.5 L/min.

Figure area sprayed per minute:

(156 m/min.) × 6 m = 936 square meters per minute

Figure minutes to spray one hectare:

$\frac{10{,}000 \text{ m}^2\text{/hectare}}{936 \text{ m}^2\text{/min}} = 10.65 \text{ min. per hectare}$

Figure spray applied per hectare:

(9.5 L/min.) × (10.65 min./hectare) = 101.18 L/ha

actual chemical application. The amount of chemical applied per acre depends on the concentration of the mixture as well as the application rate. Determining chemical concentrations will be discussed later.

Calibration with Minimum Arithmetic

One accurate method of calibrating a sprayer requires a minimum number of calculations. First, mark out an acre (an acre is 43,560 square feet), a 209-foot square or a 100 × 436 foot rectangle are two examples. Divide 43,560 by the width of the spray pattern in feet to determine the distance to spray to equal one acre. Fill the sprayer tank with water and spray the measured acre at the throttle setting and transmission speed you plan to use in the field. Measure the water needed to refill the tank to find the rate per acre. For instance, if it takes 9.5 gallons to refill the tank, you were spraying at the rate of 9.5 gallons per acre.

Use this same method to calibrate using metric measurements. First mark out a hectare (a hectare is 10 000 square meters), a 100-meter square or a 50 x 200-meter rectangle are two examples. Divide 10 000 by the width of the spray pattern in meters to determine the distance to spray to equal one hectare. Fill the sprayer tank with water and spray the measured hectare at the throttle setting and transmission speed you plan to use in the field. Measure the water needed to refill the tank to find rate per hectare. For instance, if it takes 90 liters to refill the tank, you were spraying at the rate of 90 liters per hectare.

If not satisfied with the rate, adjust pressure, change nozzle tips or change travel speed (slow down to increase the rate, speed up to decrease it) and repeat the process. Small changes can be made by adjusting pressure; large adjustments can be made by changing nozzle tips. This procedure may have to be repeated several times. After obtaining the proper application rate with water, repeat the procedure one more time using the pesticide mixture to make sure the actual delivery rate is satisfactory. Be sure test runs with chemicals are made where they will not contaminate crops or pose a hazard to humans. Avoid applying more than the approved rate per acre (hectare).

TABLE 12

CHANGING APPLICATION RATES

Change	Increase Application Rate (gal/acre or L/ha)	Reduce Application Rate (gal/acre or L/ha)
Pressure	Increase pressure	Reduce pressure
Nozzles	Use larger orifices, more nozzles	Use smaller orifices, fewer nozzles
Speed	Reduce speed	Increase speed

Changing Application Rate

Suppose the application rate you determined is unacceptable. How do you change it? You can make three kinds of changes (Table 12):

Pressure influences flow rate, as discussed earlier. Lower the pressure and you lower the flow rate. Raise pressure and flow rate is increased. However, this is not a good method to use. You must increase pressure by a factor of 4 to double the flow. High pressure increases the number of small spray particles which can cause drift problems. Pressure too high or too low also distorts nozzle distribution patterns.

Nozzle Size can be changed to alter the application rate. Use larger tips to increase the rate. The main advantage of this method, over changing the pressure, is that using the proper pressure helps to control drift and maintain the nozzle pattern. This is usually the preferred method of changing application rates.

Speed changes alter application rate. This method is practical for small changes in application rate. However, excessive speeds should be avoided for safety reasons. Low speeds increase the time needed to spray a given field. This increases labor costs and ties up equipment for a longer period of time.

Calibrating for Band Application

If the chemical is to be applied in a band application, some changes must be made in the procedure. However, the principles remain the same.

Herbicides are sometimes applied in bands (see Fig. 63) centered on the crop row. Band application reduces the amount of chemical required because only a fraction of the soil surface is treated. The space between the treated bands rows can be mechanically cultivated. Band applications can be made with equipment mounted on the planter or with postemergence equipment after the crop comes up.

Special care must be taken when calibrating for a band treatment to assure that the correct application rate is used. Herbicide recommendations are usually given as amount of chemical or product formulation to be given per acre (per hectare). The same application rate must be used per **treated acre (hectare)** — that is the actual area of soil or crop on which chemical is applied — for a band application to be effective, but the application rate per total field acre (hectare) should be reduced. Calculate the proper application per total field acre (hectare) as follows:

$$\frac{\text{Band width (in)}}{\text{Row spacing (in)}} \times \text{broadcast application rate (gpa)}$$

$$\frac{\text{Band width (cm)}}{\text{Row spacing (cm)}} \times \text{broadcast application rate (L/ha)}$$

EXAMPLE

The recommended broadcast application rate is 20 gpa. You need to apply the same mixture in 12 inch bands on 40 inch rows. You need to apply:

$$\frac{12}{40} \times 20 = 6 \text{ gpa.}$$

If the band width is 16 inches, use:

$$\frac{16}{40} \times 20 = 8 \text{ gpa.}$$

If the metric application rate is 185 L/ha to apply in 30-cm bands on 100-cm rows:

$$\frac{30}{100} \times 185 = 55.5 \text{ L/ha}$$

If the band width is 40 cm, use:

$$\frac{40}{100} \times 185 = 74 \text{ L/ha}$$

Next determine the pumping rate (gal/min or L/min) and effective boom width (ft or m) using the procedures described previously. Calculations will then be required to estimate the proper ground speed.

- Figure crop area to be sprayed per minute:

$$\frac{\text{Pumping rate (gal/min)}}{\text{Application rate (gal/ac)}} = \text{acres/min}$$

$$\frac{\text{Pumping rate (L/min)}}{\text{Application rate (L/ha)}} = \text{ha/min}$$

- Figure proper ground speed:

$$\frac{\text{Area sprayed per minute (ac/min)} \times 43{,}560}{\text{Effective boom width (ft.)}} = \text{ft/min}$$

$$\frac{\text{Area sprayed per minute (ha/min)} \times 10\,000}{\text{Effective boom width (m)}} = \text{m/min}$$

Speed can be converted to familiar units of miles per hour by simply dividing by 88 (for km/h, divide by 16.7). If the speed is too fast or too slow for the equipment being used, change the pumping rate by changing nozzle tips or operating pressure and repeat the calculations. The entire procedure should be repeated until a speed is obtained which is acceptable for the equipment being used.

Tractor speedometers are not accurate enough to use to operate the equipment at the calculated speed. Consequently, the unit should now be calibrated, using one of the procedures outlined previously to determine the actual application rate (gal/ac or L/ha) at a speed as close as possible to the estimated speed.

LOADING THE SPRAYER

There are two problems to consider when loading a sprayer. You must determine:

- *The quantity of pesticide to add to the tank (Fig. 82)*
- *Procedures to follow while mixing the pesticide*

To determine the amount of pesticide to add to the tank, you will need two more facts:

- *Sprayer-tank size (gallons or liters)*
- *Amount of pesticide needed per treated acre (hectare)*

Recommended rates are shown on the label and in extension bulletins.

Fig. 82—Transfer Pump Measures Amount of Pesticide to Add to Tank

EXAMPLE

Suppose a sprayer has a 200-gallon tank and it has been calibrated to apply 10 gallons per acre. You wish to apply one pint of a pesticide formulation per acre.

First, find the acres sprayed per tankful.

$$\frac{\text{200 gal/tankful}}{\text{10 gal/acre}} = \text{20 acres/tankful}$$

Next calculate the amount of pesticide.

$$\frac{\text{20 acres}}{\text{tankful}} \times \frac{\text{1 pint}}{\text{acre}} = \text{20 pints/tankful}$$

The final mix would consist of 2.5 gals. of pesticide formulation and 197.5 gals. of water per 200 gal. tankful.

EXAMPLE (METRIC)

Suppose a sprayer has a 750-liter tank and it has been calibrated to apply 90 liters per hectare. You wish to apply 1.25 liters of a pesticide formulation per hectare.

First, find the acres sprayed per tankful.

$$\frac{\text{750 liters/tankful}}{\text{90 liters per hectare}} = \text{8.3 hectares}$$

Next calculate the amount of pesticide.

$$\frac{\text{8.3 hectares}}{\text{tankful}} \times \frac{\text{25 liters}}{\text{hectare}} = \text{10.4 liters/tankful}$$

You may find the recommendation given in terms of active ingredient per acre. In that case refer to the next example.

EXAMPLE

Suppose you want to use the sprayer in the preceding example to apply ½ pound of active ingredient per acre (0.5 kg/ha). The formulation to be used contains 4 pounds of active ingredient per gallon (6.9 kg/L).

First, find how much formulation is needed to obtain ½ pound (0.5 kg) of active ingredient.

$$\frac{4\ \text{lbs/gal}}{\frac{1}{2}\ \text{lb}} = 8\ \text{Half-lb/gal} = \frac{1\ \text{pint}}{\frac{1}{2}\ \text{lb}}$$

$$\frac{0.5\ \text{kg/L}}{0.5\ \text{kg}} = 1\ \text{Half-kg/L} = \frac{1\ \text{L}}{0.5\ \text{kg}}$$

Next figure the amount of pesticide.

$$\frac{20\ \text{acres}}{\text{tankful}} \times \frac{1\ \text{pint}}{\text{acre}} = 20\ \text{pints/tankful}$$

$$\frac{8\ \text{hectares}}{\text{tankful}} \times \frac{1\ \text{L}}{1\ \text{ha}} = 8\ \text{liters/tankful}$$

If you are using powders, the procedure is similar to the second example.

Mixing the pesticide thoroughly and carefully is one of the most important steps in good sprayer operation:

- Incomplete mixing results in varied application rates, too heavy at times, too low otherwise
- Some chemicals can form invert emulsions if mixed improperly. An invert emulsion is a thick, mayonnaise-like mixture that will not spray properly and is very difficult to clean out of a sprayer.
- The operator is most likely to be exposed to dangerous amounts of pesticide during mixing because he is handling the material in concentrated form. Pesticides can be mixed in the tank or in a pre-mix container. Specific instructions are given on the label of each pesticide. Follow them carefully. Adding chemicals in the wrong sequence can prevent otherwise compatible materials from mixing properly.

To mix some chemicals in the tank, add the pesticide to one-half tank of water.

Turn on the agitator and mix thoroughly. Then finish filling the tank with water. For other materials, agitation must be started before adding pesticide and continued until all chemical-water mix has been used.

If a premix container is used, fill it about one-half full of water (Fig. 83) then add pesticide. Stir the mixture until it is smooth and uniform, then add it to the water in the sprayer tank. Premixing an emulsifiable concentrate with water to form an emulsion, or premixing a wettable powder with water to form a slurry, and then adding these to a partially-filled well-agitated spray tank can help reduce mixing problems with some pesticides.

When adding pesticide or premix to the tank, use a sturdy ladder or steps on the sprayer. Do not stand on a tire, parts of the frame, boxes or other supports. If you slip, you could spill the pesticide on yourself. (See Chapter 9 for further discussion of safe practices).

Fig. 83—Premixing Pesticide before Adding It to the Tank

TRANSFER PUMP

A transfer pump (see Fig. 82) eliminates much of the hazard and inconvenience involved in handling liquid pesticides. The transfer pump automatically measures the liquid as it is being pumped. It can be preset to pump a specified amount (accurate to 1/10 of a gallon or 0.4 liters).

TRANSPORT

Transport loaded sprayers as little as possible. An accident could spill a load of chemicals on the road or in a ditch, where it could be very hard to contain and cleanup. Mix as close to the field as possible and spray the entire load immediately. If agitation of some mixtures is stopped before the tank is emptied it may be difficult or impossible to obtain adequate remixing.

Use safe transport speeds. A trailer sprayer with a 300-gallon (1135 liters) tank weighs about 2,500 pounds (1135 kg) loaded. That weight can be difficult to control with a small tractor on a hill, corner, or rough terrain.

Lock the boom or booms in the transport position (Fig. 84). Use a loop of chain if a locking pin is not provided. An unsecured boom may fall and strike a tree or other obstacle or create a traffic hazard.

Any time a sprayer is transported on a road, use accessory lights, SMV emblem, and other devices to warn operators of other vehicles (Fig. 84). Obey all rules of the road and traffic signs. Check state and local regulations.

FIELD OPERATION

There is more to operating a sprayer in the field than simply turning it on and driving. The operator must maintain constant ground speed and pressure as determined while calibrating. He must also monitor the operation continuously. Watch for:

Fig. 84—Lock Boom and Use Recommended Lighting and Marking When Transporting Equipment

- *Plugged nozzles*
- *Marker operation, if one is being used, to avoid skips or overlaps (Fig. 85)*
- *Leaks*

Fig. 85—Use Dye or Foam Markers in Large, Open Fields

Fig. 86—Watch for Fences, Trees, Gates and Other Obstacles

- *Weather conditions, especially wind which can increase drift*
- *Empty tank to prevent running pump dry*
- *Obstacles in the field such as trees and fences (Fig. 86).*

CLEANING

A sprayer should be carefully cleaned after application of each different pesticide, at the end of the season, and when repairs must be made (Fig. 87). If spray material is spilled on the machine during loading or mixing, the outside of the machine should be decontaminated immediately.

The most important step in sprayer decontamination is a thorough washing with soap (or mild detergent) and water, followed with a complete rinse with plenty of water. A steam cleaner can be used, if available. Compacted deposits can be removed with a stiff bristle brush. Dry material can be removed with a vacuum cleaner.

Use a strong soap-and-water solution to clean the inside of the sprayer after using carbarmates or organophosphates. After spraying organochlorines, substitute acetic acid (vinegar) for the soap. Finally, pump soapy water through the boom and nozzles.

Wash sprayer nozzles and strainers in the same mixture recommended for the inside of the sprayer.

Wear a rubber apron or raincoat, boots, gloves, hat and goggles when decontaminating a sprayer, to avoid contact with the pesticide.

STORAGE

At the end of the season, prepare the sprayer for storage. Proper preparation and storage mean the sprayer will be ready the next time it is needed. Sprayer will also last longer if stored properly. Follow these five steps:

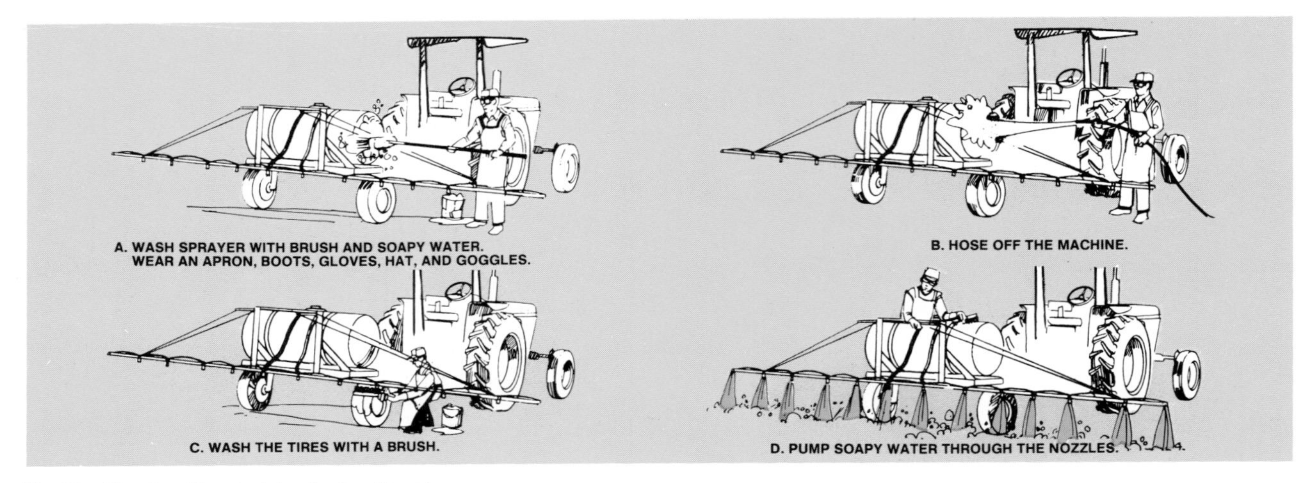

Fig. 87—Cleaning Chemical Application Machinery

- *Clean the entire unit as described above*
- *Circulate 20 gallons (75 liters) of diesel fuel through the system, including hoses and nozzles*
- *Remove, clean, and reinstall the line strainer*
- *Remove and clean nozzle tips and strainers. Store them in a jar of diesel fuel (Fig. 88). Put blank tips (without holes) in the nozzle bodies to keep dirt out of lines.*
- *For maximum life, store the unit inside. Protect polyethylene tanks from sunlight.*

Fig. 88—Store Nozzle Tips and Screens in Diesel Oil

DRIFT CONTROL

During any spraying operation some pesticide can drift away from the target area. In some cases, drifting material has caused extensive damage. The most common situation leading to serious crop or plant damage occurs when a herbicide, such as 2,4-D, drifts into a field of a sensitive crop such as soybeans, grapes, or tomatoes. In any spraying operation take all possible precautions to keep drift at a minimum. But, drift can seldom be completely eliminated. Not spraying at all is the only way to be certain of no drift.

Following are basic factors that affect drifting:

- *Every nozzle produces a range of droplet sizes, including some very fine droplets*
- *The number of small droplets produced increases as pressure is increased*
- *Small particles drift farther than larger particles in wind of the same speed*
- *The distance a particle drifts depends on droplet size and wind speed*

Controlling Drift

Since the factors that cause drift have been known for many years, "rules" for controlling drift have also been developed:

- *Use an application rate of 10 gpa (95 L/ha) or more*
- *Use a ground speed of at least 4 mph (6.5 km/h). Higher ground speed requires larger nozzle tips for the same application rate. This, in turn, increases droplet size and helps to control drift.*
- *Keep nozzles as near the ground as possible and still get good distribution*
- *Where feasible, use flooding-type nozzles, which produce mostly large particles, for drift control*
- *Spray only when winds are no more than 3 to 5 mph (5-8 km/h)*

Fig. 89—Reducing Spray Drift with Emulsion

New Approaches to Drift Control

In addition to the "standard" drift-control methods outlined above, manufacturers and researchers have been developing spraying techniques aimed specifically at preventing drift.

- **Air Emulsion** is a process of forming a light foam from the spray liquid as it leaves the nozzle. A special oil adjuvant is required. The very light foam has little tendency to float in air. As the foam breaks up, the pesticide is released (Fig. 89).

- **Invert Emulsions** can be applied by ground and air equipment. The thick "mayonnaise-like" emulsion droplets resist drifting under varying weather conditions.

- **Recirculating Sprayers** catch the excess flow and recycle it back to the tank for respraying (Fig. 90). An interesting adaptation of this technique is used to spray tall weeds in a field of a shorter crop. Streams of herbicide are sprayed horizontally over the top of the crop. The taller weeds are hit by the spray. The portion of the stream that is not intercepted by the weeds if collected for return to the tank.

- **Electrostatic charging** of droplets as they leave the nozzle enables them to be guided towards the target. A plate with an electric charge opposite to the one placed on the droplets is towed below the nozzles and guides droplets toward the ground.

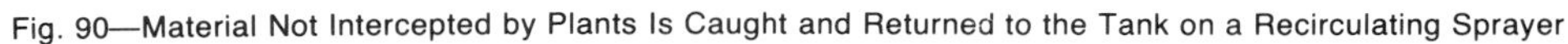

Fig. 90—Material Not Intercepted by Plants Is Caught and Returned to the Tank on a Recirculating Sprayer

SUMMARY

The liquid formulation is the most popular approach to applying pesticides while the granular and gaseous form is the most popular for fertilizers.

The common farm sprayer, while simple in design, poses a whole host of problems for the untrained operator. Many farmers have elected to hire their pesticides applied custom and leave the hassle and liability to someone else. Other farmers apply a portion of their chemicals but use custom operators for the "hot" chemicals which require extremely accurate application or are extremely dangerous to apply.

The typical farm sprayer uses a roller or centrifugal pump which removes the liquid from a polyethylene tank. The pressure is adjusted by a regulator with the unused material returning to the tank. A pressure gauge allows the operator to monitor the pump and regulator units. Certain chemicals must be constantly agitated. This is accomplished by a line bringing a portion of the chemical back to the tank where it is sprayed into the tank, generally at the bottom of the tank.

The selection of nozzles to obtain desired field coverage causes the sprayer operator to study numerous charts and push the pencil accurately.

Actual driving the sprayer in the field becomes a relatively easy task when compared to the selection and mixing of the chemical. Each pesticide used requires a careful reading of the label. In addition, the operator must pay close attention to detail and not be careless in any phase of the spraying operation.

The operator must be as careful with the last batch as he is with the first batch he applies. He is responsible for his own crops and the neighbor's. Clean up procedures are very specific and should be followed carefully. Spraying equipment is expensive and should not be abused by careless handling.

The safe application of pesticides and fertilizer over the past 25 years speaks well for the farm and custom operators. Their safety record continues to improve indicating they are capable of learning how to use the infinite number of sprayers and agricultural chemicals to produce larter quantities of more nutrition food for the entire world.

CHAPTER QUIZ

1. What are the four characteristics of a good application?

2. Name two advantages of aircraft sprayers.

3. Why do liquid fertilizers sometimes require subsurface application?

4. Name three basic functions of a sprayer.

5. What three kinds of pumps are used most frequently on sprayers?

6. Why is a pressure gauge essential?

7. What are three locations on a sprayer where strainers are used?

8. Would corn syrup or water have the higher flow rate through a nozzle?

9. Does high or low pressure produce the most small droplets?

10. Would brass or nylon tips have the longer expected life?

11. What is the application rate if your sprayer pumps 4.7 gpm and you are covering 10 acres an hour?

12. Find the application rate in gallons per acre. Average distance travelled in one minute=450 feet. Boom has 16 nozzles, 16 inches apart. Pumping rate is 3.9 gallons per minute.

13. You have a sprayer which applies 49 gal/acre at 8 mph and 40 psi. You need an application rate of 15 gpa. What could you do?

14. You are making a band application of a material which should be applied at 18 gal/acre. The bands are 14 inches wide and the row spacing is 30 inches. What application rate will you use for calibration?

15. You want to spray 14 gal/acre and apply 2 pints of pesticide per acre. How much pesticide do you add to a 350 gallon tank?

16. List four things you can do to reduce drift.

8
Application of Dry Chemicals

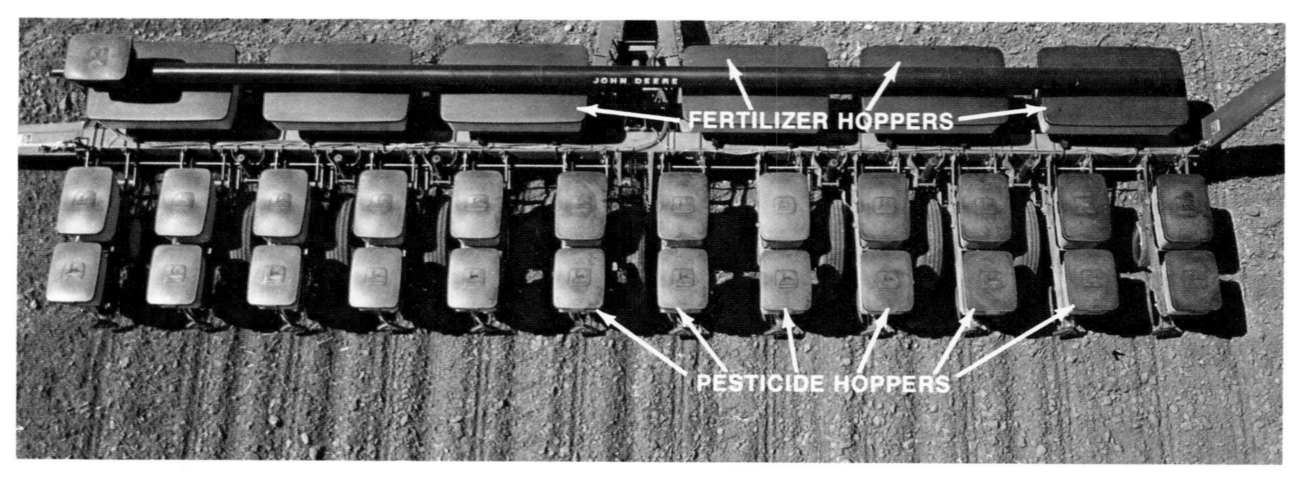

Fig. 1—Precision Chemical Application Equipment is Used on Modern Farms

APPLICATION OF DRY CHEMICALS

In recent years, many changes have been made in equipment for applying dry chemicals. There are two main reasons for these changes:

- *New chemicals have become available*
- *Application requirements have been altered*

For example, new granular herbicides and insecticides are easier to handle and control, and require smaller, lighter equipment than when materials are applied as liquids.

In all cases, read the label and follow directions before handling dry chemicals.

The primary change in fertilizer application is the increase in application rates. Instead of 100 to 200 pounds of fertilizer per acre (110 to 225 kg/ha) — which was common a few years ago — we now see applications as high as 2,000 pounds per acre (2240 kg/ha).

TYPES AND SIZES OF EQUIPMENT

Wide variations in dry-chemical-application equipment are due to the following factors:

- *Rate of application.*
- *Characteristics of the chemicals used.*
- *Placement or distribution patterns.*
- *Crop characteristics.*
- *Time of application – before, during, or after planting.*

Equipment for applying dry material may be classified in two groups:

- **Broadcast Applicators**
- **Band Applicators**

These classes may be further divided as follows:

- *Machines that apply only chemicals.*
- *Machines for planting and applying chemicals in one operation.*
- *Machines for tillage and chemical application in one operation.*
- *Machines for aerial application of fertilizer.*

BROADCAST APPLICATORS

Three common kinds of equipment used to broadcast dry chemicals are:

- **Drop-type Distributor**
- **Spin Spreaders**
- **Aircraft**

The drop-type material distributor (Fig. 2) has a trailer-mounted hopper with a series of "drops" or holes in the bottom. The operator opens or closes the drops, usually by using a lever or rope control. A simple mechanical stop controls the size of the opening. The application rate depends on travel speed as well as the rate at which chemical flows out of the hopper.

Capacity of this type of machine is usually between 1,000 and 2,000 pounds (1120 to 2240 kg/ha), and width ranges from 8 to 12 feet (2.5 to 3.5 meters). A feed shaft, driven by the ground wheels, runs the full length of the material hopper (Fig. 3). Speed of rotation of the feed shaft affects the rate of material flow out of the drops.

The spin spreader (Fig. 4) uses one (Fig. 5) or two (Fig. 6) spinning disks to distribute the material. The material is metered onto the disks by a drag chain running through the bottom of the hopper. The application rate depends on the material feed rate, width of the pattern, and the travel speed. The distribution pattern depends on the rotational speed of the disks and the point at which material is fed onto the disk.

Fig. 2—Drop-Type Material Distributor

Fig. 3—The Feedshaft Extends Across the Hopper

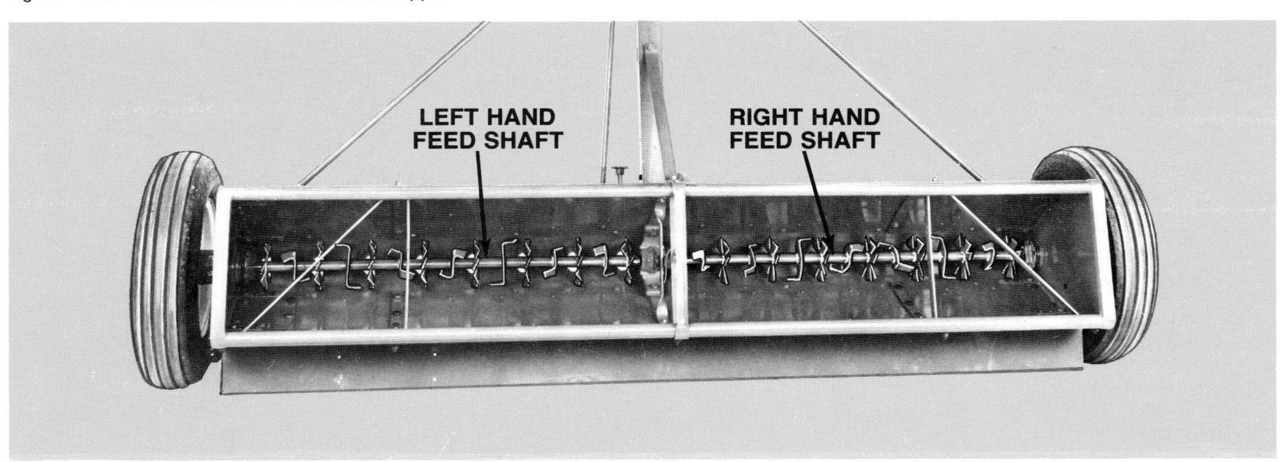

Fig. 4—Truck-Mounted Spin Spreader

Fig. 5—Single-Spinner Unit

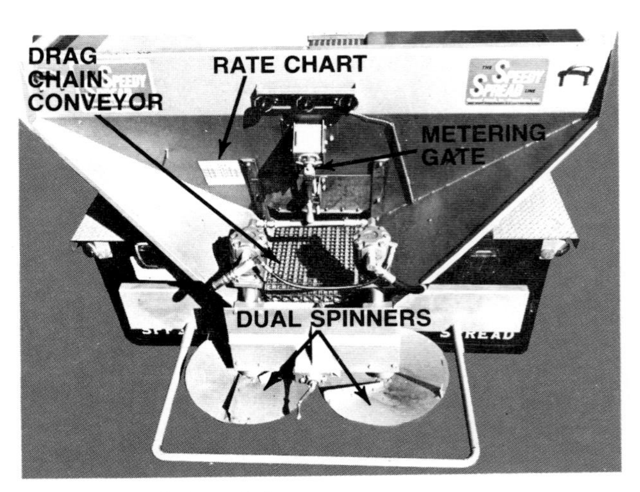

Fig. 6—Double-Spinner Unit

Fig. 7—Oscillating Tandem Axle

Hopper capacities range up to 10 tons (9 metric tons). Spread patterns may extend to 60 feet (18 meters). Application rates range from 25 to 2,000 pounds per acre (28 to 2240 kg/ha). The units may be mounted on trailers or trucks (Fig. 4). Large-capacity trailer units are usually equipped with "walking-beam" tandem axles which flex over uneven terrain (Fig. 7).

Aircraft, both helicopters and fixed-wing types, are used to apply dry chemicals. Advantages of aircraft application are that large ground areas are covered quickly without disturbing or compacting the soil and without damaging crops. Wet, soggy fields and rough terrain may be treated where it would be impossible or inconvenient to do the job on the ground. Aerial applications are used extensively on flooded rice fields in the United States.

Air application is limited to the broadcast method. High-analysis materials are usually used because aircraft can carry more chemicals per unit-weight of such chemical than with less-concentrated material. Granulated chemicals are preferred over finer types because there is less drift and greater uniformity of distribution.

On fixed-wing aircraft (Fig. 8) the chemical is metered out through a spreader under the body of the plane. The swirling air from wingtips and propeller then spreads the material. Helicopters carry centrifugal spreaders, mounted on the body of the aircraft or hung below on a cable.

BAND APPLICATORS

Band applicators are nearly always used as attachments to some other piece of equipment, usually a planter or tillage tool. They consist of the following parts (Fig. 9):

- **Hopper**
- **Metering Device**
- **Drop Tubes**
- **Openers—Chisels, Knives, Disks, or other Placement Device**

Fig. 8—Loading a Fixed-Wing Aircraft

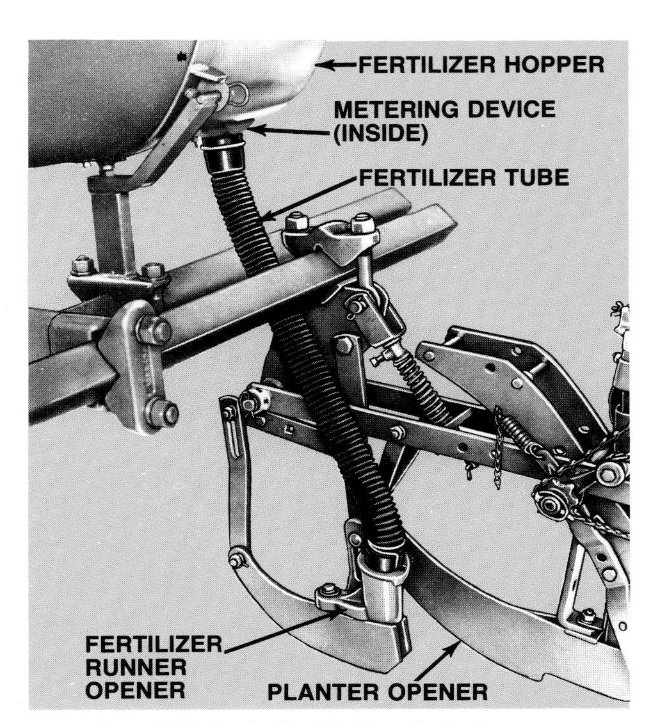

Fig. 9—Essential Parts of a Band Fertilizer Applicator

Many different combinations can be used. The equipment selected for a particular operation is often dictated by the crop and results desired.

Three types of metering systems are used:

- **Star Wheels**
- **Augers**
- **Feed Wheels**

Star wheels push a small amount of material through a gate opening each time one point on the star passes the opening (Fig. 10). Application rate depends on the rate of wheel rotation and the width of the opening.

Hoppers with a star-wheel metering system have circular bottoms. They may be made of steel or fiberglass. Capacity of each hopper is usually about 100 pounds. They may be mounted on planters, cultivators, and other machines (Fig. 11).

Augers are often used to meter material in large horizontal hoppers (Fig. 12). The rate of material application is determined by the type of auger and the opening of the flow control valve at each of the drops.

Feed wheels are used in long hoppers such as those on grain drills. The feed-wheel system is similar to the one described in the section on Drop-Type Material Spreaders (Fig. 2).

PESTICIDE EQUIPMENT

Dry pesticides are occasionally applied with spin spreaders or other broadcast equipment, but most are

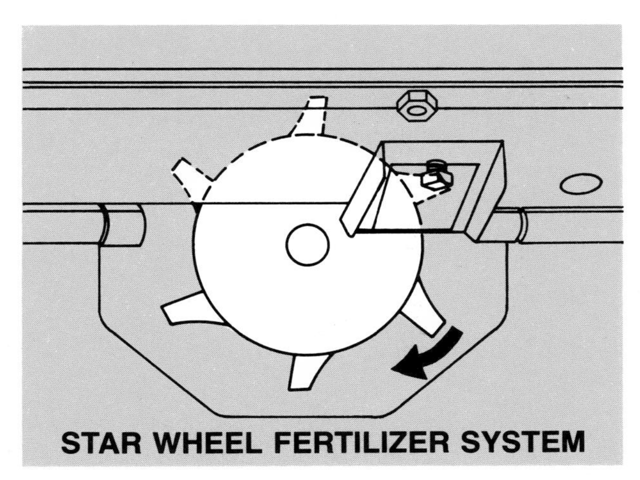

Fig. 10—Star-Wheel Fertilizer-Metering System

Fig. 11—Cultivator With Band-Fertilizer Attachment

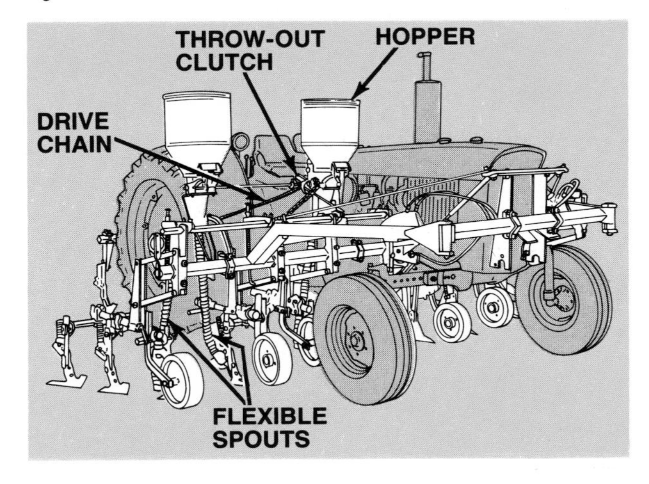

Fig. 12—Auger-Type Fertilizer Meter

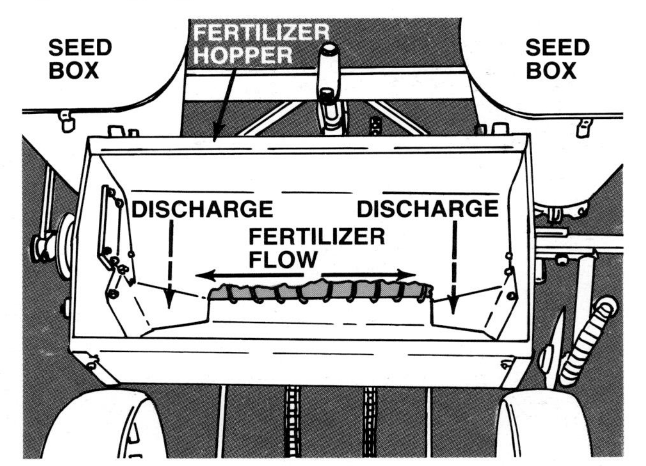

applied with specialized pesticide equipment. There are two primary types:

- **Dusters**
- **Granular Applicators**

Following are descriptions of these machines.

Dusters

There are two types of power dusters — single-outlet units (one hose and nozzle) designed for orchard dusting, and multiple-outlet units designed for field work. Dusting of field crops is no longer a common practice. Other chemical application methods are preferred.

Power dusters have these essential parts (Fig. 13):

- *Dust hopper with agitator and feed regulator.*
- *Fan.*
- *Distribution system (hoses and nozzles).*
- *Power source.*

Power dusters have hoppers with a capacity from 60 to 200 pounds (27 to 90 kg) each, depending on density of the dust. Dusting rates range from 5 to 50 pounds per acre (5.6 to 56 kg/ha).

Centrifugal fans are usually used. Delivery ranges from 500 to 1,000 cubic feet (14 to 28 m^3) of air per minute at velocities from 50 to 100 feet (15 to 30 meters) per minute.

The distribution system may have as many as 18 hoses and nozzles spaced along a boom. For field-crop dusting, a tubular boom is often used. It may have a series of holes spaced along the bottom or adjustable slotted outlets.

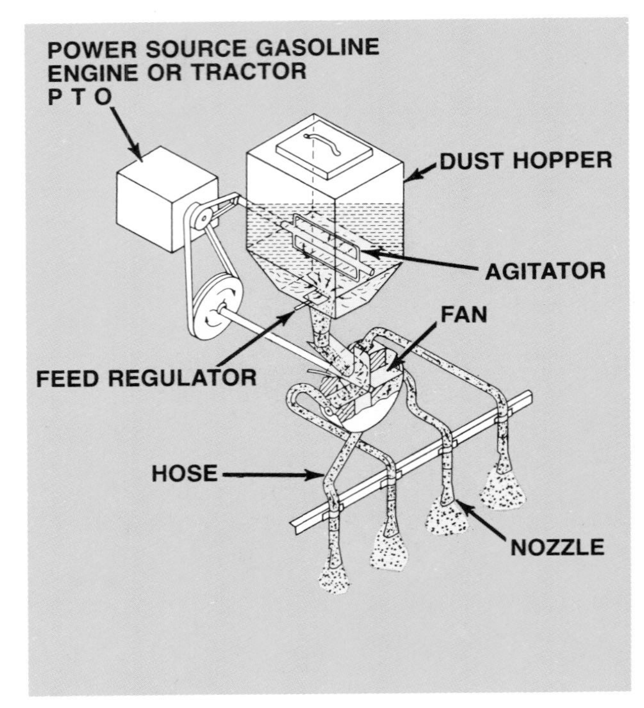

Fig. 13—Component Parts of a Typical Power Duster

Canvas drapes (Fig. 14) dragging on the ground behind the duster help confine the dust to the desired area so it can settle quickly, especially on windy days. They

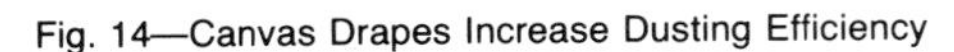

Fig. 14—Canvas Drapes Increase Dusting Efficiency

increase efficiency in the use of the dust, but should be used only when excessive mechanical injury to the crop does not result.

Several kinds of small garden-type, hand-operated dusters are also available. The nozzles are similar to those on power dusters.

Granular Applicators

Two types of granular applicators are available:

- **Gravity-Flow**
- **Positive-Feed**

The *gravity-flow* type is essentially a hopper with a metering hole in the bottom. Discharge is by gravity, but an agitator is provided in the bottom of the hopper to prevent "bridging". Rate of application is controlled to a large extent by ground speed (Fig. 15).

The *positive-feed* type has either an auger or fluted-feed metering device which rotates in the bottom of the hopper to deliver granules to discharge openings. Application rate can be adjusted by exposing more or less of the fluted-feed roller to granules in the hopper (Fig. 16), or changing speed of rotation of the auger or fluted-feed roller and size of the discharge opening.

Positive-feed applicators (Fig. 17) are more complicated and costly than the gravity-flow type. However, they are more accurate because the delivery rate is controlled by the ground speed.

The pesticide spreader is located at the end of the discharge tube. It usually has a series of inverted V's inside a plastic or sheet-metal housing (Fig. 18).

Fig. 15—Gravity-Flow Granular Applicator

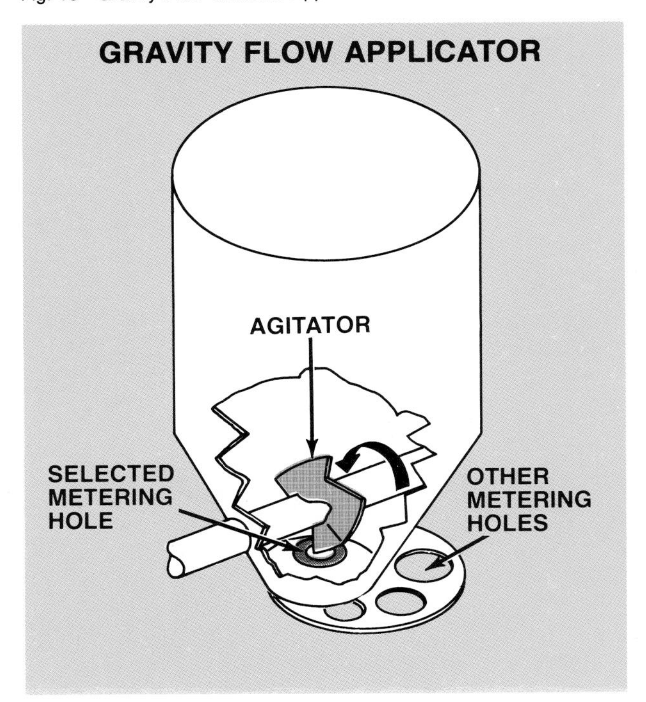

For control of some pests, such as corn rootworm, the banded pesticide should be worked into the upper one or two inches (2.5 to 5 cm) of soil. Various incorporators are used, including disks, chains, or rollers (Fig. 19).

SELECTING AN APPLICATOR

The applicator should be capable of effectively and conveniently applying the chemical to the crop. The best design depends on such factors as the crop being grown, the kind of chemical, and the terrain. The following nine features should be included in a fertilizer distributor; many of them apply equally well to applicators for other types of dry chemicals:

- *The rate of flow from the machine must be uniform to help ensure efficient use of the plant food or distribution of pesticide. Uniformity is equally important for all types of applicators. Application rate must not affect the uniformity.*
- *The mechanism to control the application rate must be easily adjustable.*
- *The machine must be able to handle many different types of fertilizers (or pesticides).*

Fig. 16—Fluted-Feed Metering System

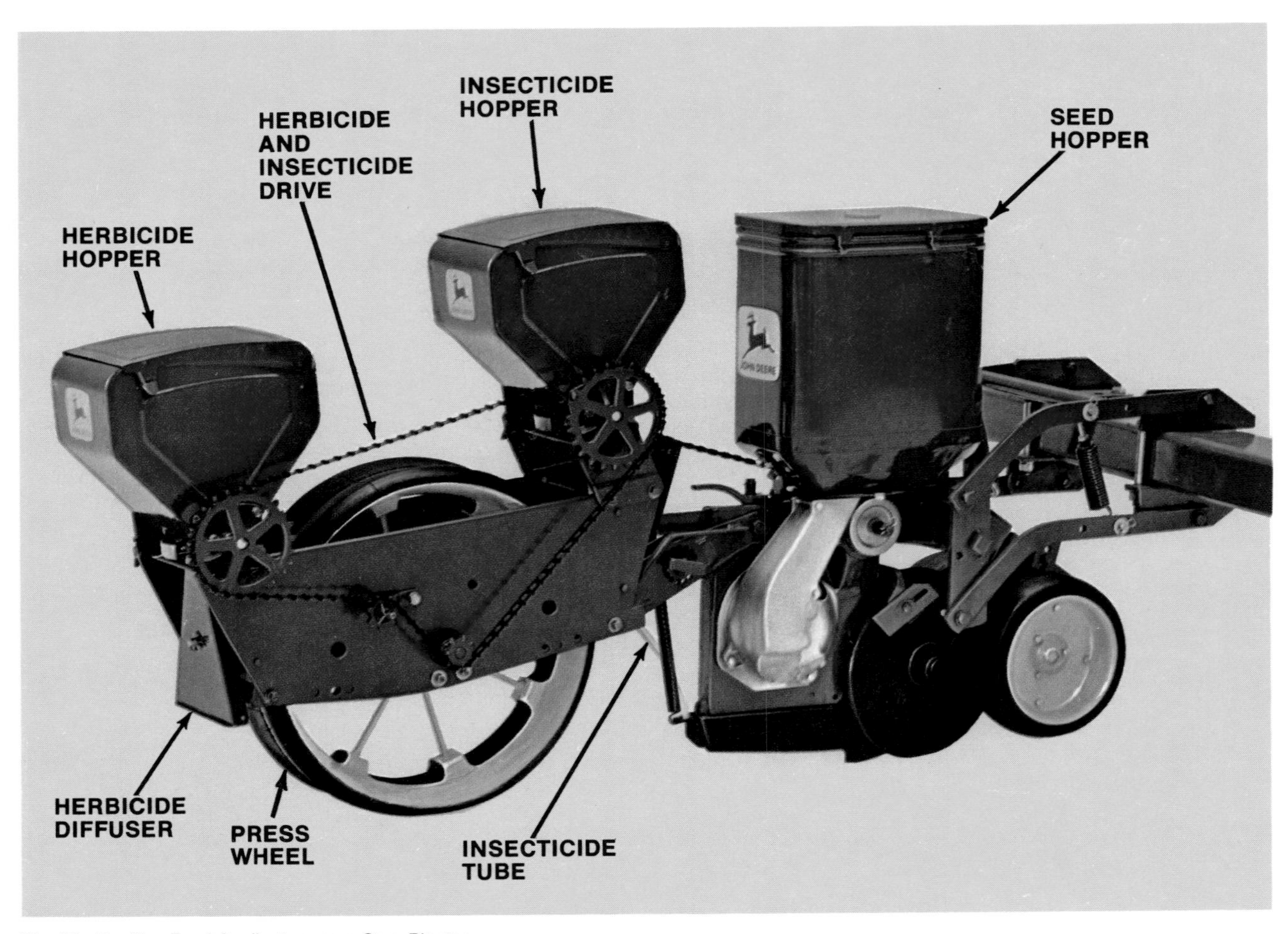

Fig. 17—Positive-Feed Applicators on a Corn Planter

Fig. 18—Pesticide Spreader

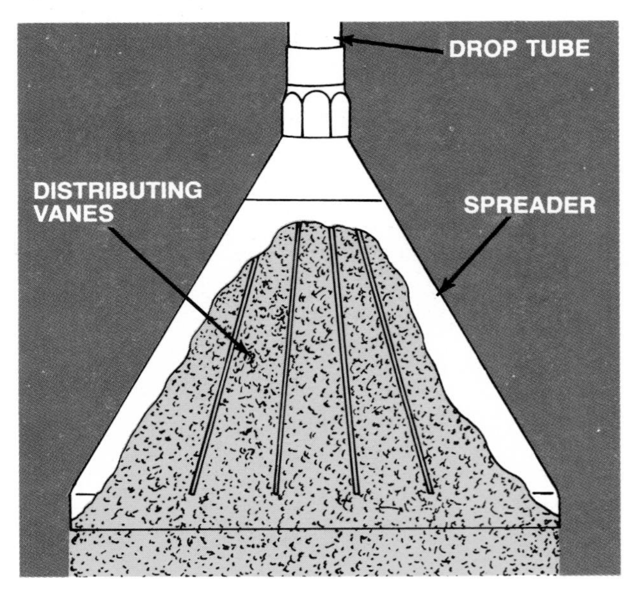

Fig. 19—Attachment for Incorporating Granular Pesticide

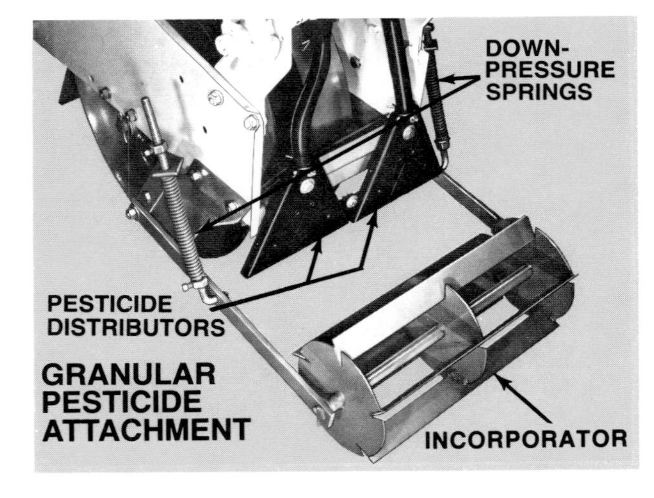

- *The distributor must be able to apply a wide range of application rates with the same degree of efficiency.*
- *Hoppers must be large enough to hold a sizeable quantity of fertilizer (or pesticide) to eliminate frequent stops.*
- *The flow of fertilizer (or pesticide) should be visible to the operator for ease in detecting flow stoppages. This is most important on machines which distribute material through tubes.*
- *The agitator should be designed to prevent bridging of material in the hopper.*
- *The mechanism must function equally well with a full or nearly-empty hopper.*
- *The machine must be easy to clean and maintain.*

APPLYING DRY CHEMICALS

We frequently think of chemical application as the operations that occur in the field where the actual distribution takes place. However, effective application requires careful checking and attention to other details both before and after the actual application.

PLANNING AND PREPARATION

Plan and prepare for chemical usage well in advance of the season. This planning should include choosing proper chemicals and the best application rates. Prepare equipment carefully, so it can function reliably and efficiently. The following suggestions should help. Refer to equipment operator's manuals and chemical label instructions for details.

Preparing the Chemical Applicator

1. *Clean the spreader thoroughly, inside and out.* Wipe off all accumulations of old oil or grease and lubricate the machine thoroughly. Follow the operator's manual lubrication instructions. Don't miss any points that should be greased or oiled, but do not over lubricate. Unnecessary grease or oil on the machine can accumulate fertilizer, chemicals, or dirt which cause rapid wear and possibly interfere with operation of the machine.

2. *Remount tires if they were removed for storage, and inflate to proper air pressure.* Tires with wrong inflation pressure will change the application rate of spreaders with ground drives. Soft tires act like small tires, and can cause the application rate to be higher than intended. Overinflated tires reduce application rates. Have the hopper half full when checking the tire pressure (Fig. 20).

3. *Check for loose bolts, pins, setscrews, missing or broken parts, etc.* Correct any problems. This check can be made while the machine is being lubricated.

4. *Inspect material drop tubes.* Be sure they are securely fastened and are not plugged. Replace broken or damaged tubes.

Fig. 20—Effect of Tire Inflation on Application Rates

5. Check for "frozen" parts. All shafts, clutches, and drives should operate freely. Turn feed shafts *by hand* (Fig. 21) before hitching to the tractor. Free or replace any parts that are binding.

Fig. 21—Check for "Frozen" Parts Before Engaging Drive

Preparing the Tractor

Consult the operator's manuals for the tractor and applicator. Some adjustments may be required on the tractor. Items to check include:

- Tractor-tire inflation pressure
- Wheel spacing
- Drawbar positioning
- Hitching details
- Front and rear tractor weights

Material Storage

Keep chemicals dry. Do not store in a damp place. Fertilizer accumulates moisture rapidly. Damp chemicals can cause accelerated corrosion of metal and may "freeze" parts to each other.

PRELIMINARY SETTINGS

Use the operator's manual or application-rate tables on decals attached to the machine as guides to set metering-gate openings, auger speeds, and other adjustments on the machine. Pay attention to the density of the chemical. Most applicators meter volume, not weight. When using a more dense material, reduce the application rate accordingly.

Determine the density by weighing a known volume of the fertilizer. For example, if ¼ cubic foot of chemical weighs 16.5 pounds (¼ m^3 = 265 kg), the density is:

$$\frac{16.5 \text{ pounds}}{1/4 \text{ cubic foot}} = 66.0 \text{ pounds/cubic foot}$$

$$\frac{265 \text{ kg}}{1/4 \text{ m}^3} = 1060 \text{ kg/m}^3$$

Any known volume can be used.

Remember that suggested settings in the manual or charts are only preliminary settings. They should be verified and changed as needed during calibration and rechecked periodically during operation.

CALIBRATION

While the calibration procedure you choose should be suitable for the kind of machine involved, the principles are the same. The basic steps are listed below. Note: During calibration of some equipment, chemical is collected, rather than actually spread, so the calibration is not always done in the field to be treated. Surface conditions should be similar, though, to obtain comparable results.

1. Determine the test area to be covered.
2. Set the machine for the desired application rate.
3. Determine the quantity actually applied.
4. Calculate the actual application rate.
5. Adjust rate settings as necessary.

Calibrating the Spin Spreader

A simple procedure has been developed for calibrating spin spreaders. It yields two pieces of information:

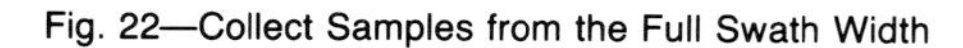

Fig. 22—Collect Samples from the Full Swath Width

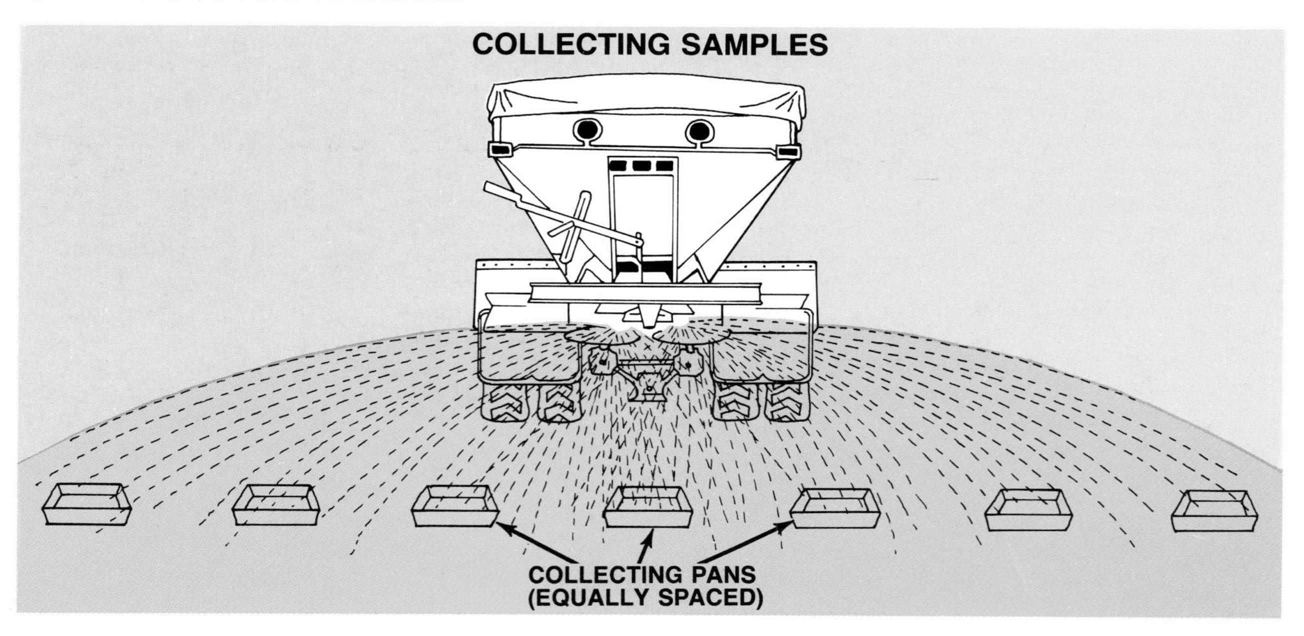

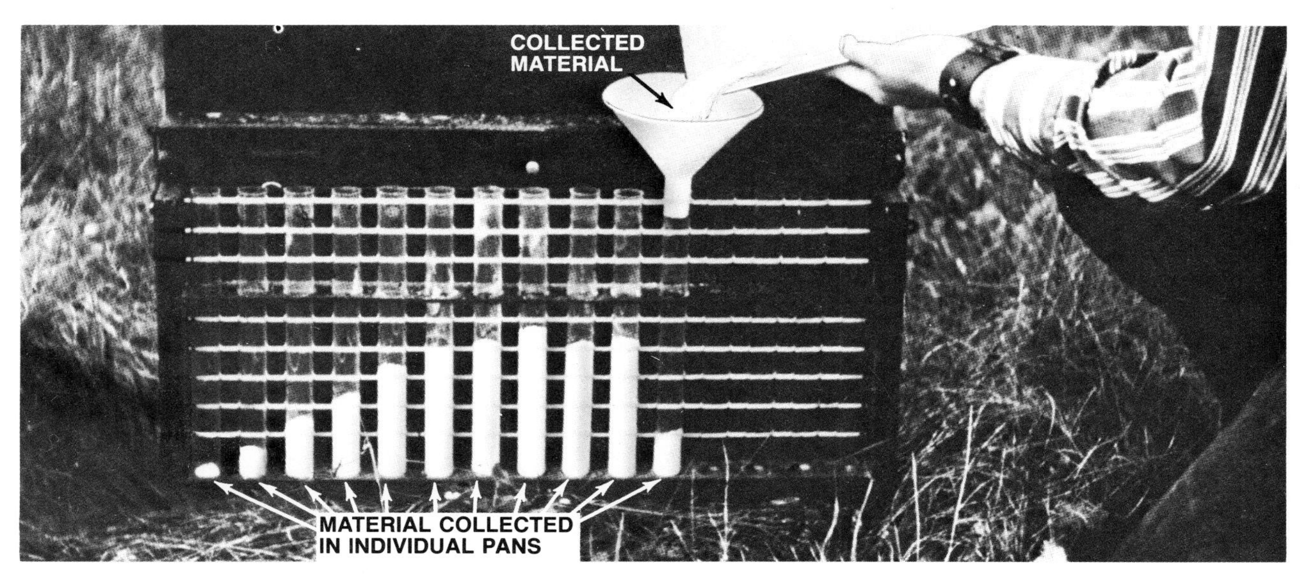

Fig. 23—Pour Material from Pans into Test Tubes

- **Application rate**
- **Application pattern**

Contact your local extension agent to see if he has the needed equipment or can find it for you. If not, supply your own. You will need the following:

- *A 10-foot (3-meter) measuring tape*
- *A funnel*
- *15 test tubes* (½ inch (1.25 cm) I.D., 4½ *inches (11.5 cm) tall)*
- *A test-tube rack*
- *15 metal baking pans (11¾ x 16¾ x 2¼ inches or 30 x 42 x 6 cm) with a baffle in each. (The diffusing lens with ½-inch (1.25-cm) squares from some fluorescent light fixtures can be cut to the same size as the pans and makes a satisfactory baffle.)*

The pans are spaced equidistant on level land. The applicator should be driven over the pans at normal field speed, straddling the center pan in a direction perpendicular to the line of pans (Fig. 22). Material collected in each pan is then poured into the corresponding test tube in the rack (Fig. 23). The amount of material in each tube gives a quick visual check on the spread pattern. The effective swath width may be found by locating the points on the sides of the swath where the height of material in the tube is approximately half of the fertilizer height in the center tube.

The approximate rate of application at each collecting point can be determined from the height of material in the corresponding test tube (Fig. 24). The material should be settled in the test tubes by holding each tube and tapping it gently on a soft surface.

Evaluation of Spread Pattern

Several typical spread patterns have been identified. The flat-top, oval, and pyramid patterns are best (Fig. 25). Other spread patterns are usually undesirable (Fig. 26).

Calibrating Granular Applicators

Granules are usually applied in bands on or near crop rows. It is essential that each unit be calibrated separately. The procedure below also can be used for broadcast units.

Fig. 24—Application Rate Determination (For equivalent art with Metric Units, see page 203, Table 9.)

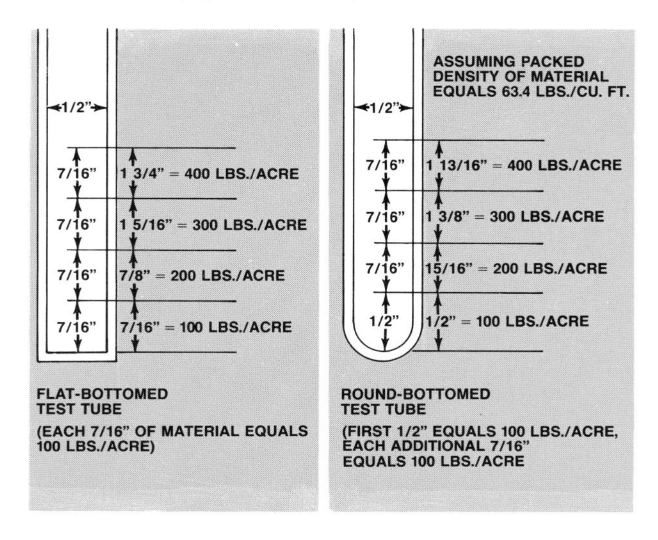

1. Measure band width in feet and calculate the distance traveled to cover an acre.

$$\frac{\text{43,560 sq. ft./acre}}{\text{Width of band (ft.)}} = \text{Distance (ft.) to cover an acre}$$

2. Disconnect drop tubes and catch the granules in a bag from each discharge point while driving a known distance. Weigh the samples individually.

3. Determine the application rate in pounds per acre.

$$\frac{\text{(Feet driven per acre - Step 1) x (Ounces collected - Step 2)}}{\text{(Feet driven - Step 2) (16 ounces/lb.)}} = \text{Pounds applied per acre}$$

Example

An applicator spreads a band of chemical 15 inches wide (1.25 ft.). Five ounces is collected from each spout while driving 500 ft.

Step 1. $\frac{\text{43,560 sq. ft. per acre}}{\text{1.25 ft.}} = \text{34,848 ft. to drive to cover one acre}$

Step 2. 5 ounces

Step 3. $\frac{\text{(34,848 ft.) x (5 ounces)}}{\text{(500 ft.) (16 ounces/lb.)}} = \text{21.8 pounds per acre}$

4. Remember, the application rate just determined is for the area actually covered. However, only a fraction of the area is covered when making a band application. To determine the amount applied per acre of land, multiply the application rate by the fraction of land being treated.

$$\frac{\text{Band width (in.)}}{\text{Row spacing (in.)}} \text{ x Application rate on bands} = \text{Pounds per field acre}$$

Example

Continuing the previous example, and assuming a 40-inch row spacing, the application rate per acre of land is:

Step 4. $\frac{\text{15 in.}}{\text{40 in.}} \text{ x 21.8 lb./acre} = \text{8.2 lb. per field acre}$

METRIC CALIBRATION

The following is a repeat of the above procedure using all metric measurements.

1. Measure band width in meters and calculate the distance traveled to cover a hectare.

$$\frac{\text{10 000 m}^2\text{/ha}}{\text{Width of band (m)}} = \text{Distance (m) to cover a hectare}$$

2. Disconnect drop tubes and catch the granules in a bag from each discharge point while driving a known distance. Weigh the samples individually.

3. Determine the application rate in kilograms per hectare.

$$\frac{\text{(Meters driven per hectare - Step 1)} \times \text{(Grams collected - Step 2)}}{\text{(Meters driven - Step 2) (1000 grams/kg)}} = \text{Pounds applied per acre}$$

Example

An applicator spreads a band of chemical 40 centimeters wide (0.4 m). Five ounces is collected from each spout while driving 150 meters.

Step 1. $\frac{\text{10 000 m}^2\text{ per ha}}{\text{0.4 m}} = \text{25 000 m to drive to cover one hectare}$

Step 2. 140 grams

Step 3. $\frac{\text{(25 000 meters)} \times \text{(140 grams)}}{\text{(150 meters) (1000 g/kg)}} = \text{23.3 kilograms per hectare}$

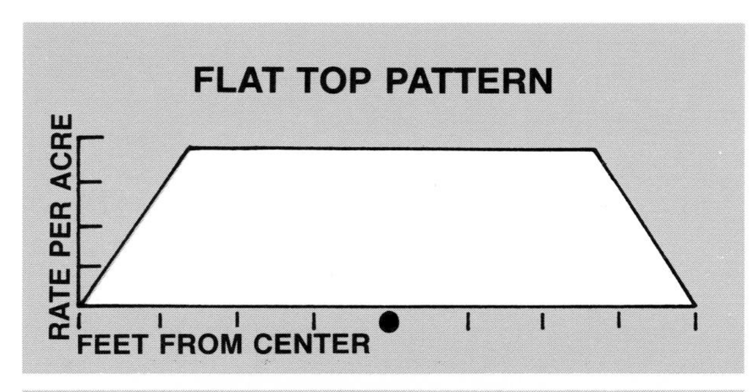

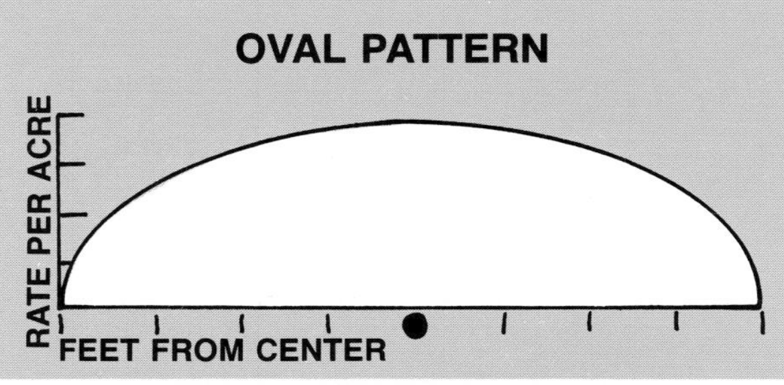

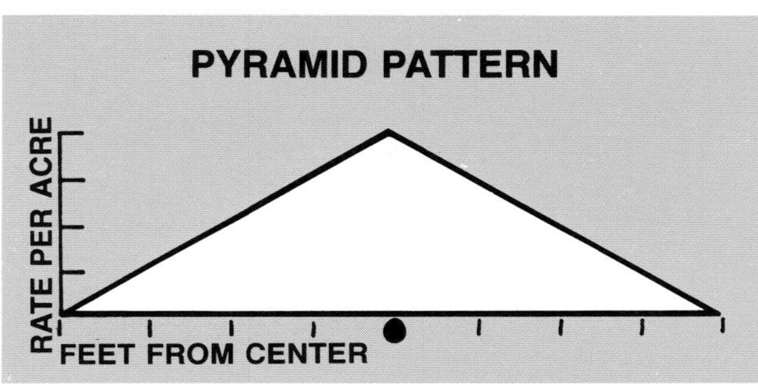

Fig. 25—Desirable Material Spread Patterns

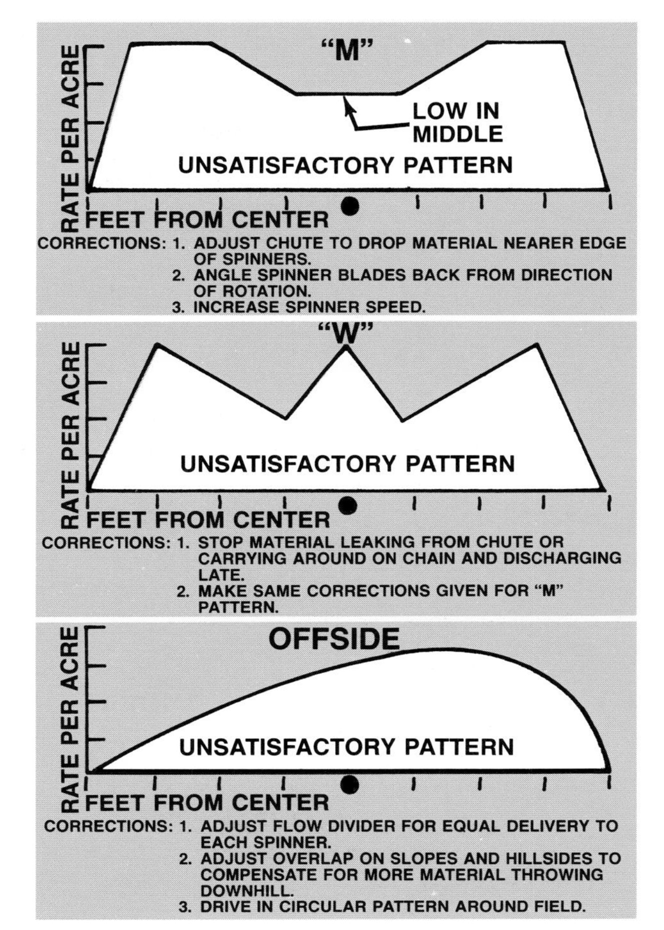

Fig. 26—Correcting Unsatisfactory Spread Patterns

METRIC CALIBRATION (Continued)

4. Remember, the application rate just determined is for the area actually covered. However, only a fraction of the area is covered when making a band application. To determine the amount applied per hectare of land, multiply the application rate by the fraction of land being treated.

$$\frac{\text{Band width (cm)}}{\text{Row spacing (cm)}} \times \text{Application rate on bands} = \text{Kilograms per field hectare}$$

Example

Continuing the previous example, and assuming a 100-cm row spacing, the application rate per hectare of land is:

Step 4. $\frac{40 \text{ cm}}{100 \text{ cm}} = 23.3 \text{ kg/ha} = 9.3 \text{ kg per field hectare}$

The preceding discussion treated a single applicator or drop tube from a hopper. Check and adjust the application rate of each discharge so that all apply the same amount.

Granules of different chemicals vary greatly in size and density. Always recalibrate equipment when changing materials.

FIELD OPERATION

The primary function of the operator in the field is to monitor and guide the equipment. Be sure chemical is coming out of all tubes. If each tube is not visible, stop periodically and check.

When using a broadcast spreader, drive carefully to obtain the proper amount of overlap. Overlap is not as important with spins spreaders (Fig. 27) because the

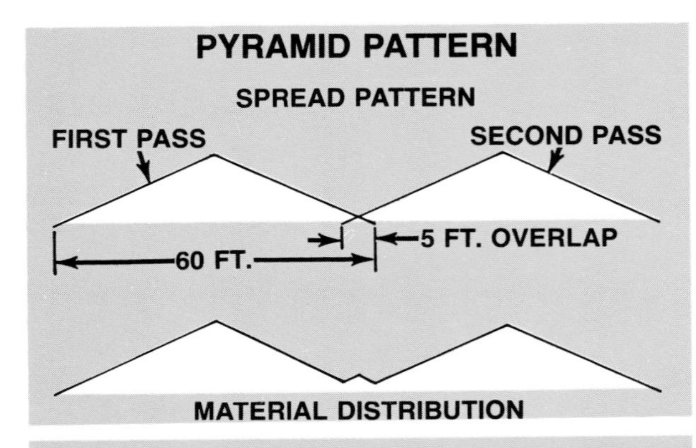

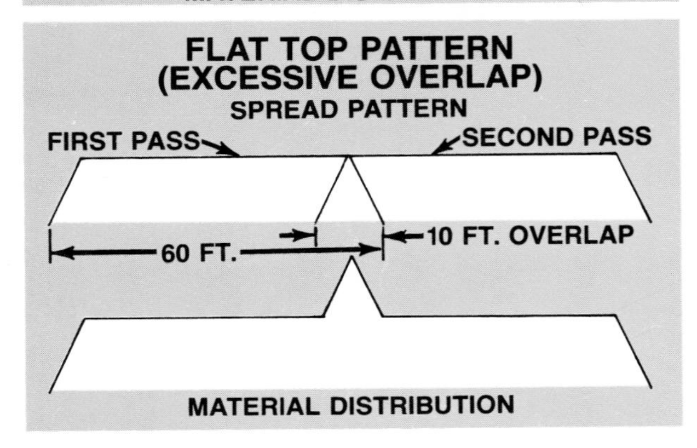

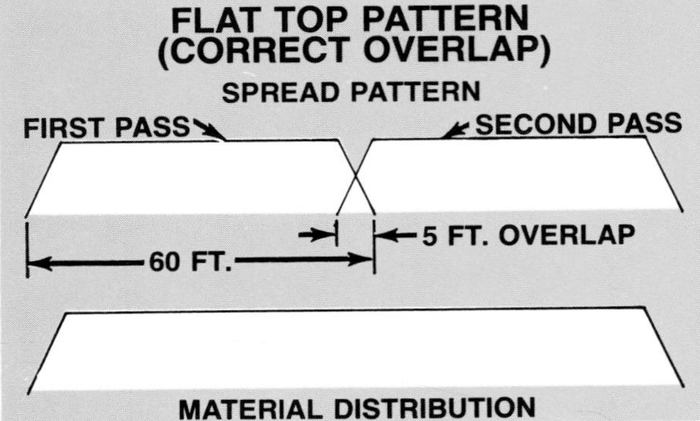

Fig. 27—Effect of Overlapping Spread Patterns

application rate tapers off at the edges. However, if the application rate is uniform across the swath, be careful not to overlap or leave skips.

Speed control is very important. As ground speed is increased, the application rate per acre decreases for power-operated and gravity-fed equipment. The drop-type distributor is less sensitive to speed changes because the feed shaft is ground-driven.

Watch both the tachometer and ground speed when using a spin spreader. As the spreader is unloaded, tractor speed tends to increase, because of the lighter load. If the engine speeds up, the spreader disk also speeds up, which widens the swath and can change the spread pattern. Widening the swath, of course, also decreases the average application rate.

The accuracy and uniformity of distribution is generally more critical for pesticides than for fertilizer application.

TRANSPORT

Transporting a spreader between farmstead and fields can be a special problem. To avoid trouble when transporting chemicals in most spreaders is simple — **don't do it.** Material may sift out of hoppers or pack extremely tight from bouncing and shaking.

Transport chemical distributors, planters, and similar equipment empty. Use accessory lights and other warning devices to alert operators of other vehicles. Check local and state regulations. See your machinery dealer for the proper safety equipment.

TRANSPORTING SPIN SPREADERS

Before transporting the spreader, make sure that it is properly and securely hitched and drives are disconnected. Don't dump a stream of chemicals along the road. Use safety chains on the hitch (Fig. 28).

If the spreader is equipped with hydraulic brakes, install a brake safety chain (Fig. 29).

When on the road, keep to the right as far as possible. Observe normal traffic rules. Be alert, and pay attention to your driving. Cover the load with a tarp to prevent blowing of material (Fig. 30). *Do not carry riders!*

STORAGE

If at all possible, store chemical-application equipment indoors when it is not being used. Clean all equipment thoroughly before storing.

Fig. 28—Hitch Safety Chain

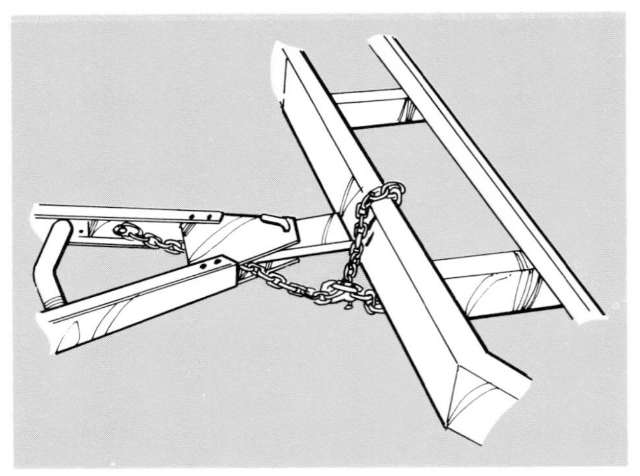

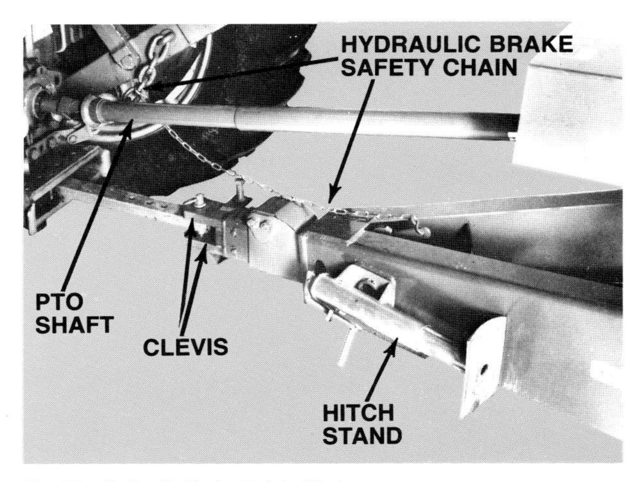

Fig. 29—Hydraulic Brake Safety Chain

CLEANING

Clean thoroughly with water. If fertilizer or pesticide has built up on the machine, use a wire brush or scraper. If that is not effective, soak fertilizer accumulations with diesel fuel. Work outdoors and keep all fires, including tobacco smokers, out of the area.

Wear appropriate protective gear when cleaning chemicals from the machine.

PREPARING FOR STORAGE

After the machine has been thoroughly cleaned, paint all parts which show paint wear.

When the machine is dry, lubricate it as recommended and oil vital parts to protect against corrosion.

If the machine has tires, set it up on blocks to take weight off the tires. Do not deflate the tires.

Check the machine carefully and note any repairs that are needed. Order the required parts and make repairs during the off-season.

Fig. 30—Cover the Load With a Tarp

SUMMARY

Dry-chemical application equipment is used to apply fertilizers and pesticides. There are many types of specialized equipment. Frequently, dry-chemical applicators are used as attachments on other machinery.

Fertilizing equipment includes broadcast and band applicators. Broadcast applicators include spin spreaders, aircraft, and drop-type fertilizer distributors.

Pesticide equipment includes dusters and granular applicators.

When planning chemical applications, consider three items:

- **The applicator**
- **The tractor**
- **The chemical**

Like other applicators, dry-chemical equipment must be correctly calibrated. The principle of calibration is to measure the machine output and adjust the application rate to the desired amount. There are three steps;

1. Determine the area treated.
2. Determine the quantity of chemical applied.
3. Calculate the application rate.

Spin-spreader calibration includes measuring the distribution pattern and output. A simple procedure is available for calibration of these spreaders.

In the field, the operator must monitor the machine operation, drive accurately to control overlap, and control the speed.

Avoid transporting loaded applicators except spin spreaders. When on the road, keep to the right, obey traffic laws, and use warning devices to alert drivers of other vehicles. Do not carry riders.

Before putting the machine in storage, clean it. Set it on blocks to take the load off the tires. Check it completely to find whether repairs are needed.

CHAPTER QUIZ

1. Recent developments have caused a major change in fertilizer application. What is this change?

2. What is the main advantage of air application of fertilizer?

3. (True or false.) Dry chemical application requires very little equipment maintenance.

4. (True or false.) Operator's manuals contain the correct settings for dry chemical application and should be used on every field.

5. (True or false.) Farmers usually apply the same chemical and at the same rates with granular band applicators over the entire field. The applicators located on the toolbar are the most accurate and should be used to calibrate the planter or other machine being used.

6. Why is a positive feed system more accurate than a gravity feed system for metering dry chemicals?

9
Safety

Fig. 1—Apply Chemicals Safely

Introduction

To anyone using chemicals and machinery, safety is extremely and personally important.

Here's why:

- *Accidents hurt–and kill*
- *Accidents cost time and money*
- *Accidents can be avoided*

An accident is possible any time you work with a machine. And the risk is greater when you are using chemicals—because *three kinds of hazards* are involved:

- *Hazards common to all machines*
- *Hazards specific to chemical equipment*
- *Hazards involved in handling chemicals themselves*

AVOIDING ACCIDENTS

Who has accidents with chemicals? There are two groups:

- *People who are not aware of the hazards*
- *People who are aware of the hazards but, because of carelessness, hurry, or some other reason, do not use safe procedures*

Fig. 2—Pesticide Labels Are Your Best Source of Information

To avoid accidents, you must first recognize the hazards. Experience and information are basic to recognizing hazards. Begin by studying equipment, operator's manuals, and recommended procedures to help identify potential hazards. The information in this chapter will help.

SAFETY INFORMATION

Information on hazards and safe working procedures are available from many sources. Five of the best are:

1. *National Safety Council Bulletins*
2. *National Agricultural Chemicals Association*
3. *FMO: Agricultural Machinery Safety*
4. *Extension bulletins distributed by the agricultural safety or pesticide specialist in your state*
5. *Chemical labels*

Labels

When you work with a chemical, your single most important source of information is the container label. Read it *twice* before handling the material the first time (Fig. 2). Read it each time you use the product.

The label includes the following information:

- *Name and address of the manufacturer*
- *Trade name of the product*
- *List of all active ingredients (both common and chemical names)*
- *Type of pesticide*
- *Formulation*
- *EPA registration number*
- *Storage and disposal instructions*
- *Hazard statements (Fig. 3)*

TOXICITY	KEY WORDS	GROUP	TYPICAL WARNING
HIGH	DANGER POISON	1	DANGER POISON KEEP OUT OF REACH OF CHILDREN POISONOUS IF SWALLOWED OR ABSORBED THROUGH SKIN FLAMMABLE • VAPOR HARMFUL Do Not Get in Eyes, on Skin or on Clothing Do Not Breathe Vapors or Spray Mist Do Not Take Internally Keep Away from Heat and Open Flame Keep Container Closed
MODERATE	WARNING	2	WARNING IRRITATING TO SKIN AND EYES CAUSES BURNS ON PROLONGED CONTACT HARMFUL IF SWALLOWED Avoid Breathing Spray Mists Keep Out of the Reach of Children Do Not Get in Eyes, on Skin or on Clothing or Shoes
LOW	CAUTION	3	CAUTION KEEP OUT OF REACH OF CHILDREN ABSORBED THROUGH SKIN CAUSES EYE IRRITATION HARMFUL IF SWALLOWED FLAMMABLE LIQUID
	NONE	4	NO SIGNAL WORDS

Fig. 3—Toxicity-Rating Key Words

Fig. 4—Read and Follow Directions

- *Directions for use*
- *Net contents statement*
- *Recommendations for use*

A chemical user who ignores the label may be headed for trouble (Fig. 4). He can permanently injure his health or the health of others. He can harm livestock, wildlife, or crops. A chemical company may spent $8 or $10 million to obtain the information required for registration of a new pesticide. That information is summarized for you on the container label. **Read and follow all instructions on the label.**

Fig. 5—Wear Protective Clothing when Applying Fertilizer

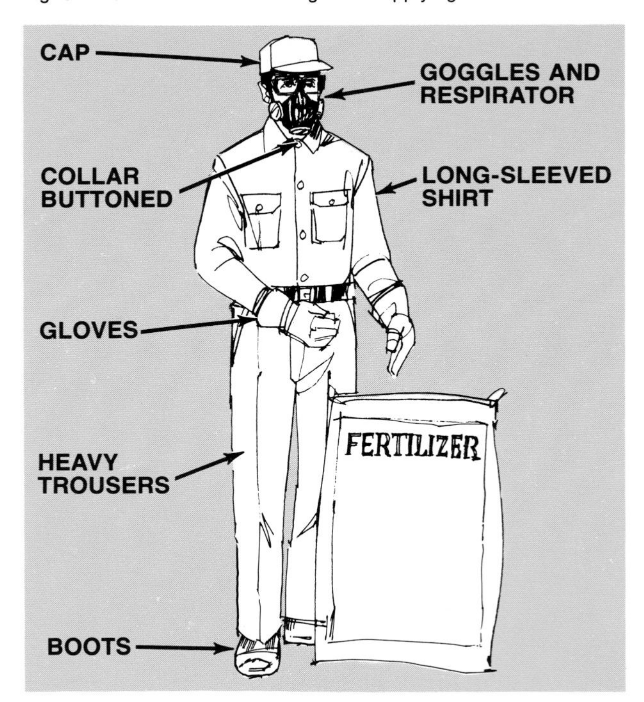

The remainder of this chapter discusses factors involved in the safe use of chemicals, including:

- *Hazards*
- *Safe working practices*
- *Procedures to follow if you have an accident*

APPLYING DRY FERTILIZERS

Dry fertilizer is hygroscopic—it attracts moisture. It can draw water out of your skin and leave it red and sore. These "skin burns" cause only minor discomfort, unless you have especially sensitive skin.

To prevent fertilizer burns on *your* body:

1. *Wear protective clothing* (Fig. 5)

2. *Wash exposed skin areas frequently*–do not leave fertilizer on your skin

3. *Let the wind blow fertilizer dust away from you.* Drive crosswind in the field and stand upwind when filling hoppers (Fig. 6)

4. *Wear a filtered respirator if you cannot avoid the dust.* It keeps the dust out of your lungs

5. *Wear goggles* to protect your eyes

Spin Spreaders

Spin spreader hazards fall into two groups:

- **Mechanical**
- **Flying particles**

Avoid accidents with spin spreaders by keeping away from flying particles and moving parts of the machine.

1. *Don't get behind the spreader when it is operating* (Fig. 7)

2. *Stop the machine and shut off the engine before starting any maintenance, repairs, or adjustments*

3. *Drive safely.* Loaded spreaders are heavy and can veer out of control on hills or turns.

ANHYDROUS AMMONIA

Anhydrous ammonia can cause painful skin burns. Even mild exposure can irritate the eyes, nose, and lungs. Permanent injury or blindness may result if ammonia gets into your eyes. Prolonged inhalation can cause suffocation and death. And, because of its low boiling point, ammonia can cause "burns" by freezing as well as by caustic action.

First Aid For Ammonia Exposure

If ammonia strikes the skin or eyes, rinse the area immediately with clean water and continue for at least 15 minutes (Fig. 8) then go directly to a doctor!

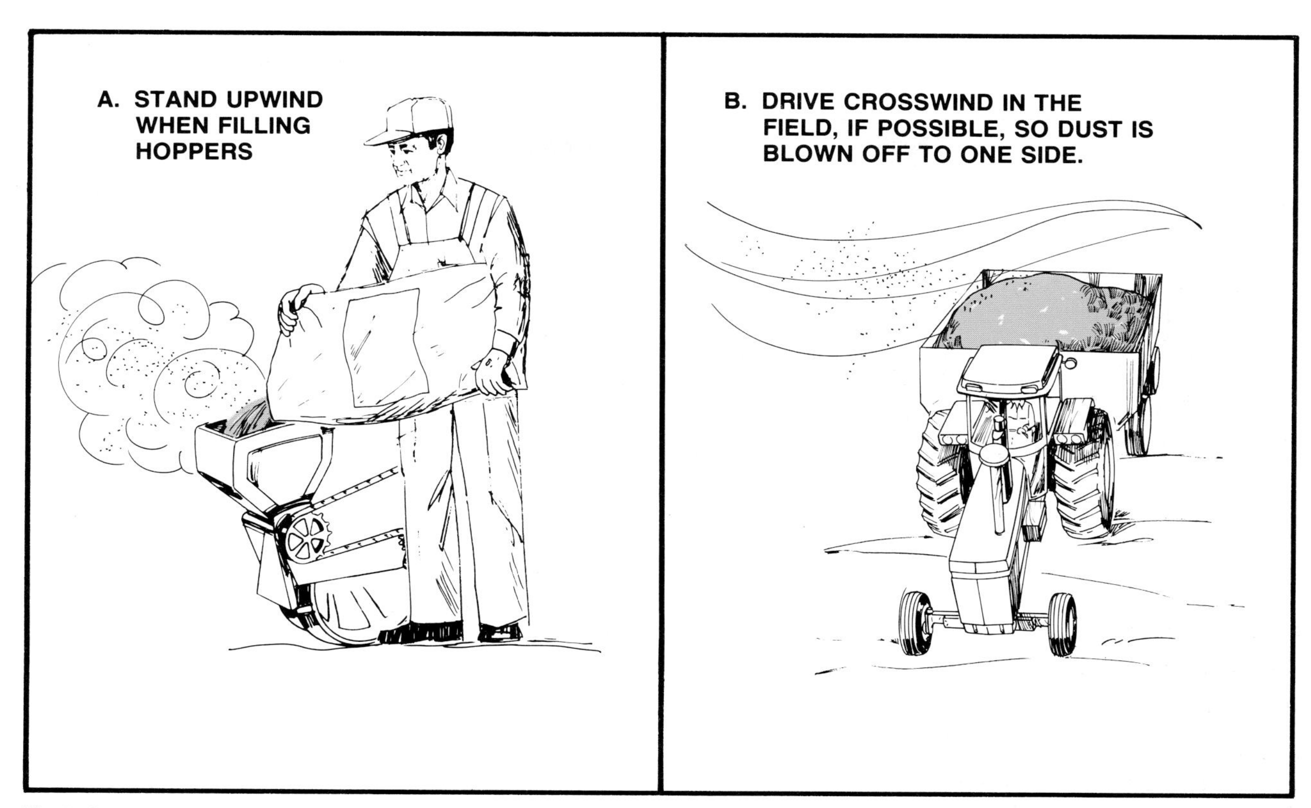

Fig. 6—Let the Wind Blow Dust from Fertilizer or Pesticide Away from You

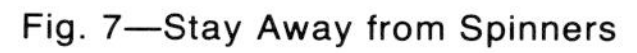

Fig. 7—Stay Away from Spinners

Fig. 8—Wash Off Ammonia with Plenty of Fresh Water

Fig. 9—Use Equipment Designed to Handle Anhydrous Ammonia

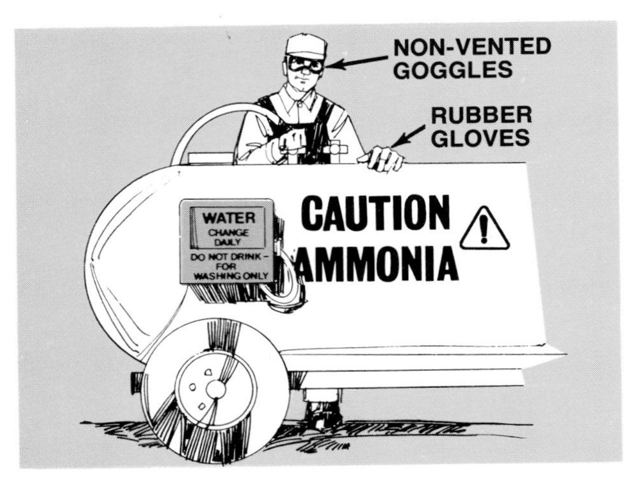

Fig. 10—Wear Goggles and Rubber Gloves when Transferring Ammonia

If a person has lost consciousness due to ammonia fumes, get him to fresh air and administer artificial respiration. Get medical help as soon as possible.

Handling Anhydrous Ammonia

Remember three important things when using anhydrous ammonia:

- *Use proper equipment*
- *Take care of the equipment*
- *Follow safe practices*

How do you do these things? Here are 12 specific recommendations:

1. *Use equipment specifically designed for use with ammonia* (Fig. 9). Be sure there are no copper or brass parts, as ammonia is corrosive to copper and copper alloys.

2. *Take care of equipment.* Check it before use to make sure all joints are tight, valves are working, and fittings are in good shape. Replace weak hoses or other parts not in top condition.

3. *Replace safety valves that do not work.* Don't attempt to repair them.

4. *Keep tanks coated with light-colored paint.* The best paint to use is a high-reflectance white paint made especially for ammonia tanks. This helps reflect heat and retards tank rusting.

5. *Take care of hoses.* They can burst unexpectedly. Check hoses and connections at least once a year. Store hoses in a cool dry place when not in use. Hang them with open ends down over an old wheel rim or some similar object. Avoid sharp bends or kinks. Tape hose ends to keep out dirt and insects. Replace any hoses that are cracked or worn.

6. *Keep heat and fire away from ammonia.* Treat ammonia like a flammable gas. Welding or cutting on anhydrous tanks should be done only by experts. Do not let anyone do it unless he has had special training. Do not allow smoking around ammonia equipment.

7. *Wear rubber gloves and tight-fitting, unvented goggles when transferring anhydrous ammonia* (Fig. 10). Goggles designed for use with grinders or other tools are not satisfactory because the air vents may allow ammonia to get behind the lenses.

8. *Immediately remove clothes contaminated with ammonia and thoroughly wash the skin area beneath.*

9. *Close valve wheels by hand.* If a wrench is used, you may damage the valve and let ammonia escape.

10. *Know how to operate the equipment.* Get training from the ammonia dealer or other competent individual. Read the operator's manuals for your equipment. Many accidents happen because equipment operators do not know how to operate the equipment correctly.

11. *Keep a supply of fresh water handy for washing.* Carry at least five gallons on each piece of equipment. Change the water daily to be sure it does not become contaminated by ammonia fumes (Fig. 11).

12. *Carry a plastic squeeze bottle of water for flushing eyes in case of an accident* (Fig. 12).

LIQUID FERTILIZERS

Aqueous ammonia and other liquid fertilizers which contain ammonia present hazards much the same as anhydrous ammonia. Eyes are most often affected in an accident. Vision is nearly always impaired and blindness can result.

Many accidents occur when the operator is transferring liquid from a nurse tank to an applicator tank. For protection during the transfer operation, wear protective clothing, including non-vented goggles or a full face shield (Fig. 13). Keep plenty of clean water available for washing in case of an accident.

Fig. 11—Keep Fresh Water Available On Ammonia Equipment

Fig. 12—Keep a Pocket-Size Container of Water Handy for Flushing Eyes

ANIMAL MANURES

Four toxic gases, ammonia, methane, carbon dioxide, and hydrogen sulfide, are produced when manure is stored without oxygen, as in liquid-manure tanks (Table 1).

Follow these rules to be safe from manure gases:

1. *Know the effects of the gases.* If you *think* you detect one of them, get to fresh air immediately.

2. *Provide maximum ventilation whenever a tank is agitated or emptied* (Fig. 14)

3. *Do not allow smoking or any other fire near a liquid-manure tank.* Methane-air mixtures can be explosive.

4. *Never go into an empty or partially-filled liquid-manure tank unless it is well ventilated and a second person is standing by with a rope and harness attached to you* (Fig. 16). This person can pull you out if

Fig. 13—Wear Protective Clothing When Transferring Aqueous Ammonia (Liquid Fertilizer with Ammonia)

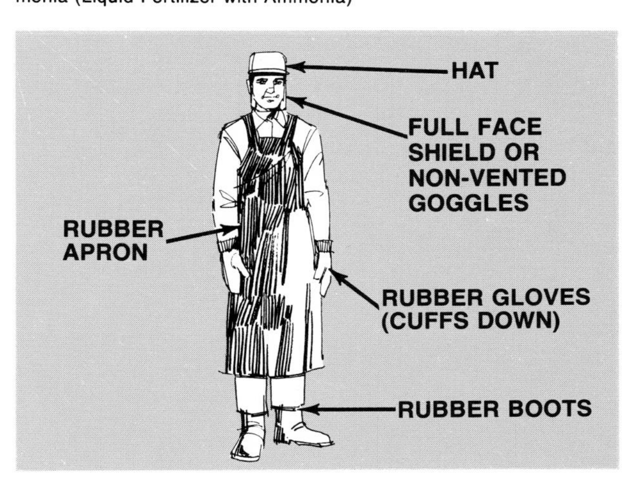

TABLE 1

MANURE GAS CHARACTERISTICS

Gas	Concentration	Odors	Typical Effects And Symptoms
AMMONIA	0.5%	Sharp, pungent	Throat and eye irritation Harsh coughing, severe irritation of eyes, nose and throat Can be fatal
CARBON DIOXIDE	3 - 6%	Odorless	Labored breathing, drowsiness, headaches Death by suffocation
METHANE	30%	Odorless	Non-Toxic Highly flammable, explosive when mixed with air
HYDROGEN SULFIDE	0.5% 1%	Rotten egg	Headaches, dizziness, nausea, excitement Can cause unconciousness and death

Fig. 14—Use Maximum Ventilation When a Liquid-Manure Tank is Being Agitated

you are overcome by gases. The best procedure is to wear a self-contained breathing apparatus while in the tank.

PESTICIDES

Pesticides are poisons. They have to be to kill pests. They may also be toxic (poisonous) to man and desirable plants and animals. Exposure to a significant amount of almost any pesticide can make a person ill.

Fig. 15—Wear a Self-Contained Breathing Apparatus if You Must Enter a Liquid-Manure Tank

Many pesticides are so toxic that exposure to even a small quantity can kill a person. When you work with pesticides, be aware of the dangers.

ACUTE AND CHRONIC EXPOSURE

If a person is accidentally splashed or sprayed with a pesticide, or swallows some pesticide and immediately loses consciousness, there is little doubt about the cause of the illness and proper medical treatment can be sought. "One-Time" cases like this are examples of acute exposure.

Chronic exposures, over several days or weeks, can be as dangerous as acute cases. The greatest risk is a combination of the two. Even limited daily exposure because of faulty equipment or not wearing proper protective clothing, reduces your body's ability to deal with an acute exposure.

KINDS OF EXPOSURE

Before a pesticide can harm you, it must enter your body. There are three means of entry.

- *Oral* (swallowing)
- *Dermal* (contact with skin)
- *Inhalation* (breathing)

ORAL EXPOSURE

Swallowing a pesticide is called oral exposure. Oral poisoning may occur during an accident such as a spill or equipment failure. But it is more likely to be caused by carelessness. Oral exposure can be either acute or chronic. The seriousness of a particular exposure depends on the oral toxicity of the material and the amount swallowed.

Acute oral toxicity is expressed by an LD_{50} value (Fig. 16). The LD_{50} is the number of milligrams of a pesticide per kilogram of body weight that kills 50 percent of test animals (usually rats or rabbits). The smaller the LD_{50}, the more toxic (dangerous) the chemical.

There is some uncertainty about applying LD_{50} values from tests on rabbits or rats to predict toxicity of a material to humans. However, most experts agree that the *relative* toxicities are about the same for humans as for the test animals. The amounts of material that would probably kill people are shown in Fig. 17.

Avoiding Oral Exposure

Most oral exposure is the result of carelessness. Proper precautions can eliminate nearly all cases.

1. Always keep pesticides in their original labeled containers.

2. Never store pesticides in containers that originally held food or beverages, such as soda-pop bottles.

3. Keep drinking-water containers and coffee-break equipment away from the pesticide working area.

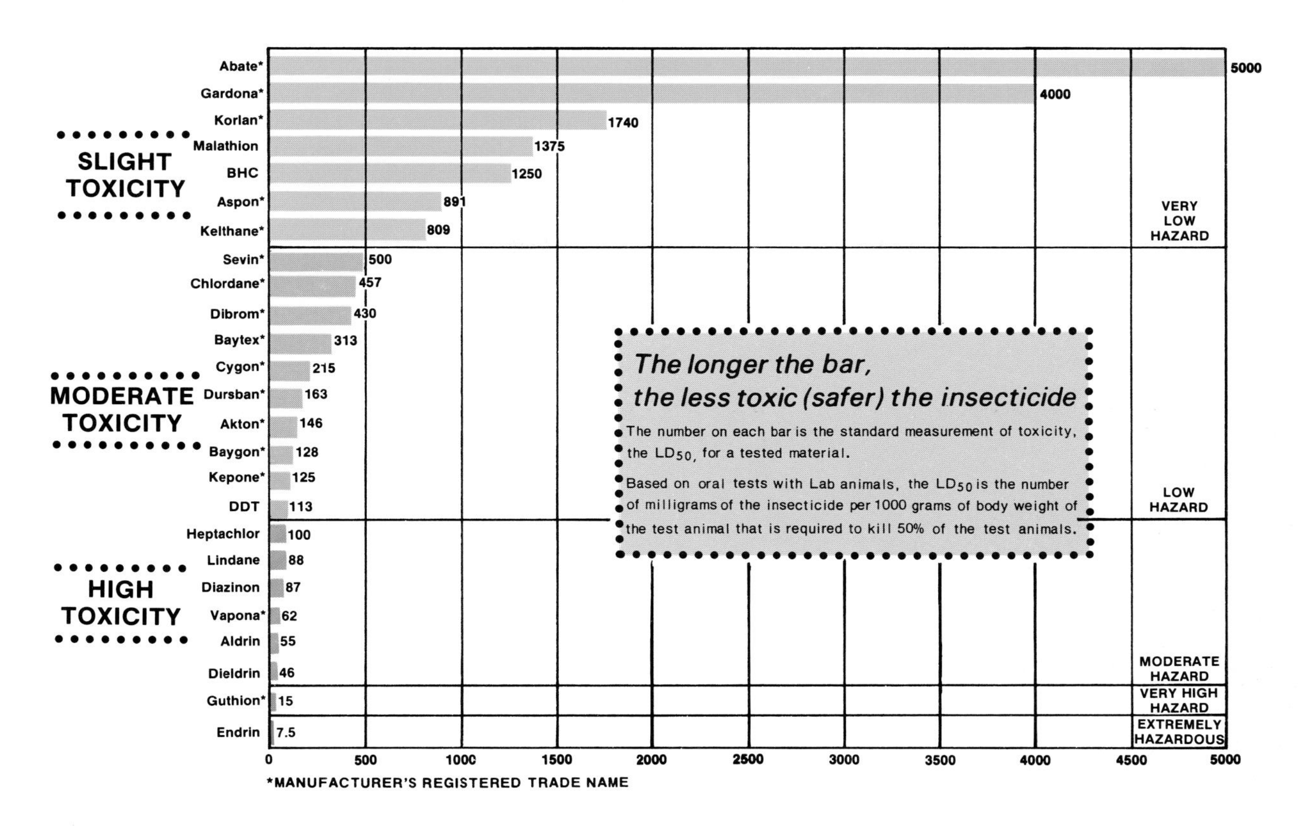

Fig. 16—Oral Toxicities of Some Insecticides

Fig. 17—The "Probably Lethal" Dose of a Pesticide Depends on the Toxicity of the Chemical and the Weight of the Victim

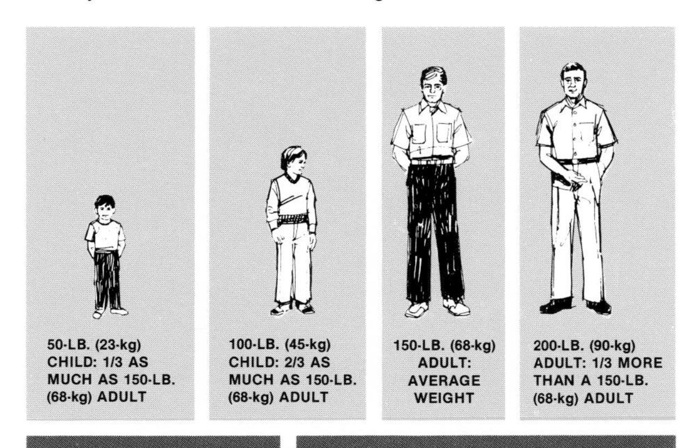

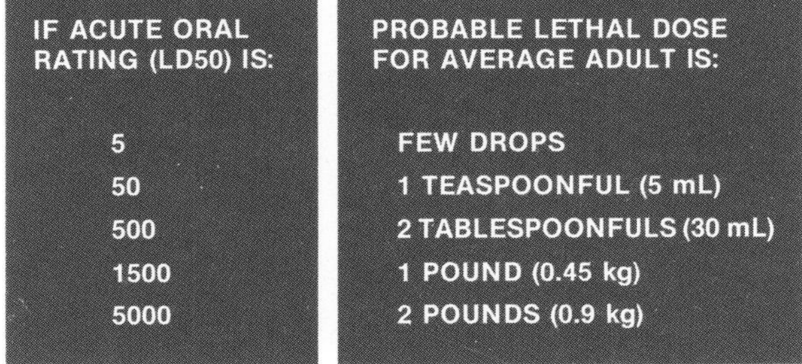

IF ACUTE ORAL RATING (LD50) IS:	PROBABLE LETHAL DOSE FOR AVERAGE ADULT IS:
5	FEW DROPS
50	1 TEASPOONFUL (5 mL)
500	2 TABLESPOONFULS (30 mL)
1500	1 POUND (0.45 kg)
5000	2 POUNDS (0.9 kg)

4. Install a backflow preventer on water lines used to fill sprayers.

5. Don't drink from or fill water containers from the hose used to fill the sprayer tank.

6. Avoid back-siphoning and contamination of water supplies by not immersing the end of the filter hose in the pesticide mixture (Fig. 18).

7. Don't eat anything that has been sprayed until the safety interval given on the label has passed.

Fig. 18—Warning Sticker to be Placed near Filler Cap of Sprayer Tank

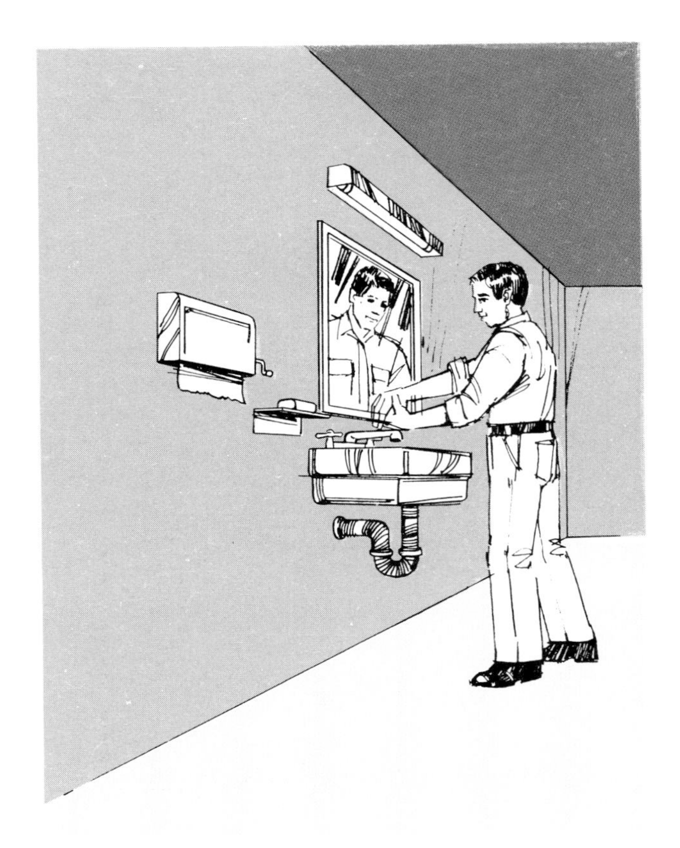

Fig. 19—Wash Carefully after Working with Pesticides

Fig. 20—Post Warning Signs in Recently-Sprayed Fields

8. Wash thoroughly with soap and water before eating, drinking, or smoking if you have been working around pesticides (Fig. 19).

9. Post signs in sprayed areas showing the earliest date that the crop can be harvested or the field entered (Fig. 20).

10. Don't smoke, drink, or eat around pesticides.

11. Clean nozzle tips with a soft brush, small wire, or low pressure (under 30 psi, 200 kPa) compressed air. Don't hold tips to your mouth and blow through them (Fig. 21).

DERMAL EXPOSURE

Dermal exposure is skin contamination. Like oral exposure, dermal exposure can be either acute or chronic. Acute exposure can be expressed as a dermal LD_{50} value (Fig. 22).

Rates of absorption through the skin are different for different parts of the body (Fig. 23). For example, absorption is more than four times faster on the forehead than on the forearm. Absorption through the skin in the scrotal, or groin area approximates the effect of injecting the pesticide directly into the bloodstream.

Absorption continues through all affected skin areas as long as pesticide remains in contact with the skin. Seriousness of exposure is increased if the affected skin area is large or if material is left on the skin for an extended period of time.

Fig. 21—Clean Clogged Nozzles with a Soft Brush, Small Wire or Air Hose — NOT YOUR BREATH

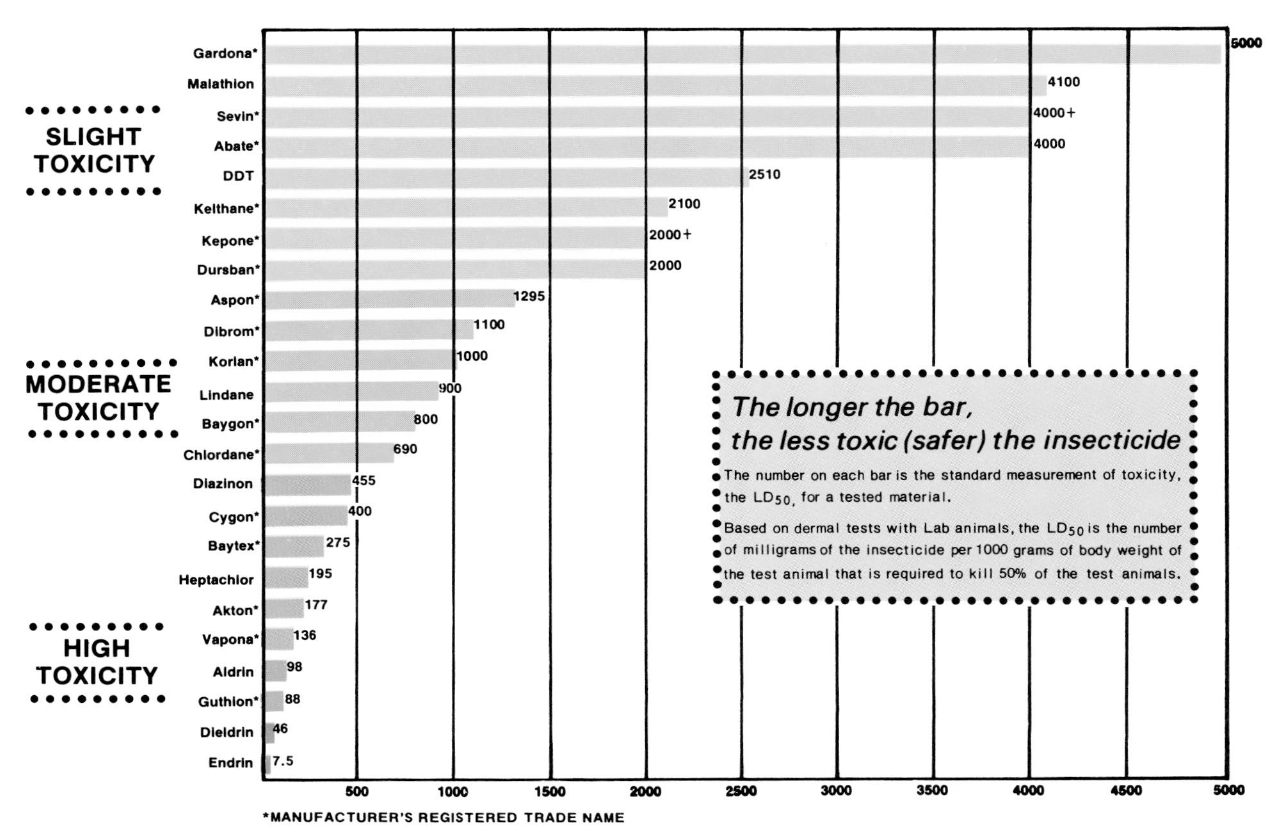

Fig. 22—Dermal Toxicities of Some Insecticides

Avoiding Dermal Exposure

Avoid dermal exposure by keeping pesticide off the skin. Protective clothing and equipment are your first line of defense. Take extra care to avoid chronic exposure. The pesticide can build up in your body and the effects may be difficult to recognize.

1. Wear protective clothing (see later section for details). Protection is particularly important if the dermal LD_{50} of the pesticide if below 250.

2. Stand on steps or a sturdy platform when filling a sprayer tank (Fig. 24).

3. Clean up and dispose of spilled pesticides promptly. If you walk through the spill area, your footwear may become contaminated and subject you to chronic exposure.

4. Avoid drift. Don't depend on the tractor cab to protect you. Small particles and vapors may pass through the air-filtering system. Wear full protective gear if you are spraying highly-toxic material. See Chapter 7 for ways to reduce drift (Fig. 25).

5. Wear clean clothes. Clothing may be contaminated by splashing, drift, and in other ways. Don't wear clothing that you suspect may be contaminated. Wash it with detergent and hot water first. Wear clean, fresh

Fig. 23—Dermal Absorption Rates Are Higher for Some Parts of the Body

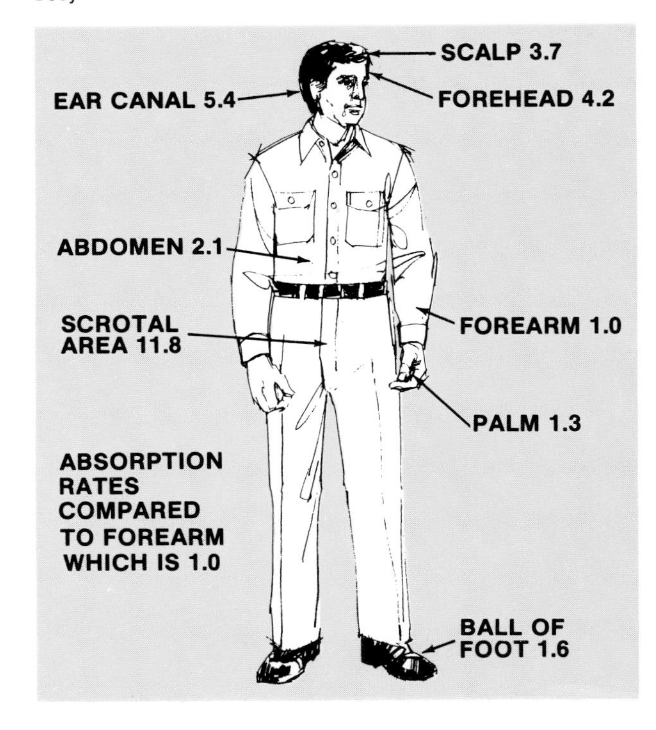

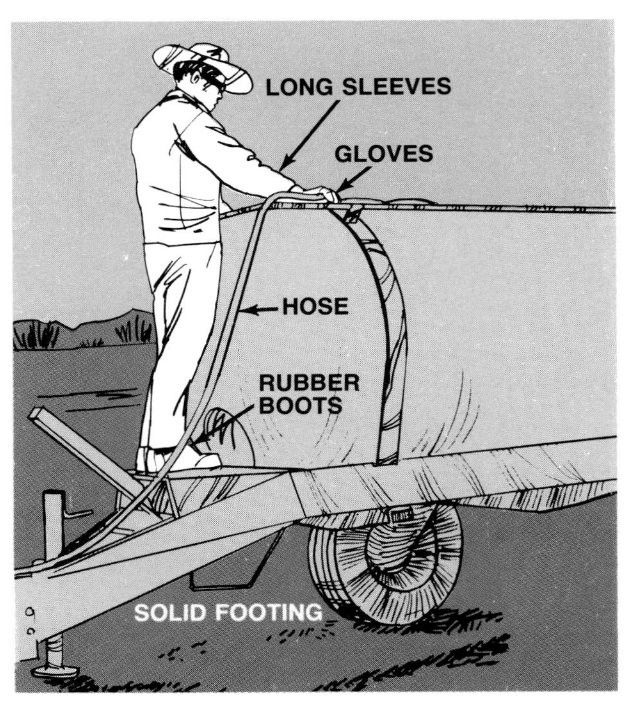

Fig. 24—Stand on Solid Footing when Filling Sprayer Tanks

clothing every day and change during the day if you think clothes have absorbed any pesticide.

6. Wash tractor seat-cushion covers frequently.

7. Observe the re-entry interval on the container label. Post signs around the area until the re-entry interval has expired (Fig. 20).

8. Follow other precautions listed under: Avoiding Oral Exposure.

INHALATION EXPOSURE

Inhalation exposure results from breathing in pesticide vapors, dust, or spray particles. Like oral and dermal exposure, it is more serious with some pesticides than others. There is no good measure of the LD_{50} type to characterize respiratory toxicity. The safest procedure is to avoid breathing in any pesticide.

Poisoning through the lungs is more likely to occur in confined areas such as greenhouses or storage sheds than it is outdoors, because the pesticide is kept inside the structure. It can occur outdoors if the concentration of material is high or if a highly-volatile material is used.

Avoiding Inhalation Exposure

1. Don't carry smoking materials when working with pesticides. If the tobacco is contaminated, the pesticide may be vaporized and drawn into your lungs with the smoke.

2. Stay out of the smoke from burning containers (Fig. 26). The smoke contains vapors from the burned pesticide. Stand upwind from the fire.

3. Avoid drift. Small drifting particles can be inhaled directly into your lungs. Use a chemical cartridge respirator for protection (Fig. 27).

4. Follow other precautions listed under: Avoiding Oral Exposure and Avoiding Dermal Exposure.

DRY PESTICIDES

GRANULAR PESTICIDES

Hazards associated with dry granular pesticides are similar to those for granular fertilizers, except that pesticides are generally more toxic.

Fig. 25—Small Drifting Particles Can Cause Dermal Exposure

Fig. 26—Stay Away from Smoke from Burning Containers

Because of their form, granular pesticides *seem* safer than liquid pesticides. This leads users to relax their safety precautions. Don't do it. Unless proper safeguards are followed, the possibility of chronic exposure from dry pesticides is great. Keep the dust away from your skin and hair. Follow the procedures outlined in the section on applying dry fertilizers.

Mixing Powdered Pesticides

When mixing or pouring powdered pesticides, avoid as much dust as possible. Keep the container low when pouring to avoid dust clouds. Try to have wind blowing dust away from you.

WETTABLE POWDERS

Wettable powders are dustier than granular pesticides. Wear a filtered respirator when working with them and always wear rubber gloves when handling toxic powders.

Open dust and powder packages from the top. When pouring, hold the container low, close to the receptacle. Pour slowly to avoid billowing dust.

After a wettable powder has been mixed in the spray tank, continue spraying until the tank is empty. Calculate needs as accurately as possible to avoid having excess material left in the tank when spraying is completed or must be stopped. Stopping for lunch or some other reason will permit the powder to settle out. The agitator cannot be relied upon to put all of the spray mix back in suspension. Variation in strength of the application can result. This can be hazardous to the operator, since he could then be exposed to more concentrated material than anticipated. Also, the sprayer will usually be more difficult to clean. The extra cleaning effort involved also exposes the operator and environment to a higher risk of accidental contamination.

Fig. 27—Use a Recommended Chemical Cartridge Respirator for Protection against Drift

SYNERGISM

Synergism occurs when the combined action of two compounds (pesticides) produces an effect greater than the additive effects of the individual compounds. For example, say a unit of pyrethrin alone would kill 10 insects, while a unit of piperonyl butoxide alone would kill five insects. If these two units of chemicals were combined, you might expect them to kill 15 insects. But, due to synergism, as many as 50 may be killed.

This increased toxicity affects all three types of exposure — oral, dermal, and inhalation. Use particular caution when using combination sprays. Never make up your own mixtures. Use only those combinations recommended by your state cooperative extension service or other reputable agency *and approved on the pesticide label* (Fig. 28).

PESTICIDE APPLICATION EQUIPMENT

To avoid hazards in using sprayers:

1. Inspect hoses and hose connections daily (Fig. 29). They should be tight enough to not leak. Wear rubber gloves while tightening joints. Relieve system pressure before doing any work on the machine (Fig. 30).

2. Use the proper application rate. The wrong application rate or a pesticide concentration that is too high may expose the operator and the environment to danger when he thinks he is safe. Calibrate sprayers frequently.

3. Avoid spray residues on machine parts when the sprayer is dismantled for repair. Wear rubber gloves. Wash the machine before doing any disassembly work.

4. Keep spare nozzle tips for field replacements. Don't try to clean a nozzle tip in the field.

PROTECTIVE CLOTHING

Protective clothing and devices should always be worn when handling, mixing, or applying pesticides (Fig. 31). Protective clothing covers the body completely except for the face. Water-repellent clothes give the best protection. Next best is a coverall-type garment that is thick enough to prevent the pesticides from penetrating easily. Full-length sleeves are necessary — don't roll them up! Keep the collar buttoned and trouser legs rolled down.

Leather absorbs some pesticides and is extremely difficult to decontaminate. Use only rubber gloves and boots. Wear a full-brimmed, water-repellent hat to protect your head and neck. Don't wear a hat made of an absorbent material such as felt. The common billed cap does not provide adequate protection.

Wash all protective gear at the end of each working day and store it in a safe place until needed. Coveralls should be washed daily. Take a shower at the end of each working day.

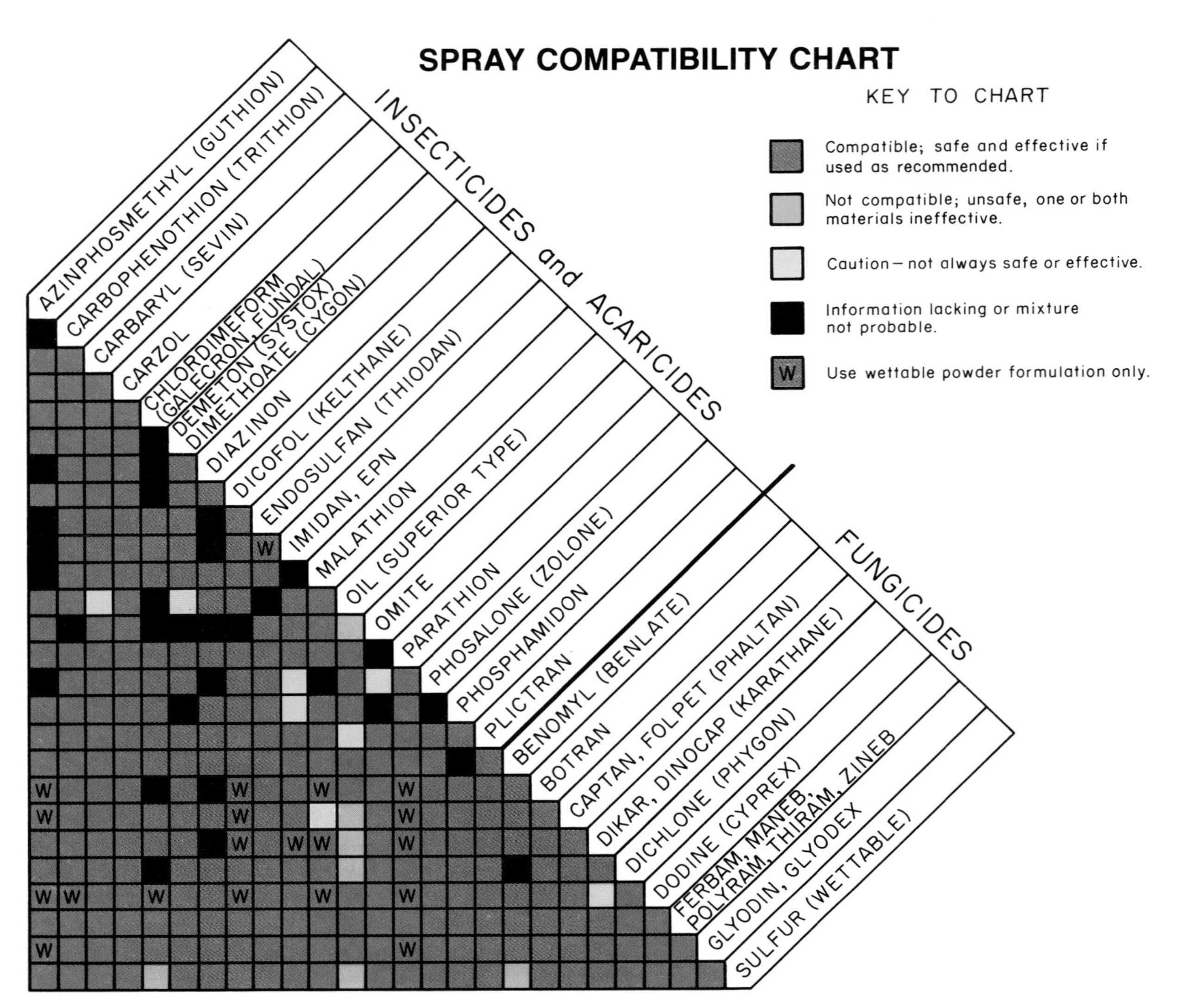

Fig. 28—Spray Compatibility Chart

Fig. 29—Look for Leaks Every Day

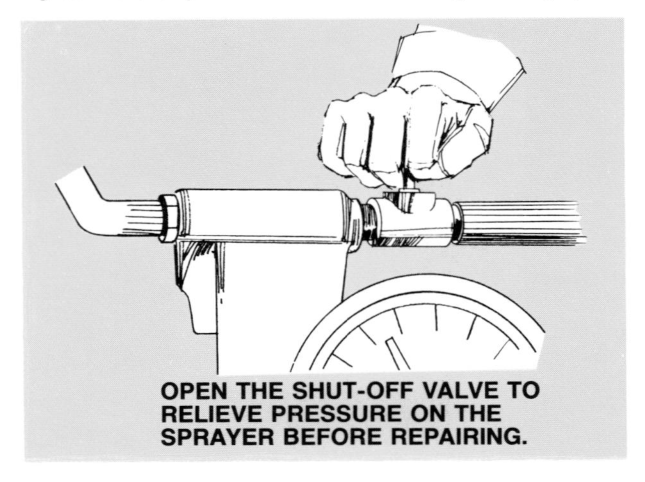

Fig. 30—Relieve System Pressure Before Working on the Sprayer

GLOVES

Do not handle or apply pesticides without gloves. The best protection for your hands is a pair of unlined rubber gauntlet gloves (extending above the wrist). Unlined gloves are easier to keep clean and the gauntlet protects the wrist. When applying organophosphates, use only natural rubber gloves. Organophosphate pesticides can penetrate some plastic gloves. Keep the upper end of the gauntlets folded down to keep liquids from running onto your arm when your hands are raised.

APRONS

Wear a rubber or plastic apron when mixing or loading the sprayer. It should reach past your boot tops. For work with organophosphates, the apron should be natural rubber.

BOOTS

Wear rubber boots when handling or applying pesticides. Keep trouser legs outside the boot tops, so that any runoff will go outside the boot.

GOGGLES

Unless a full-face respirator is being used, there is a risk to the eyes. Wear goggles when loading or mixing pesticides. Goggles should be snug but comfortable and should be the nonvented, nonfogging type.

Fig. 31—Wear Water-Repellent Protective Clothing when Spraying Pesticides

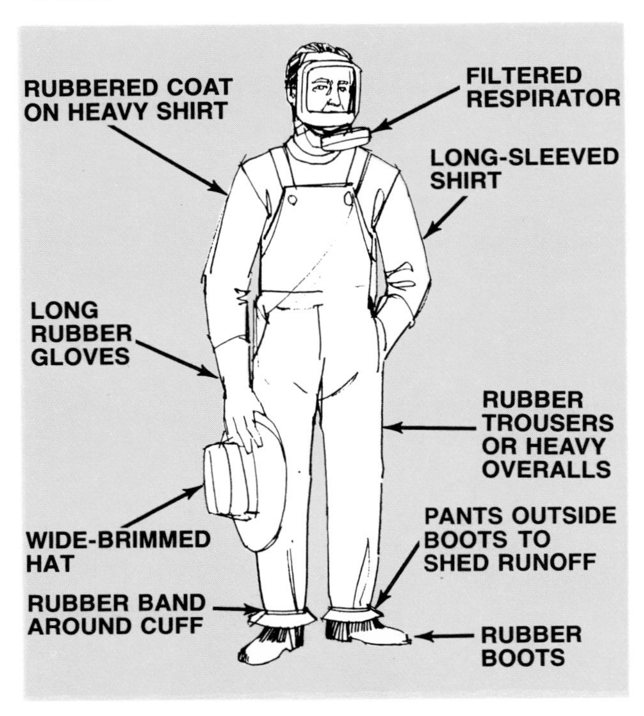

RESPIRATORS

Wear a respirator if there is a chance that dusts or vapors may be inhaled. Choose a mask or respirator approved for the particular pesticide being used. A simple gauze respirator will only filter out dust or liquid particles. It can't stop gasses and vapors and is suitable only for low-toxicity pesticides.

The respirator should cover at least your mouth and nose. There are two main types — chemical-cartridge respirators and gas masks. Cartridge respirators are light and protect against dusts and gases.

The full-face gas mask has a large canister worn on the chest. It has built-in goggles and covers the whole face. Use a gas mask when applying highly toxic materials or when mixing or application is done in an enclosed area such as a machine shed or greenhouse.

In situations where there is a lack of oxygen, use a supplied-air respirator or a self-contained breathing apparatus.

PESTICIDE STORAGE

Pesticides are usually purchased before they are needed, and must be stored between the time of purchase and use. Store them properly to prevent possible damage or hazard to:

- *People, especially children*
- *The environment*
- *Pets and domestic animals*
- *Materials and containers* (moisture can cause corrosion, extreme temperatures can deform or burst container)

Store pesticides in a separate building away from other farm structures, if possible. It should be located on high ground with good drainage and be accessible to trucks and equipment for delivery and removal under any weather conditions (Fig. 32)

Lock the storage area securely to keep children, pets, and irresponsible adults out. Costly and sometimes scarce pesticides are also a target of some thieves — another reason to keep the storage area locked. Post warning signs (Fig. 33).

Follow these tips for pesticide storage.

1. Always store pesticides in the original, tightly-closed containers (Fig. 34). Labels must remain attached and legible.

2. Stack containers only if the bottom ones are strong enough to support the stack without splitting open and spilling the material.

3. Read the label for storage instructions.

Fig. 32—Pesticide Storage Building

Fig. 33—Chemical Storage Should Be Locked and Clearly Marked

Fig. 34—Store Pesticides in Original Containers

4. Store herbicides away from the insecticides, fungicides, seed, fertilizer, and similar materials.

5. Use the material in a package only if you are confident of its contents and condition.

6. Don't store materials unnecessarily. Dispose of:

- *Materials that are no longer registered*
- *Empty or unlabeled containers*
- *Spilled materials*

7. Keep fire away from the storage area. Do not permit smoking in or near the pesticide storage area.

8. Post warnings for fire department or other emergency personnel in case of fire or accident.

9. Wear full protective clothing and equipment when working in the storage area (Fig. 35).

PESTICIDE DISPOSAL

Use extreme care when disposing of pesticides to avoid contaminating soil, air and water and to ensure that other potential hazards are minimized. Plan for the disposal of three kinds of materials:

- **Leftover Chemicals**
- **Spilled Materials**
- **Empty Containers**

Following are suggestions and guidelines.

Leftover Chemicals

Leftover spray mixtures should not be left on the farm. Storing it in odd containers is hazardous, particularly to children and animals. Calculate requirements carefully and mix only enough to cover the required area. Spread any remaining mixture on the land if possible, but avoid over-application to some areas.

Each state has approved landfills for disposal of leftover pesticides. Consult the nearest office of the Environmental Protection Agency for the location of approved landfills.

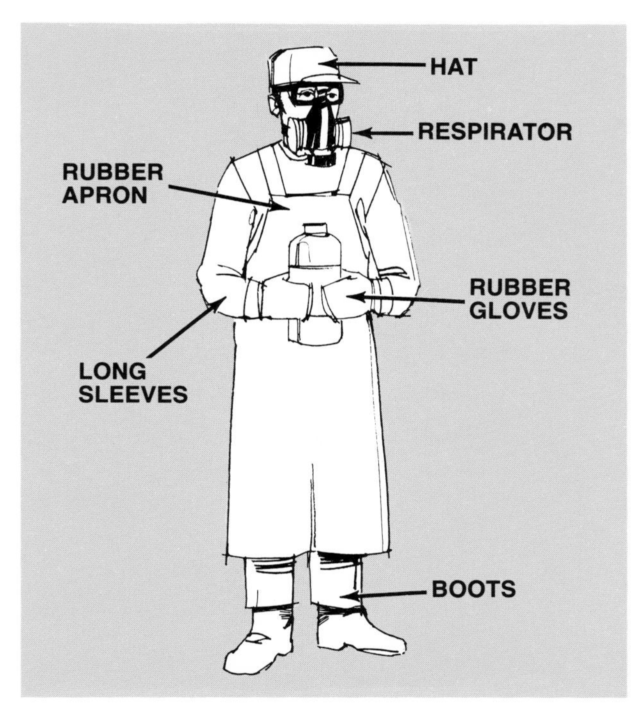

Fig. 35—Wear Protective Clothing in the Pesticide Storage Area

Spills

Keep a liquid spill from spreading by surrounding it with a ring of absorbent material such as dried clay. Then clean it up and decontaminate the area (Fig. 37).

PESTICIDE SAFETY TEAMS: If you have a large spill, or if the spilled material is highly toxic, it may be wise to call in outside assistance. Pesticide safety teams have been organized in some areas to assist in the prompt cleanup and decontamination of large chemical emergencies, such as warehouse fires or transport accidents. They also may be called upon for assistance and advice in any kind of emergency connected with pesticides.

The pesticide safety team headquarters maintains a 24-hour answering service. *Their telephone number is (800) 424-9300.* Keep this number handy. Paint it on the wall or put it on a sign posted on the pesticide storage area and near your home phone. **This number is for emergency use only. DO NOT call for information or purposes other than emergency.**

The pesticide safety teams have been set up to handle a variety of emergencies. The number can be called at any time from anywhere in the United States. It is a toll-free number.

Empty Containers

Simply pouring the contents from pesticide containers will never remove all the chemical. There always re-

Fig. 36—Dump Leftover Spray Mixture in an Approval Disposal Pit if it Cannot Be Used

mains a residue of pesticide that may be hazardous. Keep empty containers in a locked storage area and dispose of them at least once a year in a way that poses no hazard to humans, animals, or the environment.

Use a rinse-and-drain procedure to clean containers (Fig. 38). A container that has been rinsed three times with water is considered nonhazardous. Dispose of these containers along with regular garbage. Containers that have not been rinsed three times must be disposed of at a landfill approved by the EPA to accept such containers.

Check state and federal EPA regulations on disposal of pesticide containers in state-approved landfills.

Large metal containers can sometimes be returned to the supplier or sold to a firm dealing in reconditioned drums. If a drum cannot be recycled, punch it full of

Fig. 37—Cleaning Up a Spilled Pesticide

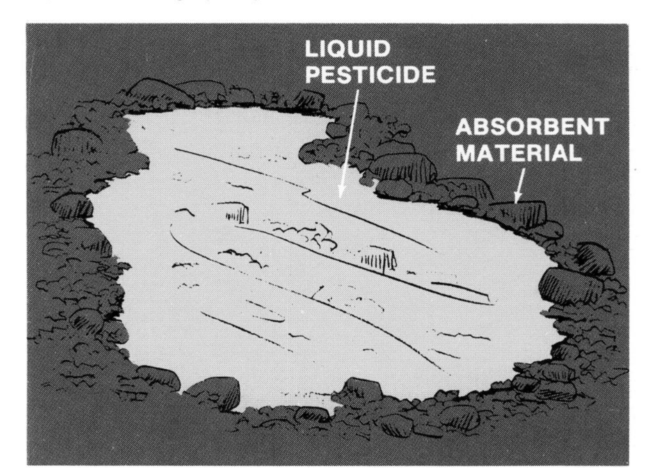

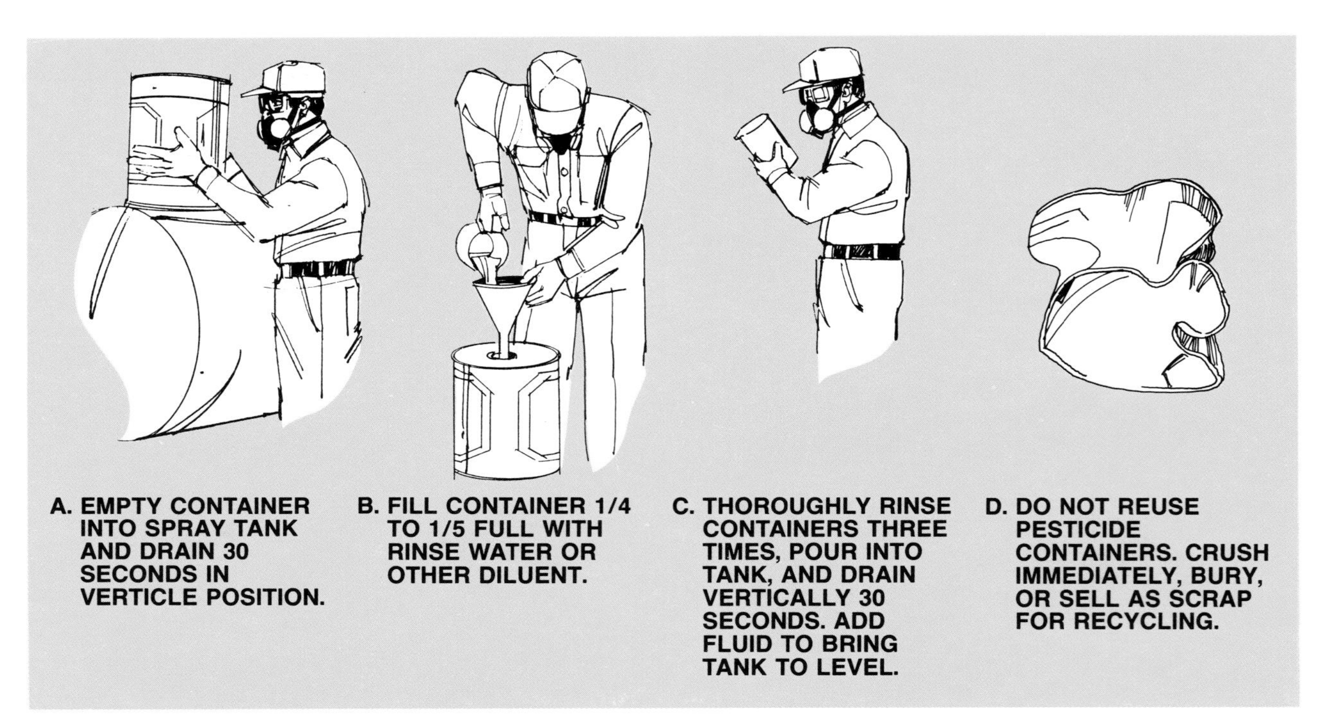

Fig. 38—Rinse and Drain Pesticide Containers before Disposal

holes and bury it. Do not use empty pesticide containers to handle feed or water for livestock, poultry, or pets.

PESTICIDE POISONING

If you or anyone using pesticides suspect or show any signs of poisoning, contact a doctor or poison-control center immediately. Poison-control centers are usually located in designated hospitals.

Poisoning symptoms depend on the chemical. For highly-toxic materials, the label lists poisoning symptoms. Read the label and know the symptoms to watch for. Following are some symptoms associated with some common pesticides:

Chlorinated Hydrocarbons

Chlorinated hydrocarbons act as stimulators to the central nervous system. They cause irritability, convulsions, or coma. Nausea and vomiting usually occur, but may not if the dose is large.

Dinitro Compounds

Early symptoms include extreme euphoria (feeling of well-being), night sweating, insomnia, restlessness, thirst, and fatigue. In more severe cases, profuse sweating, intense thirst, yellow skin, rapid pulse and respiration, fever, apathy, convulsions, and coma can all occur.

Organophosphates

Symptoms depend on the specific material, strength of the formulation, and the intensity and duration of exposure.

Initial symptoms usually include headache, weakness, blurred vision, perspiration, and nausea. Abdominal cramps, vomiting, excessive salivation, and diarrhea may occur. Breathing is difficult and the throat and chest may feel constricted. In severe cases, muscular twitching may be present along with pupil dilation. The illness can progress to loss of consciousness or convulsive seizures.

Workers who handle organophosphate insecticides should have regular periodic checks for cholinesterase activity. Organophosphate compounds, which are related to nerve gases, are toxic because they interfere with the cholinesterase enzymes that control nerve impulse transmission. Have your doctor give you a baseline cholinesterase level blood test before you begin working with these pesticides (Fig. 39).

Then have subsequent cholinesterase tests at weekly or biweekly intervals. Evidence of overexposure can thus be detected, and steps taken to correct the situation before it becomes dangerous. Follow the doctor's recommendations.

Self-diagnosis and self-treatment are extremely dangerous. Don't take drugs, even tranquilizers or aspirin, unless you discuss it with a medical authority first. Interaction between the pesticide and the drug can be very dangerous.

There is no drug that can be taken to prevent the effects of organophosphate insecticides. Some operators take atropine tablets routinely to prevent poisoning. This is dangerous. Atropine has a place in treating proven clinical cases of poisoning, but it can be dangerous if taken otherwise. Atropine is a potent drug in its own right, and if taken in the absence of organophosphate or carbamate poisoning, it may lead to serious illness. Furthermore, atropine taken in this way can mask early symptoms of poisoning, thus permitting further dangerous exposure. The consequences could be very serious.

For people who often operate away from "home base", the medic-alert system is recommended. Everyone working with a toxic pesticide should wear a tag giving this information:

- *Trade name and type of pesticide being used.*
- *Name and phone number of the physician to contact in case of an emergency.*

POISON EMERGENCIES

In spite of all precautions used in handling pesticides, poisoning can still occur. Always be ready to handle an emergency (Fig. 40). There are two things to do:

Fig. 39—Have a Blood Test for Baseline Cholinesterase before Working with Organophosphorus Insecticides

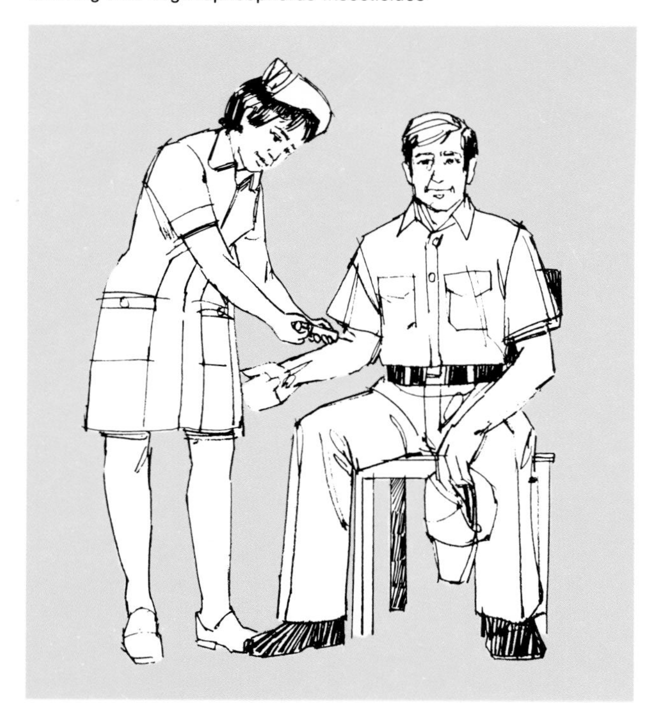

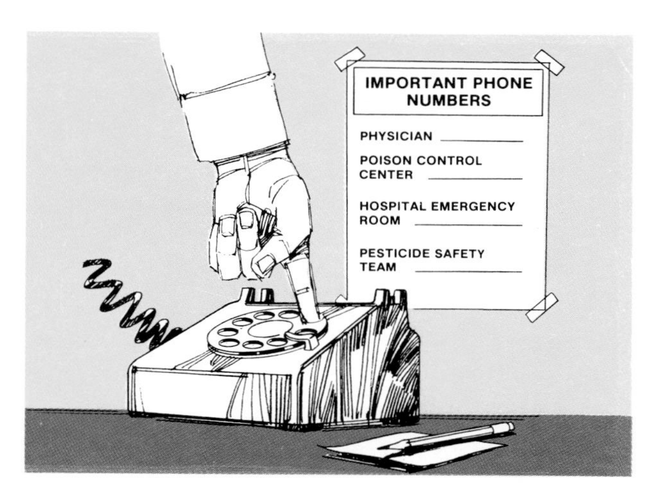

Fig. 40—Keep a List of Emergency Telephone Numbers Posted Near Your Phone

1. *Call a doctor or poison control center at once.* Supply the following information:

- **Pesticide involved**
- **Type of work the victim was doing**
- **How poisoning took place (swallowing, inhalation)**

If you cannot reach emergency help on the telephone, take the victim to the emergency room of the nearest hospital.

2. *Give the victim first aid as advised by the doctor.* Instructions are also given on the container label. If possible, send the label along when the victim is taken to the hospital.

SUMMARY

An accident is possible any time you work with a machine. And the risk is greater when using chemicals—because three kinds of hazards are involved:

- *Hazards common to all machines*
- *Hazards specific to chemical equipment*
- *Hazards involved in handling chemicals themselves*

Information on hazards and safe working procedures are available from many sources:

1. National Safety Council Bulletins
2. National Agricultural Chemicals Association
3. FMO—Agricultural Machinery Safety
4. Extension bulletins distributed by the agricultural safety or pesticide specialist in your state
5. Chemical labels

When working with chemicals, the single most important source of information is the container label.

The application of dry fertilizers, including spin spreaders, liquid fertilizers, including anhydrous ammonia, animal manures, and pesticides all require maximum safety precautions, protective clothing, proper application and correct storage and disposal.

The pesticide safety team handles a variety of emergencies and can be reached at (800) 424-9300. **This number is for emergency calls only.** Chemical poisoning can be fatal and victims should be taken to the emergency room of the nearest hospital.

CHAPTER QUIZ

1. What metals should never be used in ammonia-application equipment?

2. Name eight kinds of information found on a pesticide label.

3. What is your best source of information about a particular pesticide?

4. What hazards are associated with spin spreaders?

5. What should you do if you get ammonia in your eyes?

6. What are three paths for a pesticide to enter your body?

7. What is meant by LD_{50}?

8. (True or false.) Absorption rates differ for different parts of the body.

9. (True or false.) A tractor cab is adequate protection against drift.

10. List five items of protective clothing.

11. Why should herbicides, pesticides, seed, fertilizer, and other materials be stored separately?

12. What is the telephone number for Pesticide Safety Teams?

13. What is the reason for getting a baseline cholinesterase test?

14. What is the first thing to do if someone is poisoned by a pesticide?

15. Who would you call if you accidently broke open a drum of parathion?

Appendix

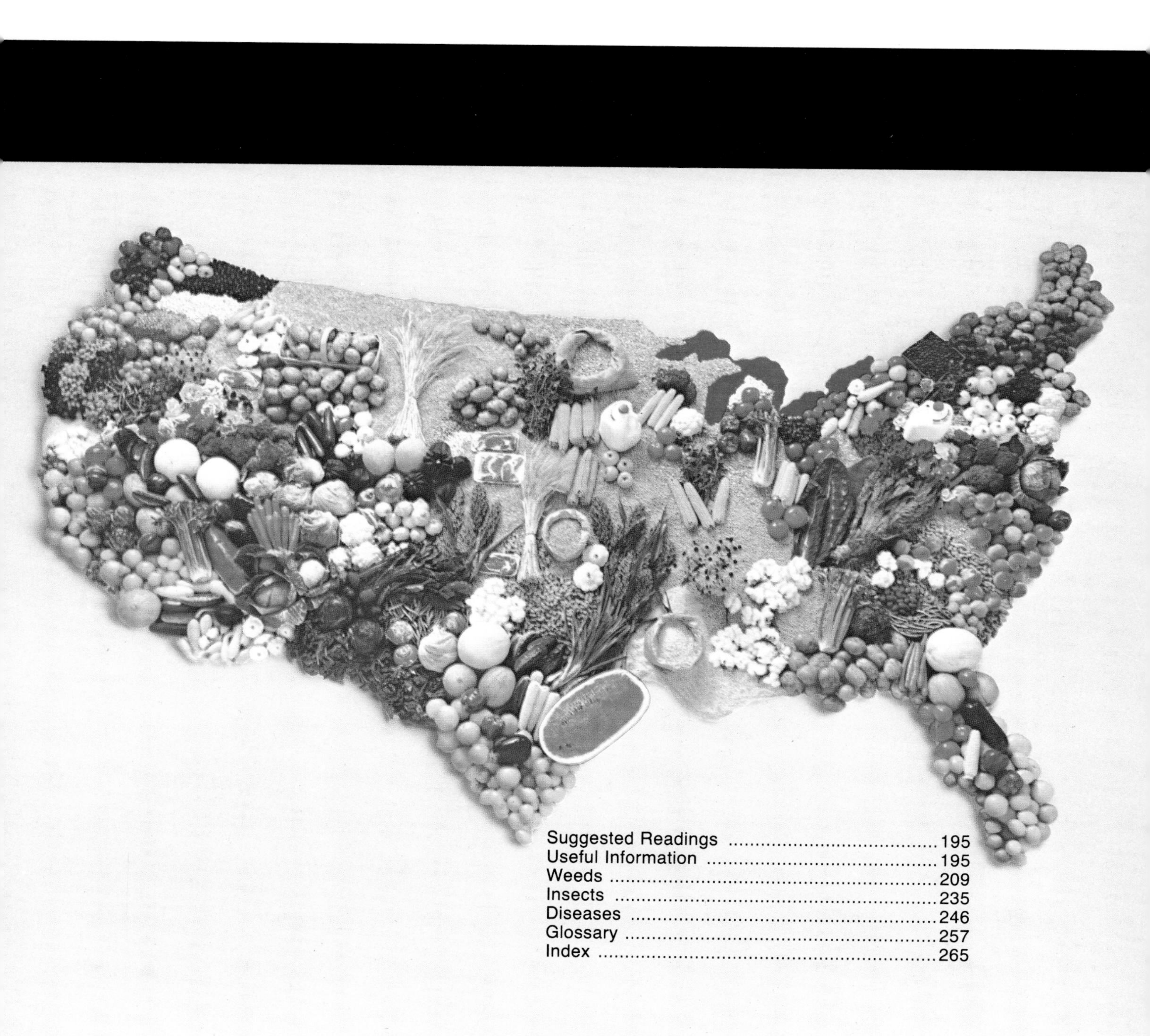

SUGGESTED READINGS

TEXTS

Biological Control of Insect Pest and Weeds; DeBach, P. and Schlinger, E.I.; Reinhold Publishing Corp., New York, 1964.

Changing Patterns in Fertilizer Use; Nelson, L.B. and others; Soil Science Society of America, Madison, Wisconsin, 1968.

Composite List of Weeds; Weed Science Society of America, Champaign, IL, 1971.

Conservation Farming; Deere & Co. Service Training Dept., Moline, IL, 1980.

Farm Chemicals Handbook; Meister Publishing Company, Willoughby, Ohio, 1982.

Fundamentals of Applied Entomology, Third Edition Pfadt, R.E. and others; Macmillan, New York, 1978.

Fundamentals of Machine Operation — Agricultural Machinery Safety; Deere & Co. Service Training Dept., Moline, IL, 1974.

Herbicide Handbook, Fourth Edition; Weed Science Society of America; Champaign, IL, 1979.

Mode of Action of Herbicides, Second Edition; Ashton, F.M. and Crafts, A.S.; John Wiley & Sons, New York, 1981.

Pesticides and Pollution; Bloom, S.C. and Degler, S.E.; Bureau of National Affairs, Washington, D.C., 1969.

Production of Field Crops; Kipps, M.S., McGraw-Hill Publications in the Agricultural Sciences, New York, 1970.

The Profitable Use of Farm Chemicals; Lavertown, S.; Oxford University Press, London, New York, 1962.

Product Guide; Mobay Chemical Corp., Kansas City, Missouri, 1982.

The Scientific Principles of Crop Protection; Martin, H.; St. Martin's Press, New York; 5th edition, 1964.

Use of Pesticides; The President's Science Advisory Committee, 1963.

Using Commercial Fertilizers; McVickar, M.H. and Walker, W.M.; Interstate Printers & Publishers, Danville, Illinois, 1961.

Weed Control; Crafts, A.S. and Robbins, W.W.; McGraw-Hill, New York, 1962.

Weed Science Principles and Practices; Klingman, G.C. and Ashton, F.M.; John Wiley & Sons, New York, 1975.

World Crop Protection; Stapley, J.H. and Gayner, F.C.H.; CRC Press, Cleveland, Ohio, Vol. 1, Pests and Diseases, 1969.

World Crop Protection; Hassall, K.A.; CRC Press, Cleveland, Ohio (1969).

VISUALS

Crop Chemicals Slide Set (FMO 13202S). 35 mm Color. Matching set of 200 slides for illustrations in FMO Crop Chemicals. John Deere Service Training, Dept. F., John Deere Road, Moline, Illinois 61265.

INSTRUCTOR'S KITS

Crop Chemicals Instructor's Kits (FMO 13702K). Looseleaf binder of activities, exercises, masters based on FMO Crop Chemicals text. Includes copy of text. John Deere Service Training, Dept. F., John Deere Road, Moline, Illinois 61265.

USEFUL INFORMATION

The following tables and charts are designed to serve as a quick reference to useful information related to crop chemicals.

Where needed, examples are provided as to how to use this information.

WEIGHTS AND MEASURES

The following two charts show conversions of weights and measures from *U.S. systems* to *metric* and vice versa.

WEIGHTS AND MEASURES—U.S. TO METRIC

U.S. System			Metric Equivalent
LENGTH			
Unit	*Abbreviation*	*Equivalents In Other Units*	
Mile	mi	5280 feet, 320 rods, 1760 yards	1.609 kilometers
Rod	rd	5.50 yards, 16.5 feet	5.029 meters
Yard	yd	3 feet, 36 inches	0.914 meters
Foot	ft. or ′	12 inches, 0.333 yards	30.480 centimeters
Inch	in or ″	0.083 feet, 0.027 yards	2.540 centimeters
AREA			
Unit	*Abbreviation*	*Equivalents In Other Units*	
Square Mile	sq mi or m²	640 acres, 102,400 square rods	2.590 square kilometers
Acre	A	4840 square yards, 43,560 square feet	0.405 hectares, 4047 square meters
Square Rod	sq rd or rd²	30.25 square yards, 0.006 acres	25.293 square meters
Square Yard	sq yd or yd²	1296 square inches, 9 square feet	0.836 square meters
Square Foot	sq ft or ft²	144 square inches, 0.111 square yards	0.093 square meters
Square Inch	sq in or in²	0.007 square feet, 0.00077 square yards	6.451 square centimeters
VOLUME			
Unit	*Abbreviation*	*Equivalents In Other Units*	
Cubic Yard	cu yd or yd³	27 cubic feet, 46,656 cubic inches	0.765 cubic meters
Cubic Foot	cu ft or ft³	1728 cubic inches, 0.0370 cubic yards	0.028 cubic meters
Cubic Inch	cu in or in³	0.00058 cubic feet, 0.000021 cubic yards	16.387 cubic centimeters
CAPACITY			
Unit	*Abbreviation*	*U.S. Liquid Measure*	
Gallon	gal	4 quarts (231 cubic inches)	3.785 liters
Quart	qt	2 pints (57.75 cubic inches)	0.946 liters
Pint	pt	4 gills (28.875 cubic inches)	0.473 liters
Gill	gi	4 fluidounces (7.218 cubic inches)	118.291 milliliters
Fluidounce	fl oz	8 fluidrams (1.804 cubic inches)	29.573 milliliters
Fluidram	fl dr	60 minims (0.225 cubic inches)	3.696 milliliters
Minim	min	1/60 fluidram (0.003759 cubic inches)	0.061610 milliliters
		U.S. Dry Measure	
Bushel	bu	4 pecks (2150.42 cubic inches)	35.238 liters
Peck	pk	8 quarts (537.605 cubic inches)	8.809 liters
Quart	qt	2 pints (67.200 cubic inches)	1.101 liters
Pint	pt	½ quart (33.600 cubic inches)	0.550 liters
MASS AND WEIGHT			
Unit	*Abbreviation*	*Equivalents In Other Units*	
Ton	tn (seldom used)		
short ton		20 short hundredweight, 2000 pounds	0.907 metric tons
long ton		20 long hundredweight, 2240 pounds	1.016 metric tons
Hundredweight	cwt		
short hundredweight		100 pounds, 0.05 short tons	45.359 kilograms
long hundredweight		112 pounds, 0.05 long tons	50.802 kilograms
Pound	lb or lb av also #	16 ounces, 7000 grains	0.453 kilograms
Ounce	oz or oz av	16 drams, 437.5 grains	28.349 grams
Dram	dr or dr av	27.343 grains, 0.0625 ounces	1.771 grams
Grain	gr	0.036 drams, 0.002285 ounces	0.0648 grams

WEIGHTS AND MEASURES—METRIC TO U.S.

Metric System			U.S. Equivalent
LENGTH			
Unit	*Abbreviation*	*Number of Meters*	
Kilometer	km	1,000	0.62 mile
Hectometer	hm	100	109.36 yards
Decameter	dkm	10	32.81 feet
Meter	m	1	39.37 inches
Decimeter	dm	0.1	3.94 inches
Centimeter	cm	0.01	0.39 inch
Millimeter	mm	0.001	0.04 inch

Metric System			U.S. Equivalent
AREA			
Unit	*Abbreviation*	*Number of Square Meters*	
Square Kilometer	sq km or km^2	1,000,000	0.3861 square mile
Hectare	ha	10,000	2.47 acres
Are	a	100	119.60 square yards
Centare	ca	1	10.76 square feet
Square Centimeter	sq cm or cm^2	0.0001	0.155 square inch

Metric System			U.S. Equivalent
VOLUME			
Unit	*Abbreviation*	*Number of Cubic Meters*	
Stere	s	1	1.31 cubic yards
Decistere	ds	0.10	3.53 cubic feet
Cubic Centimeter	cu cm or cm^3 also cc	0.000001	0.061 cubic inch

Metric System			U.S. Equivalent		
CAPACITY					
Unit	*Abbreviation*	*Number of Liters*	*Cubic*	*Dry*	*Liquid*
Kiloliter	kl	1,000	1.31 cubic yards		
Hectoliter	hl	100	3.53 cubic feet	2.84 bushels	
Decaliter	dkl	10	0.35 cubic foot	1.14 pecks	2.64 gallons
Liter	l	1	61.02 cubic inches	0.908 quart	1.057 quarts
Deciliter	dl	0.10	6.1 cubic inches	0.18 pint	0.21 pint
Centiliter	cl	0.01	0.6 cubic inch		0.338 fluidounce
Milliliter	ml	0.001	0.06 cubic inch		0.27 fluidram

Metric System			U.S. Equivalent
MASS AND WEIGHT			
Unit	*Abbreviation*	*Number of Grams*	
Metric Ton	MT or t	1,000,000	1.1 tons
Quintal	q	100,000	220.46 pounds
Kilogram	kg	1,000	2.2046 pounds
Hectogram	hg	100	3.527 ounces
Decagram	dkg	10	0.353 ounce
Gram	g or gm	1	0.035 ounce
Decigram	dg	0.10	1.543 grains
Centigram	cg	0.01	0.154 grain
Milligram	mg	0.001	0.015 grain

CALIBRATION FORMULAS

$$\text{gpm/nozzle} = \frac{7.5}{\text{sec/pt}}$$

$$= \frac{\text{gpa X s X w}}{5{,}940}$$

$$\text{gpa} = \frac{5{,}940 \text{ X gpm/nozzle}}{\text{s X w}}$$

$$= \frac{44{,}500}{\text{s X w X sec/pt}}$$

$$= \frac{43{,}560 \text{ X gal discharged}}{\text{swath width (ft) X length of run (ft)}}$$

$$\text{mph} = 0.682 \text{ X } \frac{\text{length of run (ft)}}{\text{time (sec)}}$$

gpa ....gallons per acre, broadcast basis
gpm ....gallons per minute
mph ....miles per hour
sec ....seconds
pt ....pints
ft ....feet
s ....ground speed (miles per hour)
w ....nozzle spacing on boom or band width (inches)

CALIBRATION FORMULAS (METRIC)

$$\text{L/min/nozzle} = \frac{60}{\text{sec/L}}$$

$$= \frac{\text{L/ha X s X w}}{600{,}000}$$

$$\text{L/ha} = \frac{600{,}000 \text{ X L/min/nozzle}}{\text{s X w}}$$

$$= \frac{10\,000 \text{ X L discharged}}{\text{swath width (m) X length of run (m)}}$$

$$\text{km/h} = 3.6 \text{ X } \frac{\text{length of run (m)}}{\text{time (sec)}}$$

L/ha ....liters per hectare, broadcast basis
L/min ....liters per minute
km/h ....Kilometers per hour
sec ....seconds
m ....meters
s ....ground speed (kilometers per hour)
w ....nozzle spacing on boom or band width (centimeters)

DETERMINING VOLUME

Volume of a cylinder is radius squared x 3.1416 x length of cylinder.

Volume of cone is radius squared x 1.0472 (e.g., round hopper bottom).

Volume of pyramid is area of base x 1/3 the height. (e.g., square hopper bottom).

VOLUME CONVERSION FACTORS:
(U.S. Customary Measures)

Cu. ft. x 7.48 equals gallons.

Cu. ft. x 62.4 equals pounds water.

Gallons x 8.330 equals pounds water.

Gallons x 0.1337 equals cu. ft.

Cu. in. ÷ 1,728 = cu. ft.

Cu. yd. x 27 = cu. ft.

Cu. ft. ÷ 27 = cu. yds.

DETERMINING AREA

To find circumference of a circle when diameter is known, multiply diameter by 3.1416 (approx. 3-1/7).

To find diameter when circumference is known, divide circumference by 3.1416 or multiply by 0.3183.

Area of circle = radius squared x 3.1416; or diameter squared x 0.7854.

Area of rectangle or square = length x width.

Area of right triangle = length x width ÷ 2.

Area of other triangles = baxe x height at right angle to base ÷ 2 (see diagram).

FERTILIZER BLENDING FORMULA

A blending formula is the "recipe" used for mixing fertilizer ingredients to obtain a particular analysis and may be calculated using simple arithmetic. For example, a typical blending formula for one ton of 12-12-12 will contain: (1) ammonium nitrate, (2) concentrated superphosphate, and (3) muriate of potash.

First determine how many pounds of each nutrient are needed by using the method explained in the section on fertilizer analysis. One ton of 12-12-12 contains 240 pounds of nitrogen, 240 pounds of phosphoric acid, and 240 pounds of potash (0.12 × 2,000 lb. of each plant food). (One metric ton of 12-12-12 contains 120 kilograms of nitrogen, 120 kilograms of phosphoric acid, and 120 kilograms of potash (0.12 × 1000 kg of each plant food.)

Next, find the pounds and kilograms of each ingredient needed. The amount of ammonium nitrate needed is found by dividing the pounds (kg) of nitrogen needed by the percentage of nitrogen in the ammonium nitrate. Don't forget to change the percentage to a decimal.

$\frac{240}{.34}$ = 705.9 pounds of 34% ammonium nitrate

$\frac{120}{.34}$ = 353 kilograms of 34% ammonium nitrate

Similar calculations will determine the amounts of other ingredients needed.

$\frac{240}{.46}$ = 521.7 pounds of 46% superphosphate

$\frac{120}{.46}$ = 260 kilograms of 46% superphosphate

$\frac{240}{.62}$ = 387.1 pounds of 62% muriate of potash

$\frac{120}{.62}$ = 193.5 kilograms of 62% muriate of potash

The total weight of the three ingredients comes to 1614.7 pounds (806.5 kg). An additional 385.3 pounds (193.5 kg) of filler must be added to make an even 2,000 pounds (1000 kg) of 12-12-12.

ACREAGE PER MILE OF VARIOUS WIDTHS

Width	Acres
1 foot	0.121
5 feet	0.605
8 feet	0.968
10 feet	1.21
12 feet	1.452
14 feet	1.694
15 feet	1.815
16 feet	1.936
18 feet	2.178
20 feet	2.42
24 feet	2.904
25 feet	3.025

HOW TO DETERMINE FIELD SPEED

When applying chemicals you should know how fast you are driving. You can determine your speed as follows:

1. Mark off a distance of 176 feet in the field.

2. Drive the measured distance at the speed you would like to travel.

3. Check the number of seconds required to drive between the markers with a stop watch or watch with a sweep second hand.

4. Divide the time in seconds into 120 for speed in miles per hour (mph).

5. Adjust speed, if necessary, to the recommended speed.

The chart below lists the time in seconds for speeds up to 8 miles per hour.

FIELD SPEEDS

Time To Drive 176 Feet	Speed
120 seconds	1 mph
60 seconds	2 mph
40 seconds	3 mph
30 seconds	4 mph
24 seconds	5 mph
20 seconds	6 mph
17 seconds	7 mph
15 seconds	8 mph

HOW TO DETERMINE FIELD SPEED (METRIC)

When planting you should know how fast you are driving. You can determine your speed as follows:

1. Mark off a distance of 33.4 meters in the field.

2. Drive the measured distance at the speed you would like to plant.

3. Check the number of seconds required to drive between the markers with a stop watch or watch with a sweep second hand.

4. Divide the time in seconds into 120 for speed in kilometers per hour (km/h).

5. Adjust the planting speed, if necessary, to the recommended speed.

The chart below lists the time in seconds for speeds up to 8 kilometers per hour.

(Metric chart on next page)

FIELD SPEEDS

Time to Drive 33.4 Meters	Speed
120 seconds	1 km/h
60 seconds	2 km/h
40 seconds	3 km/h
30 seconds	4 km/h
24 seconds	5 km/h
20 seconds	6 km/h
17 seconds	7 km/h
15 seconds	8 km/h

ACRES-PER-HOUR CHART

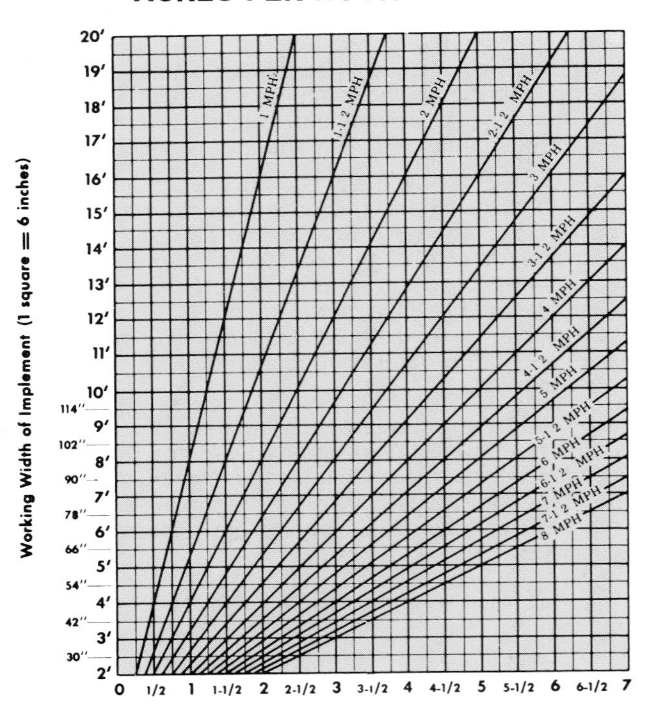

DIRECTIONS: In the left-hand column, find the line representing the working width of your equipment. Follow the line to the right until it reaches the diagonal line representing your speed of travel. From this point, follow the nearest vertical line directly to the bottom of the chart and estimate the acres per hour from the nearest figure. Note: If your equipment is wider than 20 feet, or the horizontal and diagonal lines do not meet on the chart, make your estimate using half (or one-fourth) the working width and multiplying the result by two (or four, whichever is applicable).

EXAMPLE: (Using a 3-bottom, 14-inch plow cutting 42 inches, traveling at 3¼ mph.) Find 42 inches in the left-hand column. Follow the line at 42 to the right to a point midway between the diagonal lines marked 3 mph and 3½ mph. From this point, follow nearest vertical line to bottom of chart. Note acreage per hour is slightly less than 1½.

TABLES FROM TEXT WITH EQUIVALENT METRIC VALUES

(Corresponding TABLES with U.S. Customary measurements are shown on pages listed at the bottom of each TABLE shown here).

TABLE 1

PLANT NUTRIENTS NEEDED TO PRODUCE 2540 KG/HA CORN

Substance	Kilograms per Ha	Approx. Equiv.
Water	6.7 to 9 million	48 to 61 cm of rain
Oxygen	11 300	
Carbon	8745	3.6 tonnes of coal
Nitrogen	220	
Potassium	185	
Phosphorus	37	
Calcium	63	65 kg of limestone
Magnesium	56	182 kg of Epsom salts
Sulfur	37	
Iron	3.4	
Manganese	0.5	
Boron	0.1	0.17 kg of borax
Chlorine	trace	
Zinc	trace	1 dry-cell-battery case
Copper	trace	17 m of No. 9 wire

See page 17.

TABLE 2

POUNDS OF NITROGEN IN CERTAIN CROPS

Crop	Yield	Nitrogen (kg)
Alfalfa	5.4 metric tons	152
Apples	21 m^3	23
Barley	3.5 m^3	50
Corn	5.25 m^3	61
Cotton	1450 kg of lint and seed	34
Oats	3.5 m^3	30
Oranges	800 boxes	39
Peaches	21 m^3	16
Potatoes	14 m^3	50
Soybeans	1.75 m^3	61
Sugar Beets	27 metric tons	57
Timothy	2.7 metric tons	39
Tobacco	1270 kg of stems and leaves	43
Tomatoes	27 metric tons	77
Wheat	2.1 m^3	34

See page 17.

TABLE 3

POUNDS OF PHOSPHORUS IN CROPS

Crop	Yield	Phosphoric Acid (kg)
Alfalfa	5.4 metric tons	32
Apples	21 m^3	7
Barley	3.5 m^3	18
Corn	5.25 m^3	23
Cotton	1450 kg of lint and seed	14
Oats	3.5 m^3	11
Oranges	800 boxes	14
Peaches	21 m^3	4.5
Potatoes	14 m^3	16
Soybeans	1.75 m^3	18
Sugar Beets	27 metric tons	23
Timothy	2.7 metric tons	14
Tobacco	1270 kg of stems and leaves	11
Tomatoes	27 metric tons	25
Wheat	2.1 m^3	16

See page 21.

TABLE 4

POTASSIUM CONTENT OF CROPS

Crop	Yield	Potassium (kg)
Alfalfa	5.4 metric tons	123
Apples	21 m^3	32
Barley	3.5 m^3	16
Corn	5.25 m^3	16
Cotton	1450 kg of lint and seed	18
Oats	3.5 m^3	9
Oranges	800 boxes	64
Peaches	21 m^3	30
Potatoes	14 m^3	91
Soybeans	1.75 m^3	32
Sugar Beets	27 metric tons	91
Timothy	2.7 metric tons	41
Tobacco	1270 kg of stems and leaves	86
Tomatoes	27 metric tons	163
Wheat	2.1 m^3	7

See page 23.

TABLE 5

LIME NEEDED FOR EQUAL ACID-NEUTRALIZATION

Material	Kilograms needed to equal one kilogram of ground limestone
Ground limestone	1.0
Burnt lime	0.6
Hydrated lime	0.67
Marl	1.0-1.5
Oyster shells	1.0-1.5

See page 27.

TABLE 6

LIME TO RAISE pH ONE UNIT

Soil type	*Ground limestone	*Burnt lime	*Hydrated lime
Light sandy	1680	940	1240
Sandy loams	2240	1254	1660
Loams	3360	1880	2485
Silt loams and clay loams	3920	2195	2900

*Application rate in kilograms per hectare.

See page 27.

TABLE 7

PUMP CHARACTERISTICS

Pump	Capacity (Lpm)	Speed (rpm)	Max. Press. (kPa)	Material
Gear	19-75	500-1800	690	Non-abrasive
Diaphragm	4-38	200-1200	690	Abrasive
Flexible Impeller	0-115	500-1500	345	Mildly Abrasive
Roller	0-130	600-1800	2070	Non-abrasive
Centrifugal	0-570	600-4000	345*	Abrasive
Piston	0-225	500-1000	5515	Abrasive

*Multi-stage units develop higher pressures

See page 132.

TABLE 8

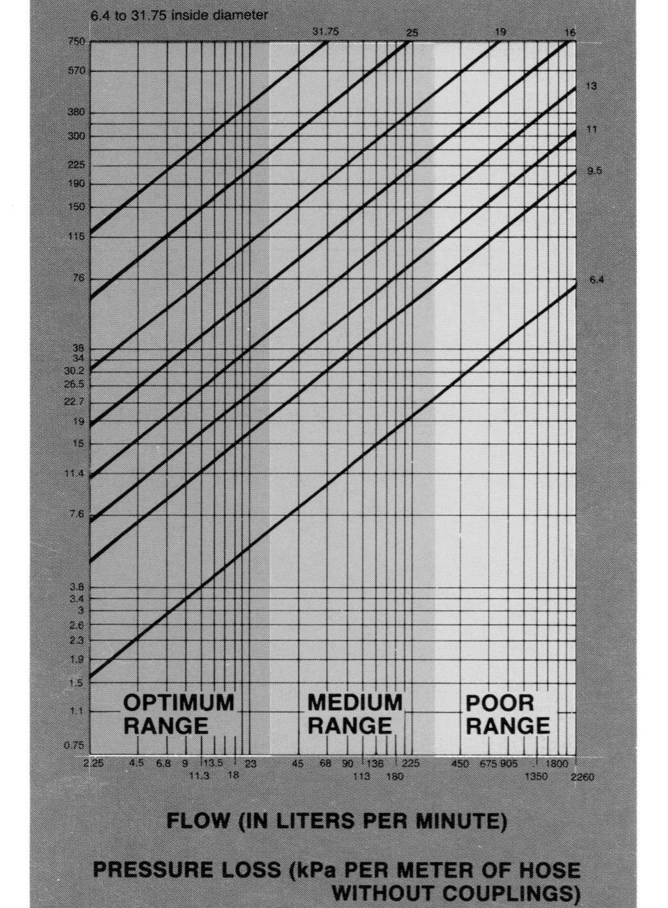

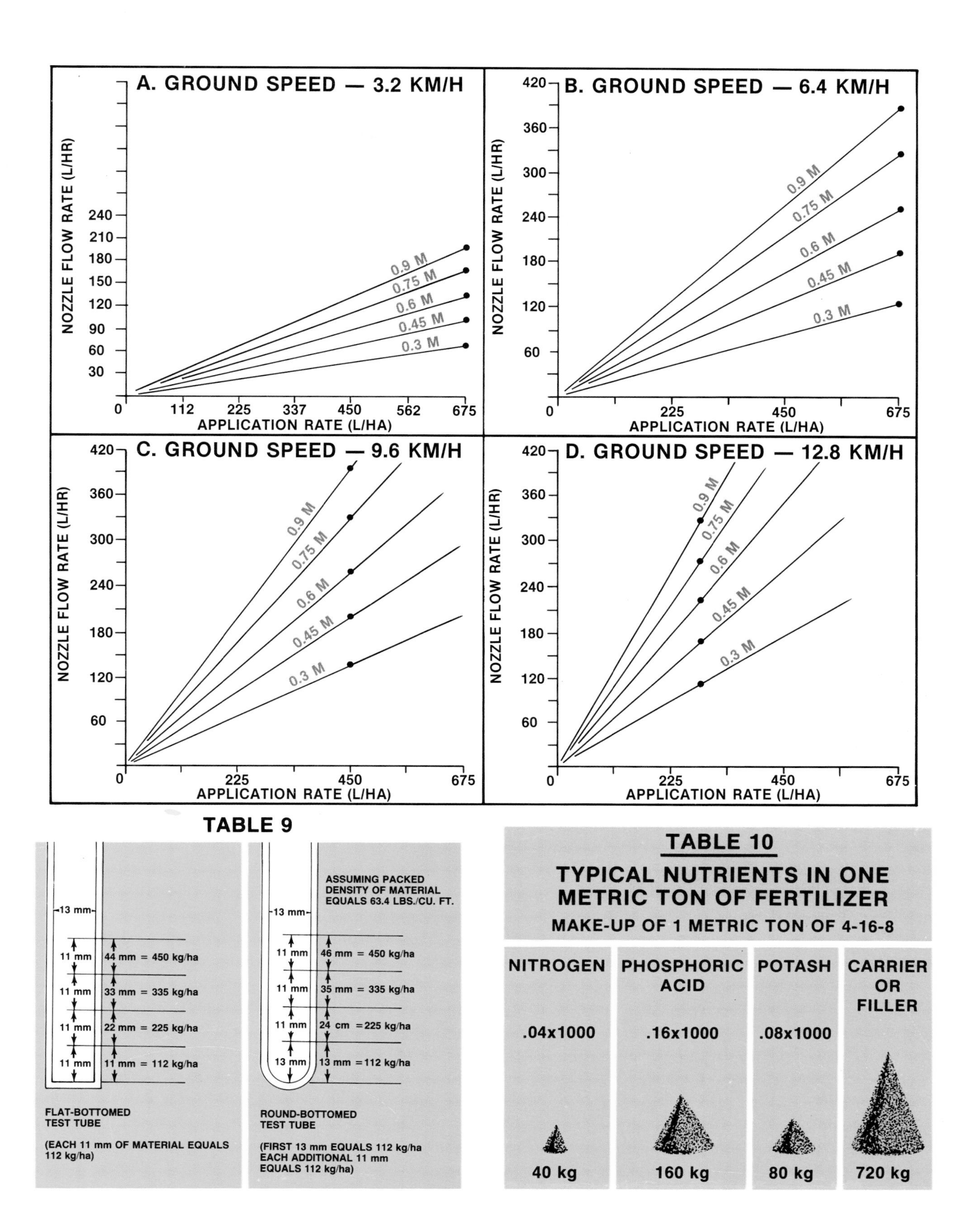
A. GROUND SPEED — 3.2 KM/H
NOZZLE FLOW RATE (L/HR)
240
210
180
150
120
90
60
30
0
112
225
337
450
562
675
APPLICATION RATE (L/HA)
0.9 M
0.75 M
0.6 M
0.45 M
0.3 M
B. GROUND SPEED — 6.4 KM/H
NOZZLE FLOW RATE (L/HR)
420
360
300
240
180
120
60
0
225
450
675
APPLICATION RATE (L/HA)
0.9 M
0.75 M
0.6 M
0.45 M
0.3 M
C. GROUND SPEED — 9.6 KM/H
NOZZLE FLOW RATE (L/HR)
420
360
300
240
180
120
60
0
225
450
675
APPLICATION RATE (L/HA)
0.9 M
0.75 M
0.6 M
0.45 M
0.3 M
D. GROUND SPEED — 12.8 KM/H
NOZZLE FLOW RATE (L/HR)
420
360
300
240
180
120
60
0
225
450
675
APPLICATION RATE (L/HA)
0.9 M
0.75 M
0.6 M
0.45 M
0.3 M
TABLE 9
13 mm
11 mm
44 mm = 450 kg/ha
11 mm
33 mm = 335 kg/ha
11 mm
22 mm = 225 kg/ha
11 mm
11 mm = 112 kg/ha
FLAT-BOTTOMED TEST TUBE
(EACH 11 mm OF MATERIAL EQUALS 112 kg/ha)
ASSUMING PACKED DENSITY OF MATERIAL EQUALS 63.4 LBS./CU. FT.
13 mm
11 mm
46 mm = 450 kg/ha
11 mm
35 mm = 335 kg/ha
11 mm
24 cm = 225 kg/ha
13 mm
13 mm = 112 kg/ha
ROUND-BOTTOMED TEST TUBE
(FIRST 13 mm EQUALS 112 kg/ha EACH ADDITIONAL 11 mm EQUALS 112 kg/ha)
TABLE 10
TYPICAL NUTRIENTS IN ONE METRIC TON OF FERTILIZER
MAKE-UP OF 1 METRIC TON OF 4-16-8
NITROGEN
.04x1000
40 kg
PHOSPHORIC ACID
.16x1000
160 kg
POTASH
.08x1000
80 kg
CARRIER OR FILLER
720 kg

WEEDS, INSECTS, & DISEASES

THE FOLLOWING PAGES ILLUSTRATE SOME OF THE COMMON WEEDS, INSECTS AND DISEASES THAT PLAGUE CROPS. A DETAILED INDEX APPEARS ON THE NEXT PAGE.

SEE YOUR LOCAL COUNTY EXTENSION AGENT FOR CONFIRMATION OF THOSE PESTS OF WHICH YOU ARE IN DOUBT. HE MAY ALSO SUGGEST RECOMMENDED TREATMENT FOR THE PEST.

WEEDS AND GRASSES ....pg 209

INSECTS OF AGRONOMIC CROPS & STORED PRODUCTS ....pg. 235

CORN INSECTS

COTTON INSECTS

FORAGE CROP INSECTS

INSECT PESTS OF STORED PRODUCTS

RICE INSECTS

SOYBEAN INSECTS

STORED GRAIN INSECTS

TOBACCO INSECTS

INSECTS OF HORTICULTURAL CROPS AND FOREST TREE CROPS ....pg. 240

FRUIT INSECTS

ORNAMENTAL PLANT INSECTS

PECAN INSECTS

SHADE AND FOREST INSECTS

VEGETABLE INSECTS

BENEFICIAL INSECTS

DISEASES OF AGRONOMIC CROPS ....pg. 246

CORN DISEASES

COTTON DISEASES

FORAGE DISEASES

PEANUT DISEASES

SMALL GRAIN DISEASES

SOYBEAN DISEASES

TOBACCO DISEASES

DISEASES OF HORTICULTURAL CROPS, TREES AND TURF AND WOOD ROTS ....pg. 250

APPLE DISEASES

CRUCIFER DISEASES

CUCURBIT DISEASES

GRAPE DISEASES

LETTUCE DISEASES

MISCELLANEOUS DISEASES

ORNAMENTAL PLANT DISEASES

PEACH AND PLUM DISEASES

PECAN DISEASES

STRAWBERRY AND BRAMBLE DISEASES

SWEET POTATO DISEASES

TOMATO DISEASES

TREE DISEASES

TURF DISEASES

WOOD ROT DISEASES

AMARANTH, Spiny — Summer Annual

ASTER, Whiteheath — Perennial

BARLEY, Little — Winter Annual

BARLEY, Wild — Annual or Perennial

BARNYARDGRASS — Summer Annual

BEGGARWEED, Florida — Summer Annual

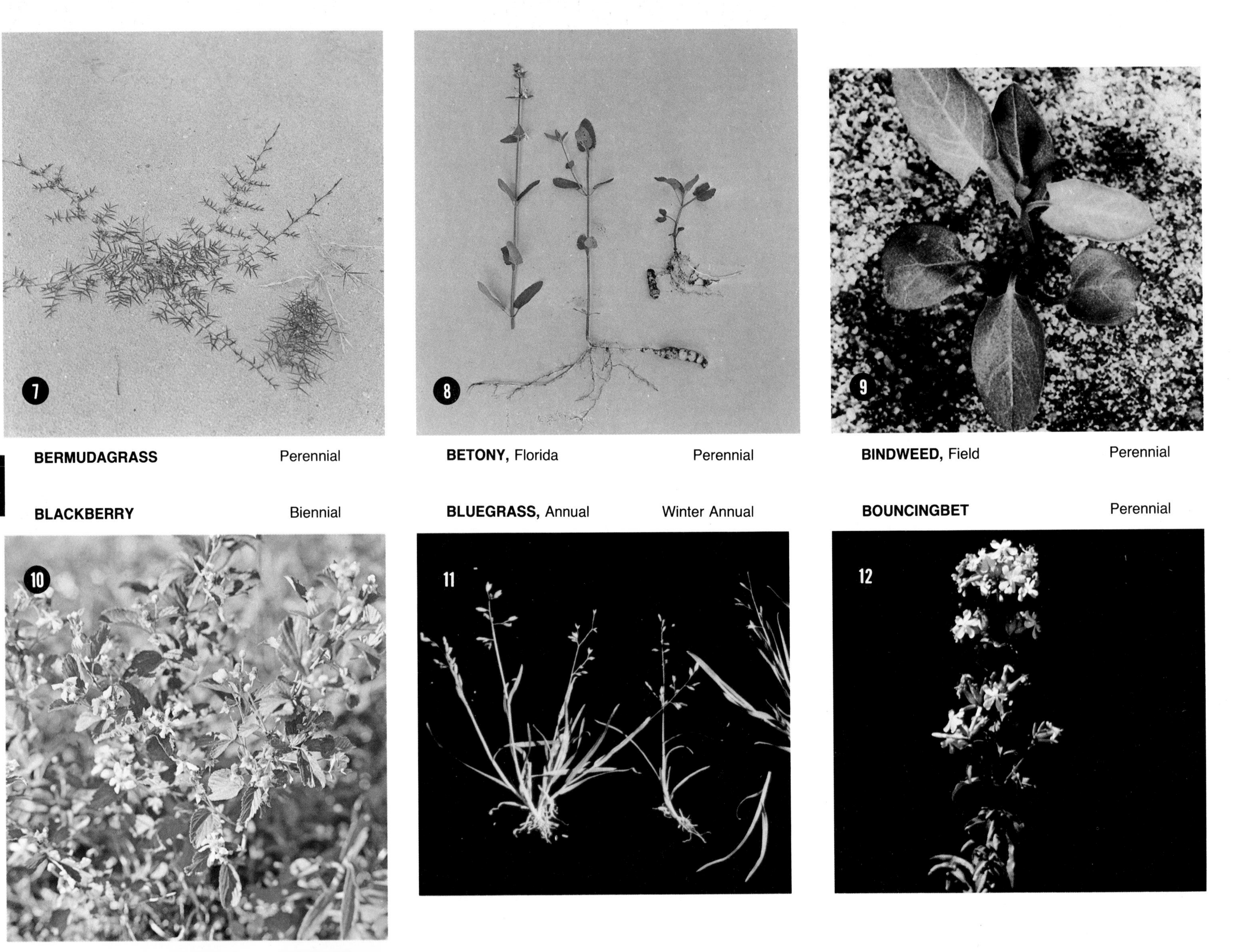

BERMUDAGRASS Perennial

BETONY, Florida Perennial

BINDWEED, Field Perennial

BLACKBERRY Biennial

BLUEGRASS, Annual Winter Annual

BOUNCINGBET Perennial

BRACKEN, Eastern — Perennial

BROME, Downy — Annual

BROOMSEDGE — Perennial

BUCKWHEAT, Wild — Annual

BUFFALOBUR — Annual

BURCUCUMBER — Summer Annual

BURDOCK Biennial

BUTTERCUP Winter Annual

CAMPHORWEED Summer Annual or Biennial

CARPETWEED Summer Annual

CARROT, Wild Biennial

CATCHFLY, Nightflowering Winter Annual

CHICKWEED Winter Annual

CHICKWEED, Mouseear Perennial

CHICORY Perennial

COCKLE, Corn Winter Annual

COCKLEBUR, Common Summer Annual

COPPERLEAF, Virginia Summer Annual

CORNFLOWER Winter Annual

CRABGRASS, Large Summer Annual

CROTALARIA, Showy Summer Annual

CROTON, Tropic Summer Annual

CROTON, Woolly Summer Annual

CROWFOOTGRASS Summer Annual

CUCUMBER, Wild — Summer Annual

CUDWEED — Biennial

DANDELION — Perennial

DAYFLOWER — Summer Annual

DICHONDRA — Perennial

DOCK, Broadleaf — Perennial

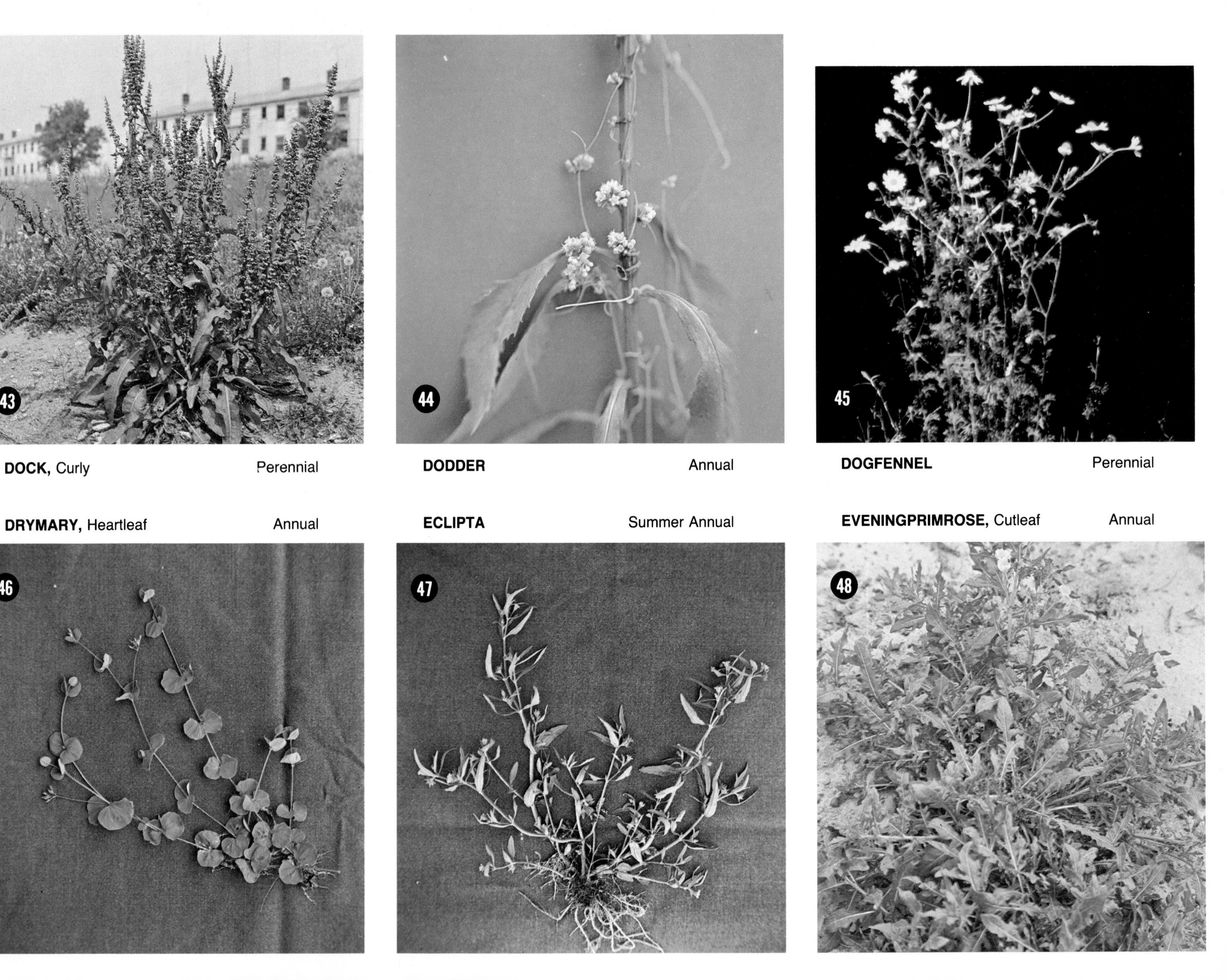

DOCK, Curly — Perennial

DODDER — Annual

DOGFENNEL — Perennial

DRYMARY, Heartleaf — Annual

ECLIPTA — Summer Annual

EVENINGPRIMROSE, Cutleaf — Annual

FALSEDANDELION, Carolina — Winter Annual or Biennial

FILAREE, Redstem — Winter Annual or Biennial

FLEABANE, Annual — Summer Annual or Biennial

FLEABANE, Daisy — Summer Annual

FOXTAIL — Summer Annual

FOXTAIL, Giant — Annual

FOXTAIL, Green — Annual

GALINSOGA, Hairy — Summer Annual

GARLIC, Wild — Perennial

GERANIUM, Carolina — Winter Annual or Biennial

GOLDENROD — Perennial

GOOSEGRASS — Summer Annual

GREENBRIAR, Common — Perennial

GROUNDCHERRY, Smooth — Perennial

GROUNDSEL — Perennial

HEMP — Annual

HEMP, Dogbane — Perennial

HEMP, Sesbania — Summer Annual

HENBIT Winter Annual

HONEYSUCKLE, Japanese Perennial

HORSENETTLE Perennial

HORSETAIL Perennial

HORSEWEED Annual

IVY, Poison Perennial

JIMSONWEED Summer Annual

JOHNSONGRASS Perennial

KNOTWEED Summer Annual

KOCHIA Annual

KUDZU Perennial

LAMBSQUARTERS, Common Summer Annual

LETTUCE, Wild — Winter Annual

MARESTAIL — Summer Annual

MAYWEED — Winter Annual

MELON, Smell — Annual

MILKWEED, Butterfly — Perennial

MILKWEED, Climbing — Perennial

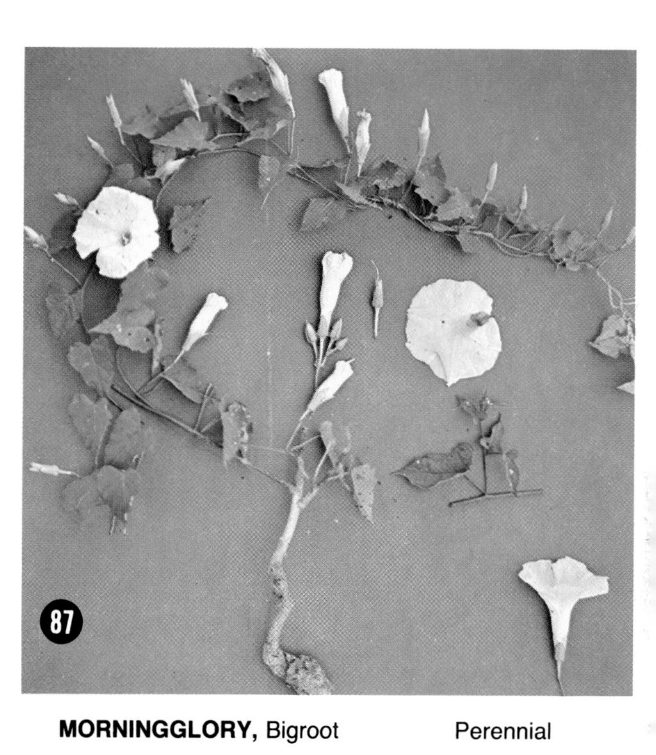

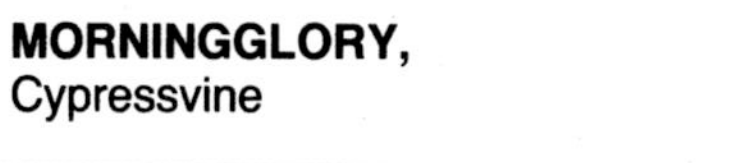

MILKWEED, Common — Perennial

MORNINGGLORY, Cypressvine — Summer Annual

MILKWEED, Honeyvine — Perennial

MORNINGGLORY, Smallflower — Summer Annual

MORNINGGLORY, Bigroot — Perennial

MORNINGGLORY, Tall — Annual

MULLEIN, Common — Biennial

MUSTARD, Hedge — Annual

MUSTARD, Wild — Winter Annual

NIGHTSHADE, Bitter — Perennial

NIGHTSHADE, Black — Summer Annual

NIGHTSHADE, Silverleaf — Perennial

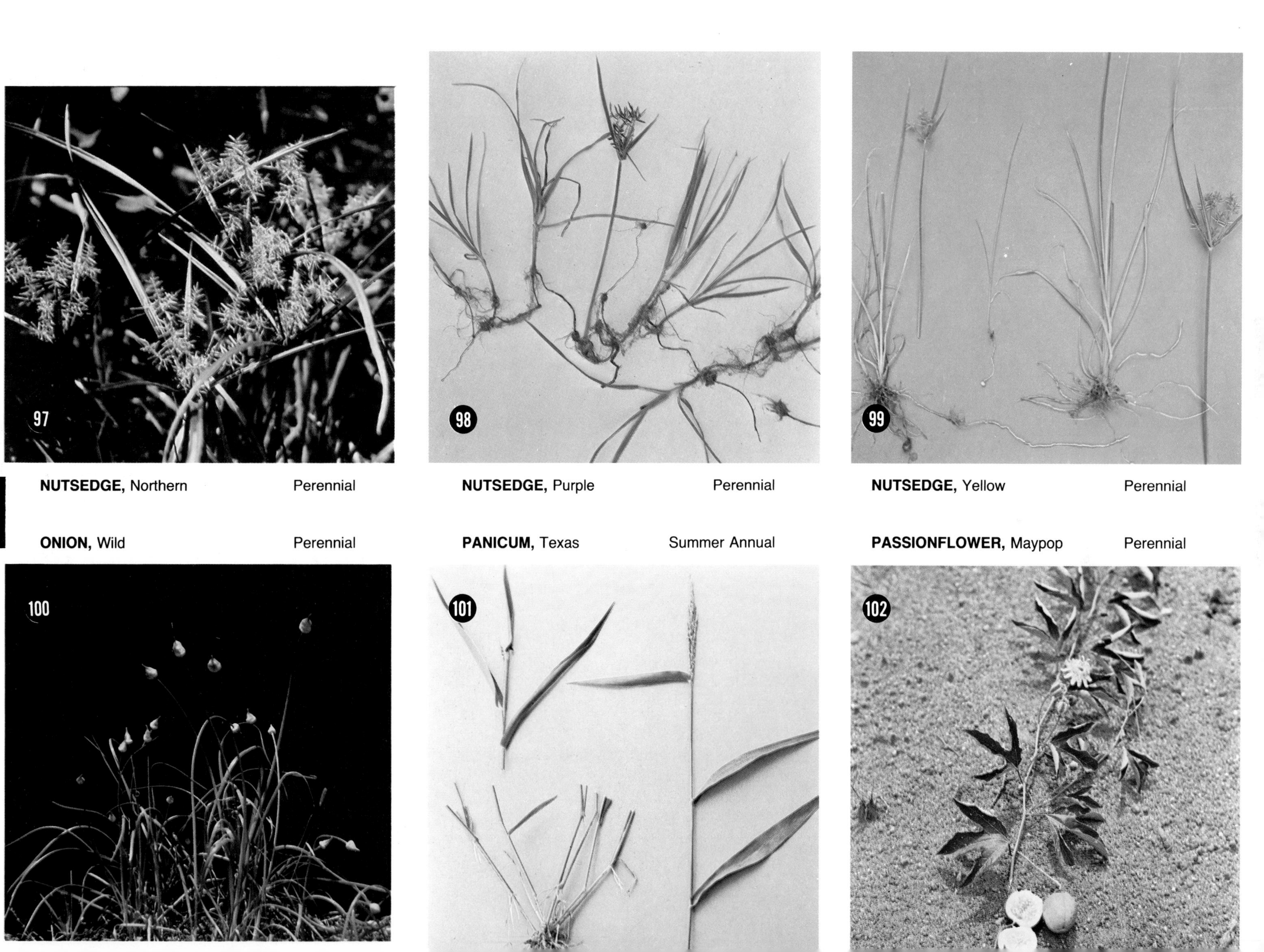

NUTSEDGE, Northern — Perennial

NUTSEDGE, Purple — Perennial

NUTSEDGE, Yellow — Perennial

ONION, Wild — Perennial

PANICUM, Texas — Summer Annual

PASSIONFLOWER, Maypop — Perennial

PEA, Partridge — Summer Annual

PEPPERGRASS — Annual or Biennial

PEAR, Prickly — Perennial

PEPPERVINE — Perennial

PENNYCRESS — Annual

PEPPERWEED, Virginia — Winter Annual or Biennial

PIGWEED Annual

PIGWEED, Prostrate Annual

PIGWEED, Redroot Summer Annual

PLANTAIN, Bracted Winter Annual

PLANTAIN, Broadleaf Perennial

PLANTAIN, Buckhorn Perennial

POKEBERRY Perennial

POORJOE Summer Annual

PURSLANE, Common Summer Annual

PUSLEY, Florida Summer Annual

RAGWEED, Common Summer Annual

RAGWEED, Giant Summer Annual

REDVINE Perennial

ROCKET, Yellow Perennial or Biennial

SANDBUR Summer Annual

SHEPHERDSPURSE Winter Annual

SICKLEPOD Summer Annual

SIDA, Prickly Summer Annual

SIGNALGRASS, Broadleaf — Summer Annual

SMARTWEED, Pennsylvania — Summer Annual

SMUTGRASS — Perennial

SNEEZEWEED, Bitter — Summer Annual

SNOWTHISTLE — Winter Annual

SORREL, Red — Perennial

SORREL, Red — Seed Head — Perennial

SPANISHNEEDLES — Summer Annual

SPURGE, Leafy — Annual

SPURGE, Prostrate — Annual

SPURGE, Spotted — Summer Annual

STARBUR, Bristly — Summer Annual

SUNFLOWER, Wild — Annual

SWINECRESS — Winter Annual or Biennial

THISTLE, Blessed — Winter Annual

THISTLE, Canada — Perennial

THISTLE, Musk — Biennial

TORPEDOGRASS — Perennial

TRUMPETCREEPER Perennial

VASEYGRASS Perennial

VELVETLEAF Annual

VERVAIN, Blue Summer Annual

VERVAIN, Prostrate Winter Annual

VETCH, Smooth Winter Annual

WATERMELON, Wild (Citron) — Summer Annual

WITCHWEED — Annual

WOODSORREL, Common Yellow — Annual

YARROW, Common — Perennial

INSECTS OF AGRONOMIC CROPS AND STORED PRODUCTS

Billbug

Corn Earworm

3 European Corn Borer

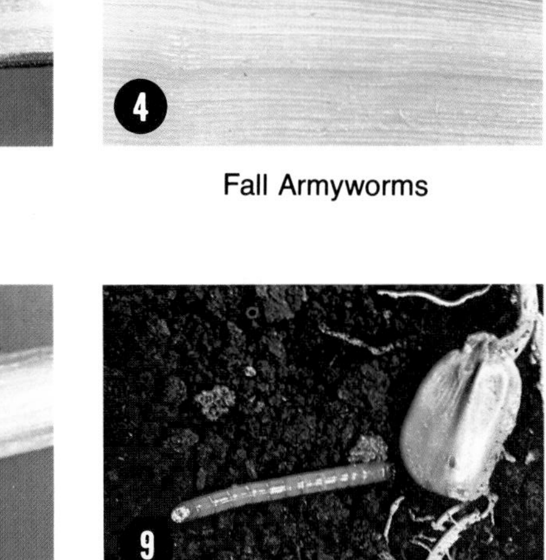

Fall Armyworms

5 Southern Corn Rootworm Adult

Southern Corn Rootworm Larva

7 Southern Cornstalk Borer

Sugarcane Beetle

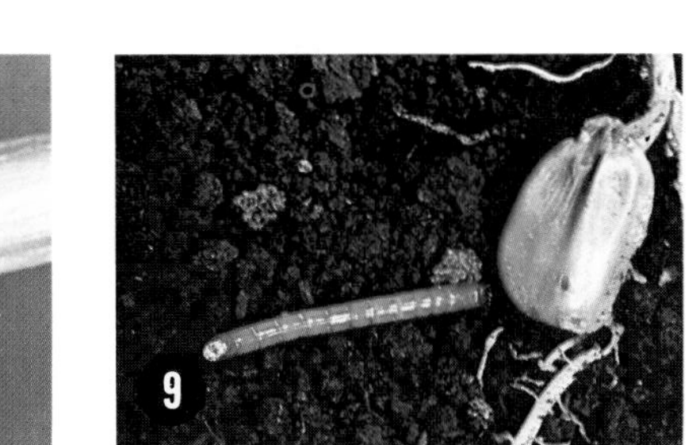

Wireworms

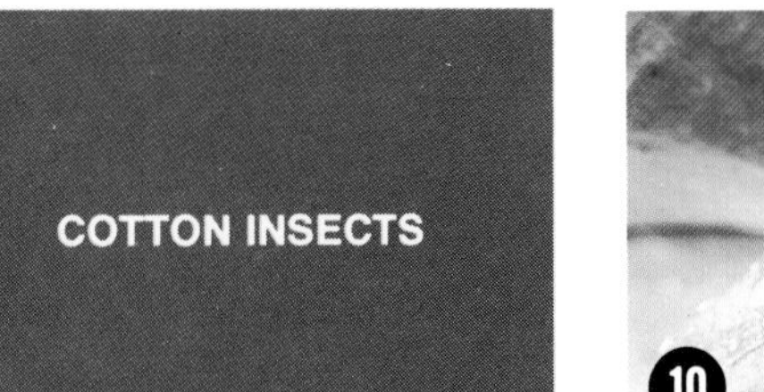

Boll Weevil

Boll Weevil
Larva and Pupa

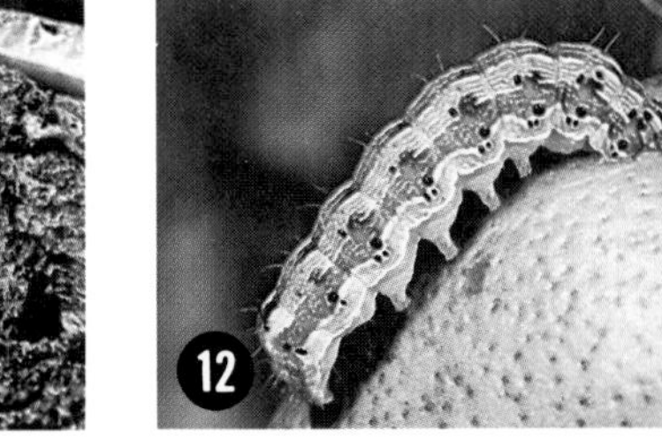

Bollworm Attacking Boll

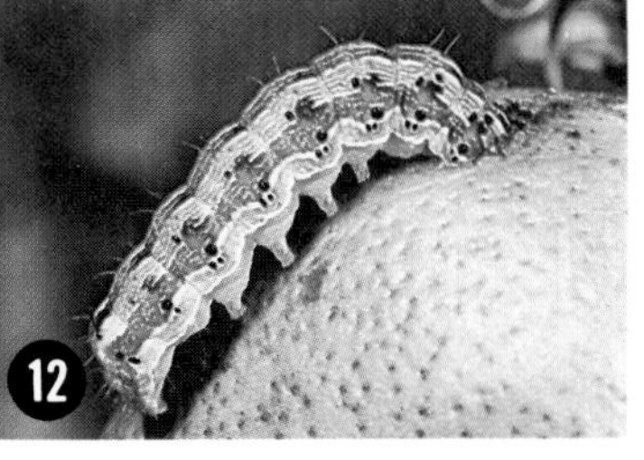

Bollworm Moths and Pupa

Cotton Aphids and Predator

15 Loopers and Damaged Leaf

Two-Spotted Spider Mite

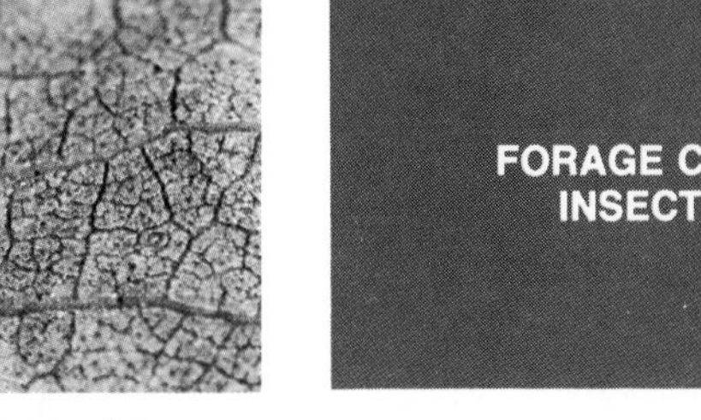

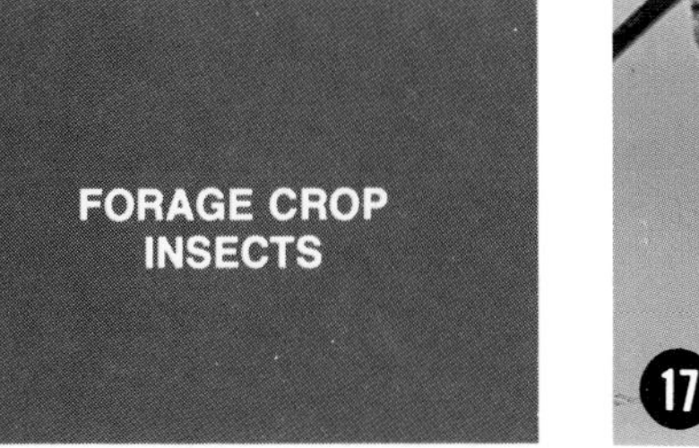

Alfalfa Weevil-Adult

Alfalfa Weevil-Larva

Corn Earworm

Fall Armyworm

Green June Beetle-Larva

Red-Legged Grasshopper

Sorghum Webworm

Sorghum Midge

Spittlebug Nymph in Spittle

Two Lined Spittlebug Adult

White Grub

Almond Moth Larvae

Black Carpet Beetle

Casemaking Clothes Moths

Cigarette Beetle

Cowpea Weevil

Dermestid (Trogoderma glabrum) (Herbst) Grain Beetle-Adult

Dermestid (Trogoderma glabrum) (Herbst) Grain Beetle-Larva

Flat Grain Beetle

Furniture Carpet Beetle

37 Indian Meal Moth Adult and Larvae

38 Rice Weevil

39 Tobacco Moth-Adult

40 Tobacco Moth-Larvae

41 Webbing Clothes Moth

42 Long-horned Grasshopper

43 Rice Weevil

44 Stinkbug

45 Bahia Grass Borer

46 Brown Stinkbug

47 Corn Earworm

48 Diseased Caterpillar

49 Green Cloverworm

50 Looper

51 Margined Blister Beetles

52 Mexican Bean Beetle

53 Southern Green Stinkbug

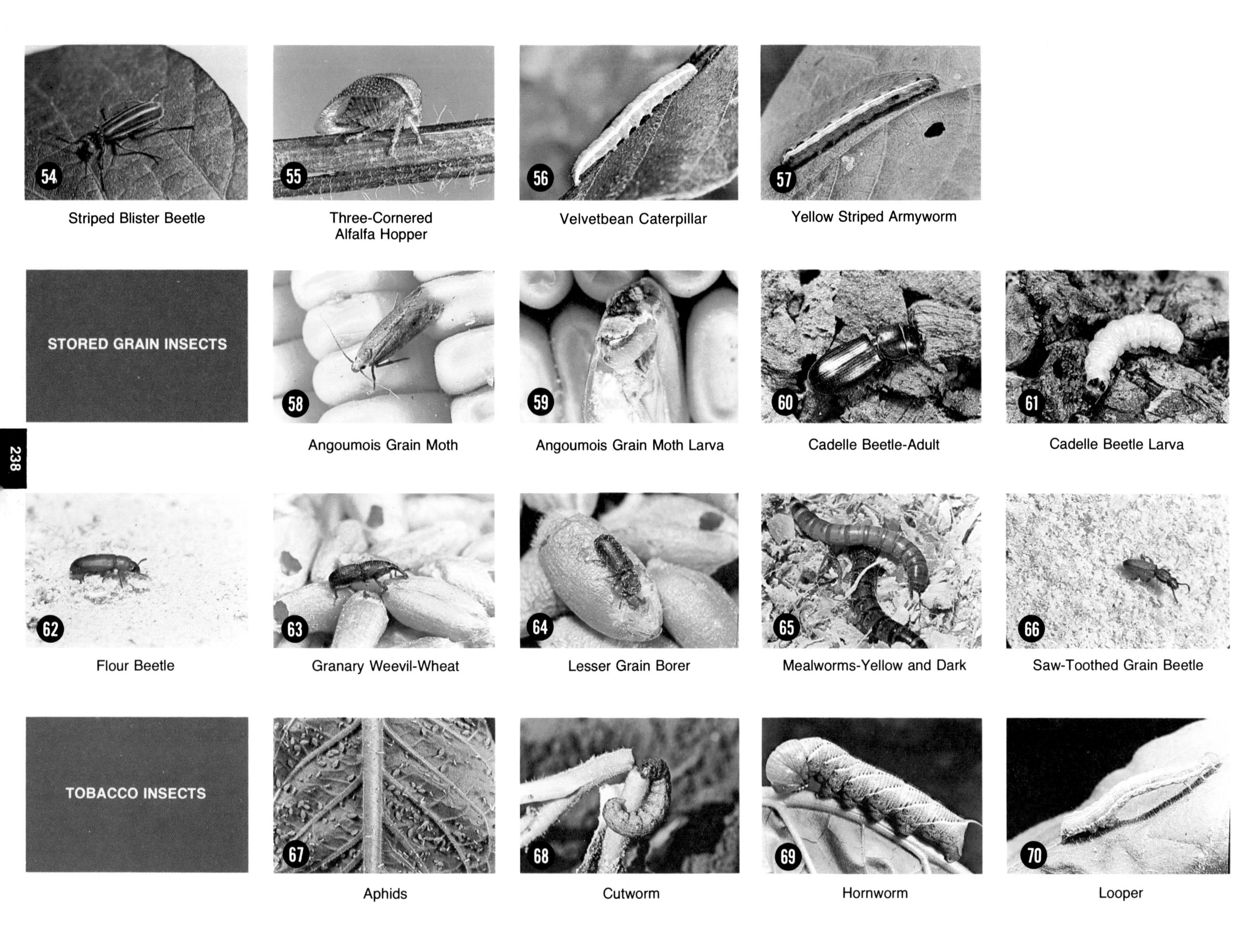

54 Striped Blister Beetle

55 Three-Cornered Alfalfa Hopper

56 Velvetbean Caterpillar

57 Yellow Striped Armyworm

58 Angoumois Grain Moth

59 Angoumois Grain Moth Larva

60 Cadelle Beetle-Adult

61 Cadelle Beetle Larva

62 Flour Beetle

63 Granary Weevil-Wheat

64 Lesser Grain Borer

65 Mealworms-Yellow and Dark

66 Saw-Toothed Grain Beetle

67 Aphids

68 Cutworm

69 Hornworm

70 Looper

Snowy Tree Cricket

Tobacco Fleabeetle-Adult

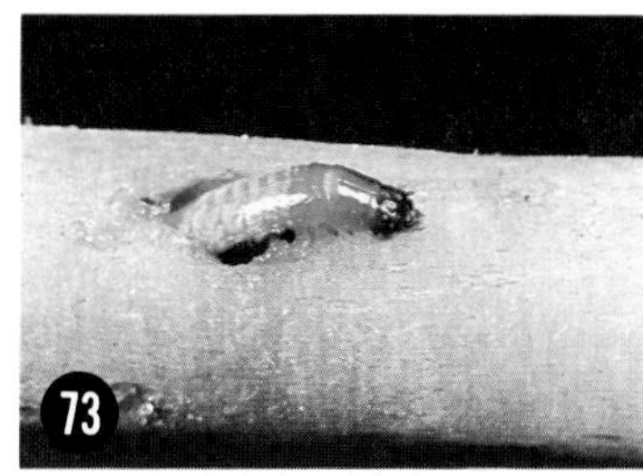

Tobacco Wireworm

Vegetable Weevil-Adult

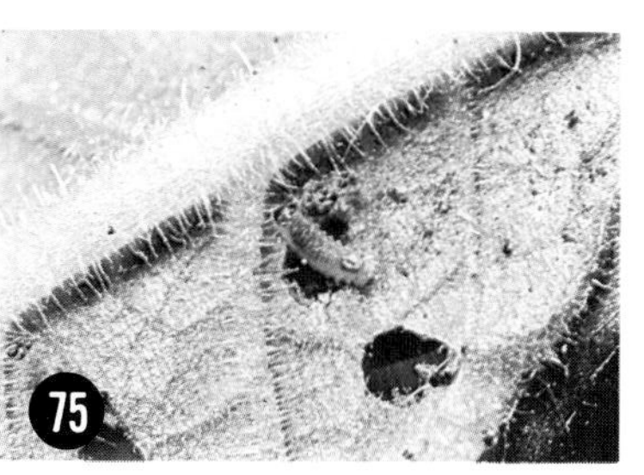

Vegetable Weevil-Larva

INSECTS OF HORTICULTURAL CROPS AND FOREST TREE CROPS

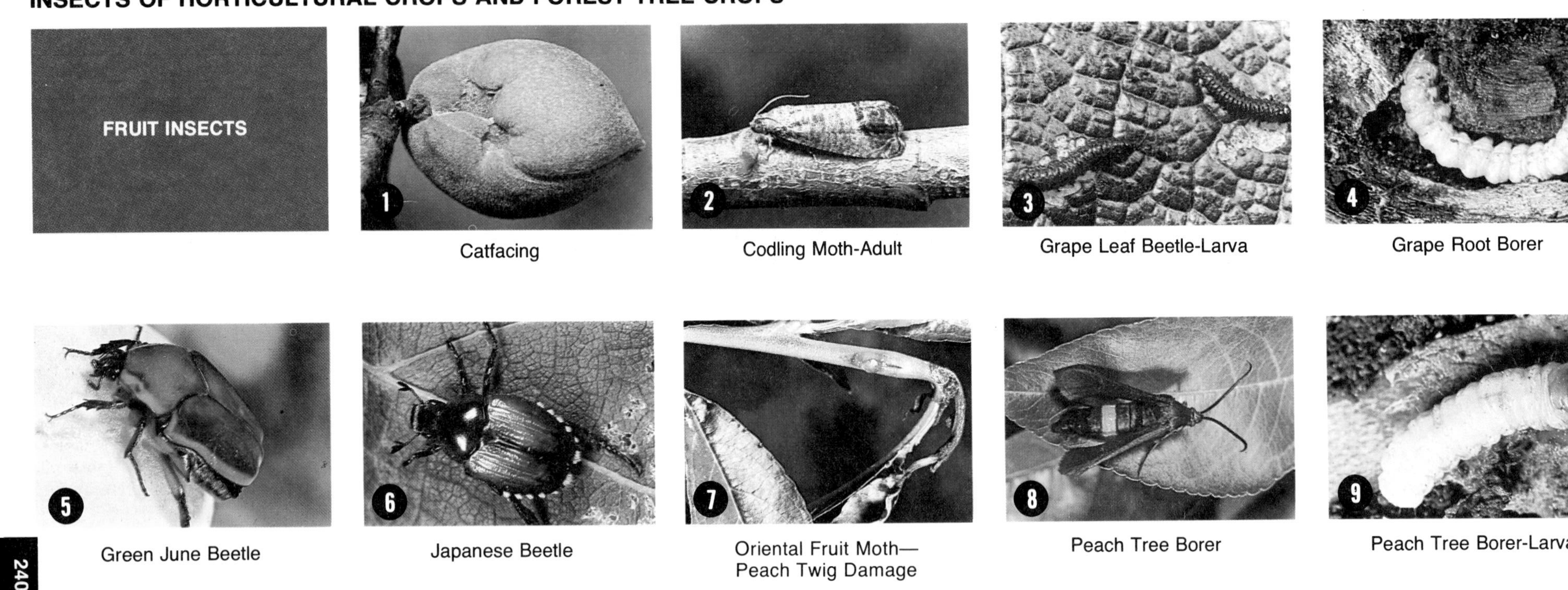

FRUIT INSECTS

1 Catfacing

2 Codling Moth-Adult

3 Grape Leaf Beetle-Larva

4 Grape Root Borer

5 Green June Beetle

6 Japanese Beetle

7 Oriental Fruit Moth—Peach Twig Damage

8 Peach Tree Borer

9 Peach Tree Borer-Larva

10 Plum Curculio

11 Plum Curculio-Larva

12 Red-Humped Caterpillar

13 Rose Chafer

ORNAMENTAL PLANT INSECTS

14 Aphids

15 Azalea Mealybug

16 Bagworm

17 Black Wooly Bear

18 Boxelder Bug

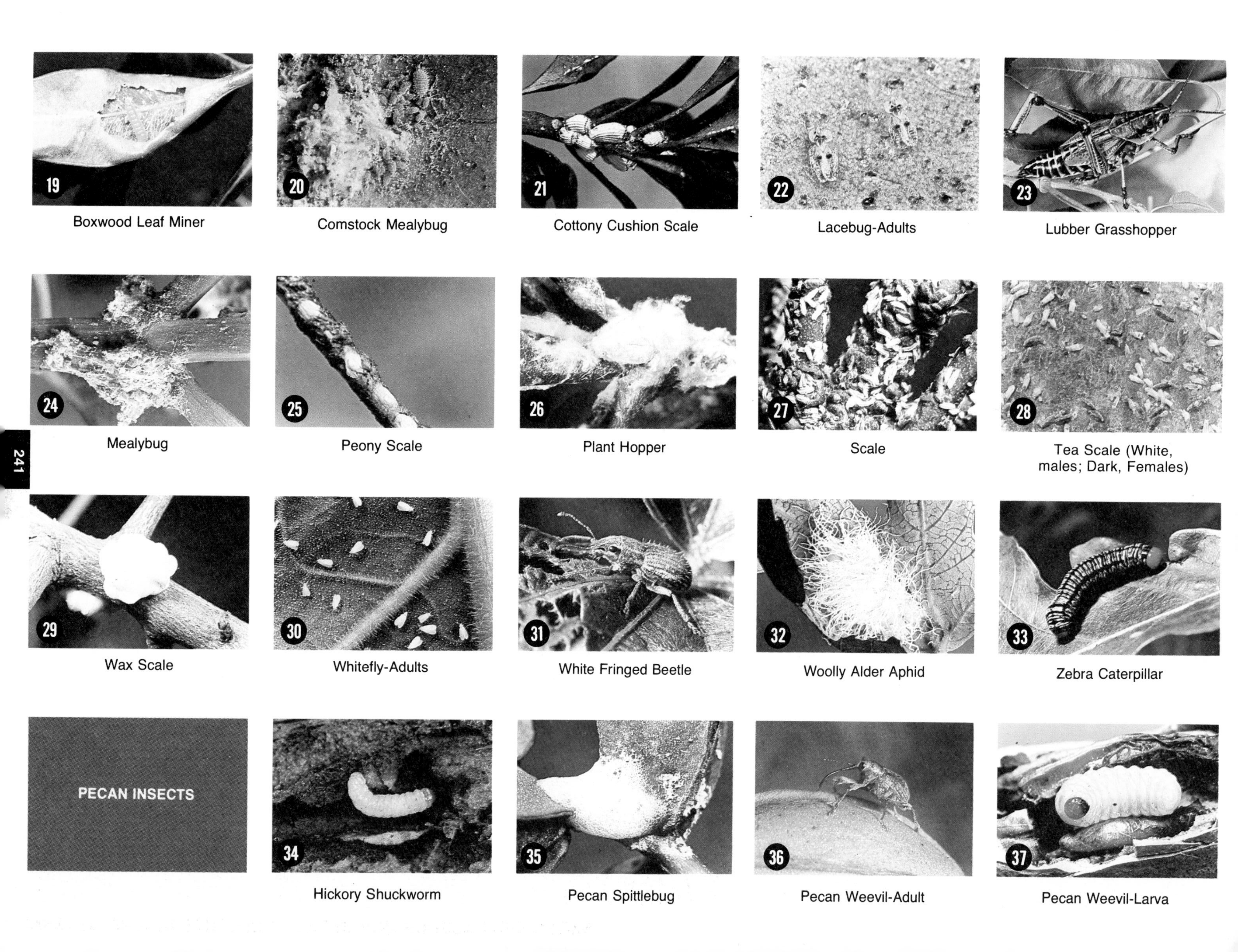

Boxwood Leaf Miner

Comstock Mealybug

Cottony Cushion Scale

Lacebug-Adults

Lubber Grasshopper

Mealybug

Peony Scale

Plant Hopper

Scale

Tea Scale (White, males; Dark, Females)

Wax Scale

Whitefly-Adults

White Fringed Beetle

Woolly Alder Aphid

Zebra Caterpillar

Hickory Shuckworm

Pecan Spittlebug

Pecan Weevil-Adult

Pecan Weevil-Larva

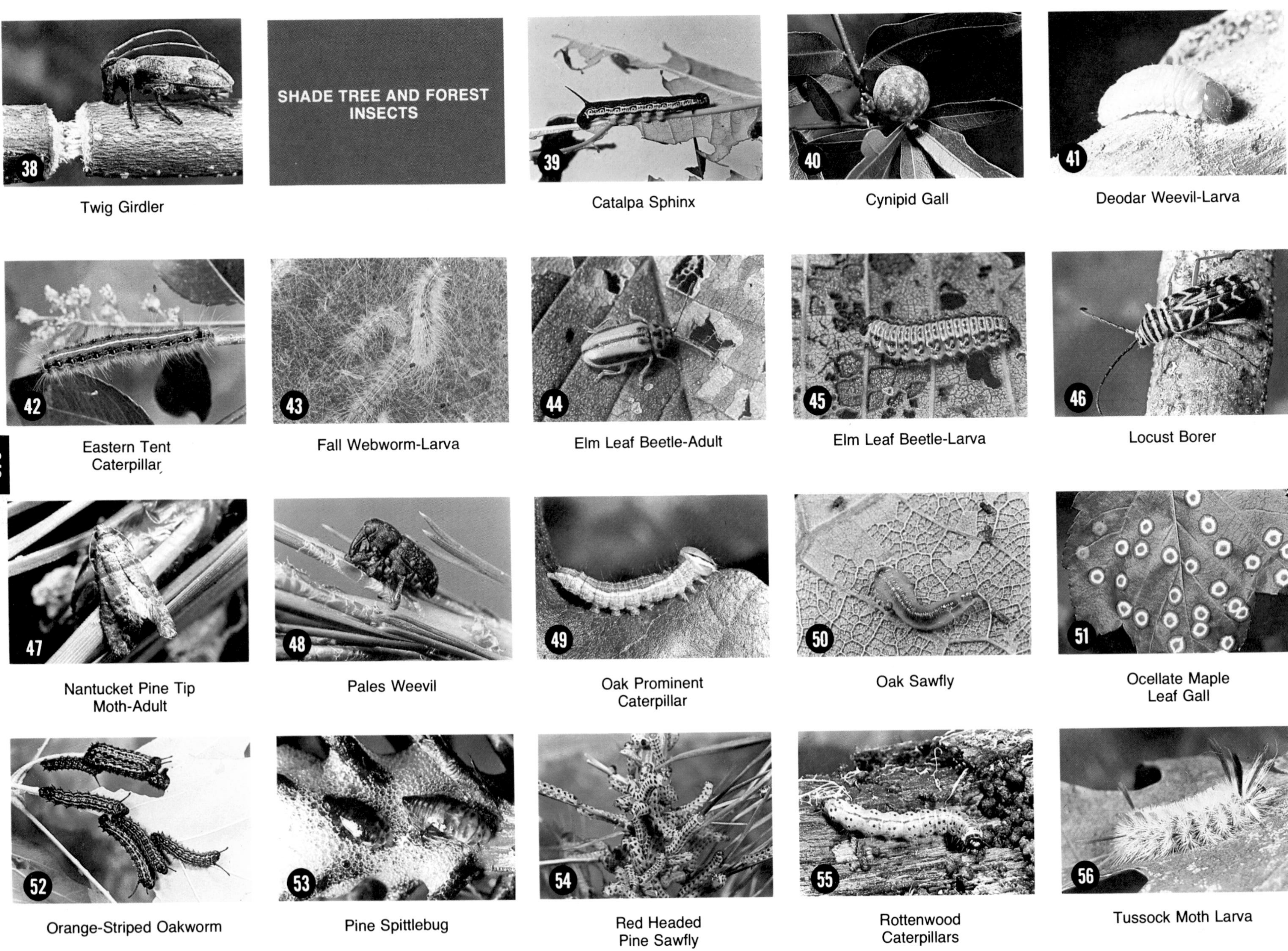

Twig Girdler

Catalpa Sphinx

Cynipid Gall

Deodar Weevil-Larva

Eastern Tent Caterpillar

Fall Webworm-Larva

Elm Leaf Beetle-Adult

Elm Leaf Beetle-Larva

Locust Borer

Nantucket Pine Tip Moth-Adult

Pales Weevil

Oak Prominent Caterpillar

Oak Sawfly

Ocellate Maple Leaf Gall

Orange-Striped Oakworm

Pine Spittlebug

Red Headed Pine Sawfly

Rottenwood Caterpillars

Tussock Moth Larva

VEGETABLE INSECTS

57 Aphids or Plant Lice

58 Asparagus Beetle-Adult

59 Asparagus Beetle-Larva

60 Banded Cucumber Beetle

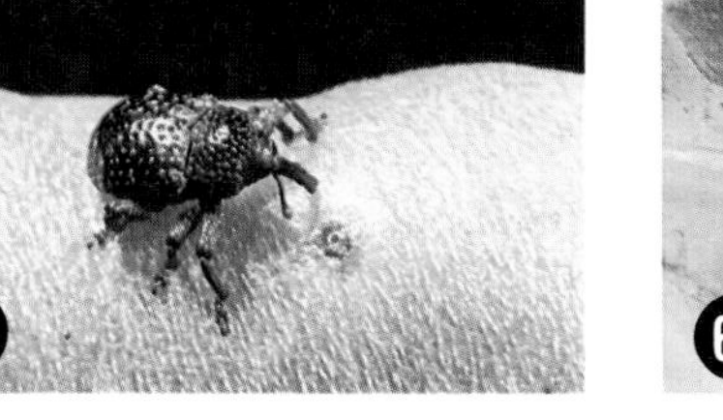
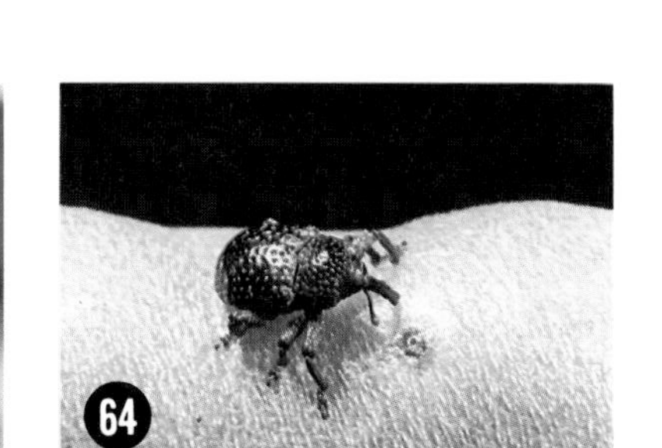
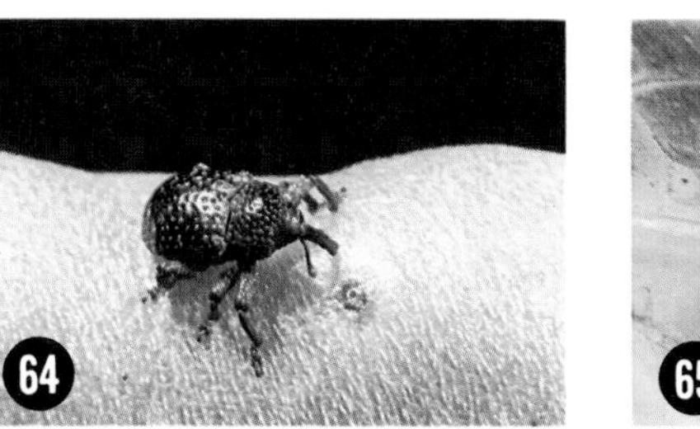

61 Black Cutworm

62 Colorado Potato Beetle-Adult

63 Colorado Potato Beetle-Larva

64 Cowpea Curculio

65 Cross Striped Cabbage Worm

66 Diamond Back Moth-Larva

67 Grub Chapin (Plectris aliena)

68 Harlequin Bug

69 Imported Cabbage Worm

70 Leaf-Footed Bug

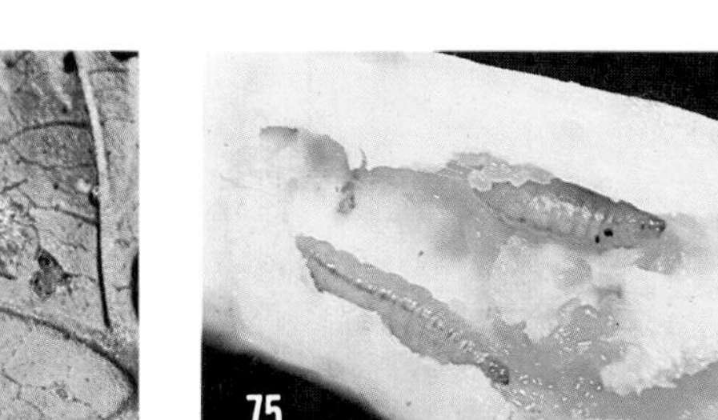

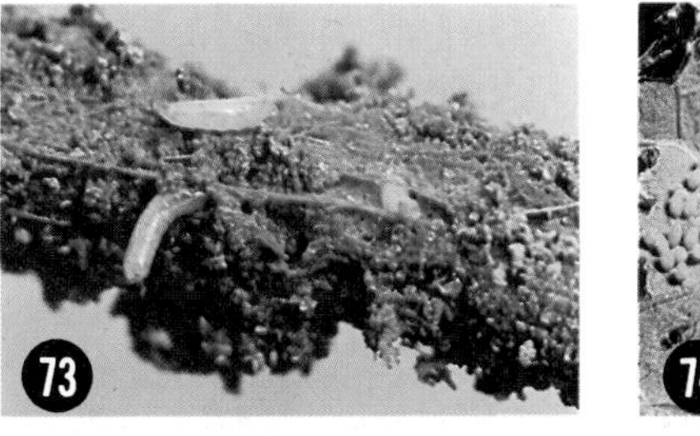
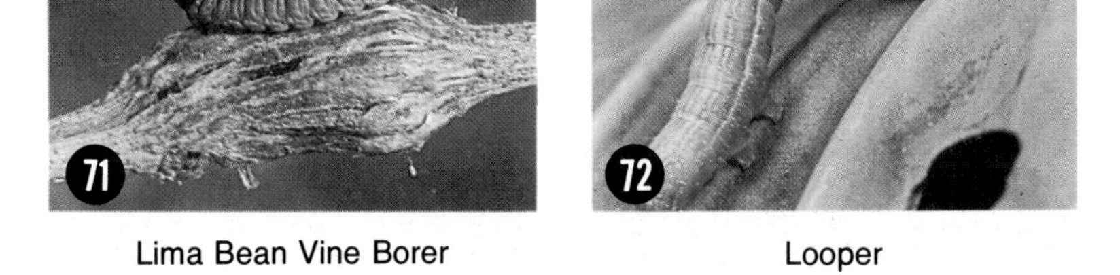

71 Lima Bean Vine Borer

72 Looper

73 Maggot

74 Mexican Bean Beetle

75 Pickleworm-Older Larvae

Pickleworm-Young Larvae

Silver-Spotted Skipper

Spider Mites

Striped Cucumber Beetle

Sweet Potato Weevil

Squash Beetle

Tobacco Hornworm

Tomato Fruitworm

Tortoise Beetle (Gold Bug)-Two Species

BENEFICIAL INSECTS

1 Ant Lion (Doodle Bug)

2 Assassin Bug

3 Braconid Parasite Cacoons

4 Ground Beetle Fiery Hunter

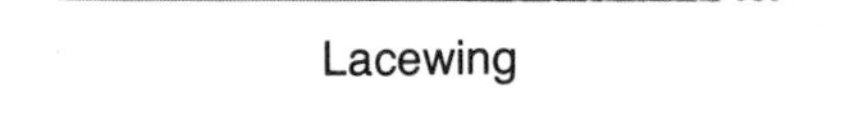

5 Lacewing

6 Ladybird Beetle-Adult

7 Ladybird Beetle-Larva

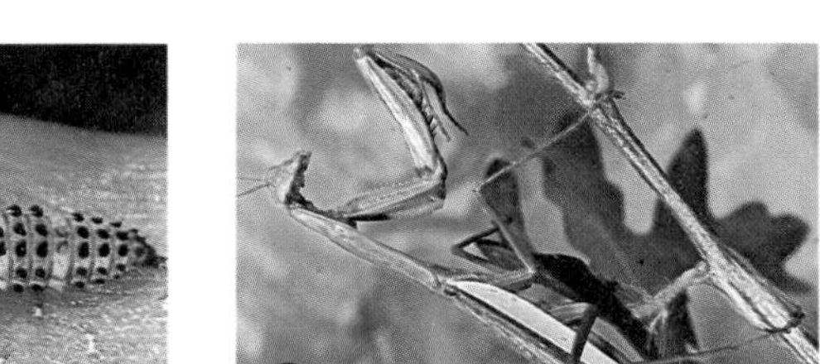

8 Praying Mantis

9 Predaceous Stinkbug

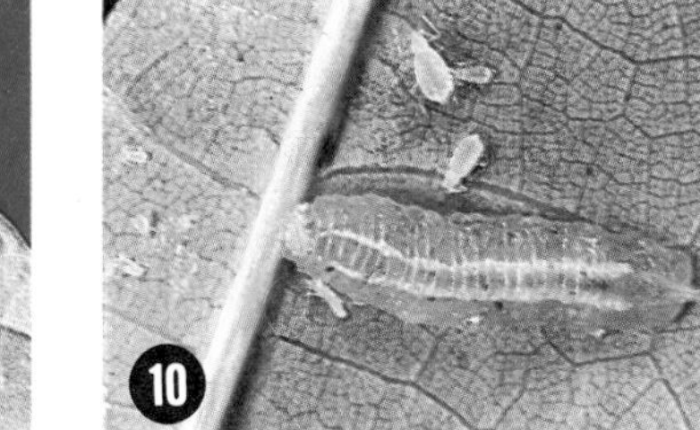
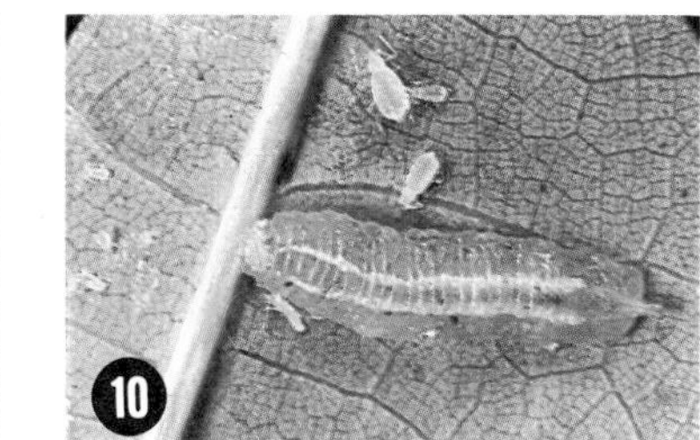

10 Syrphid Fly-Larva

11 Tiger Beetle

DISEASES OF AGRONOMIC CROPS

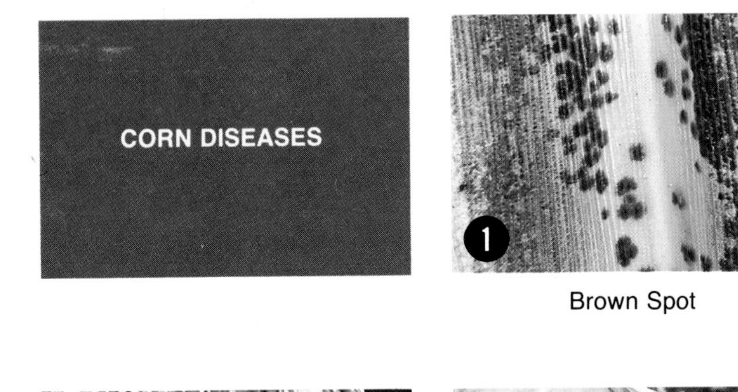

CORN DISEASES

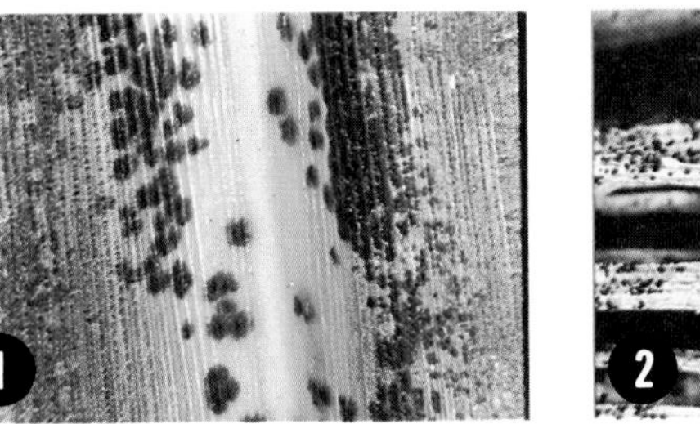

1 Brown Spot

2 Charcoal Rot

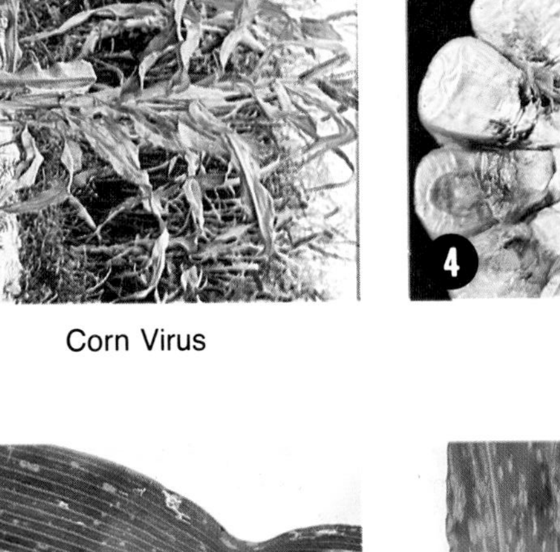

3 Corn Virus

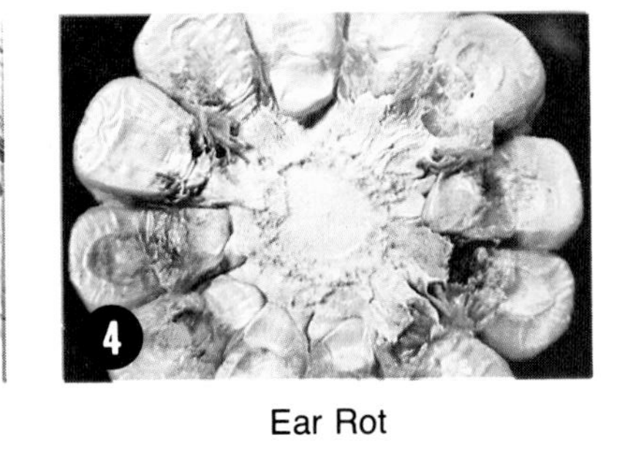

4 Ear Rot

5 Fusarium Rot

6 Lodging

7 Nematode Damage to Roots

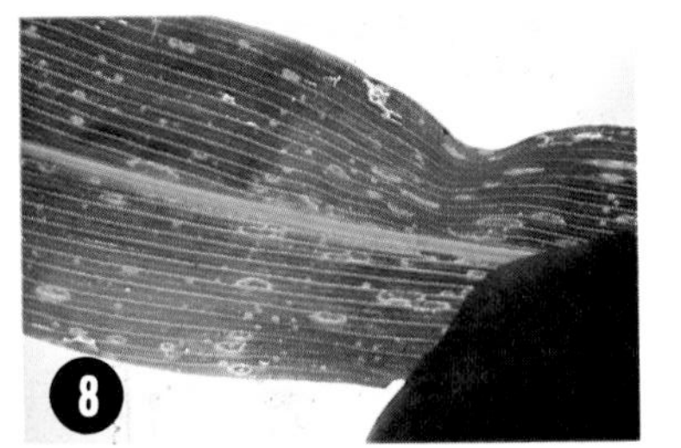
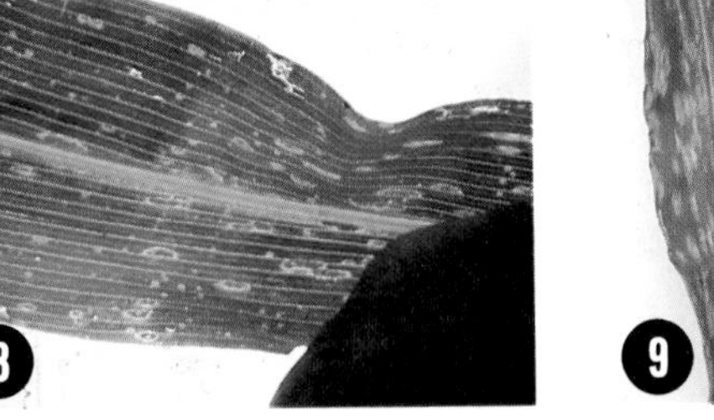

8 Corn Resistant to Southern Leaf Blight

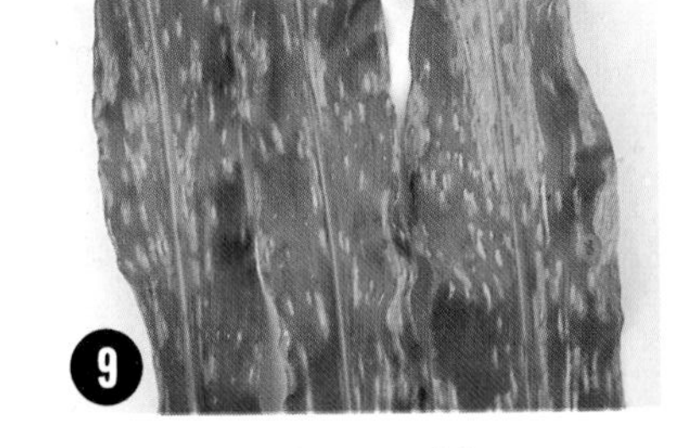
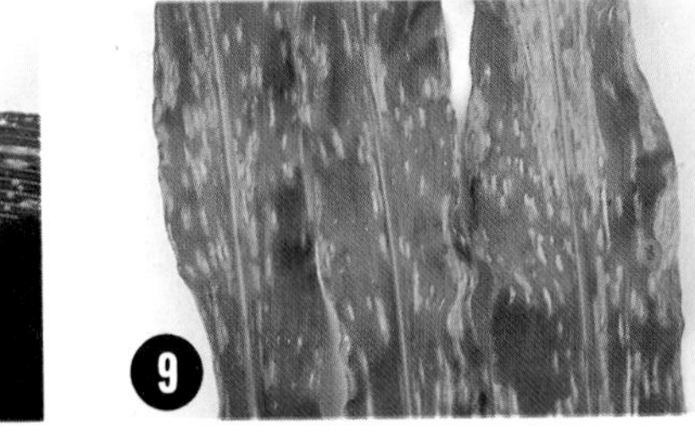

9 Corn Susceptible to Southern Leaf Blight

10 Smut

11 Virus Damage

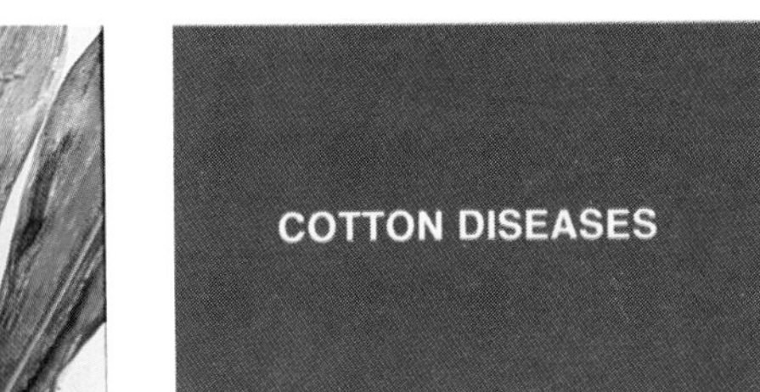

COTTON DISEASES

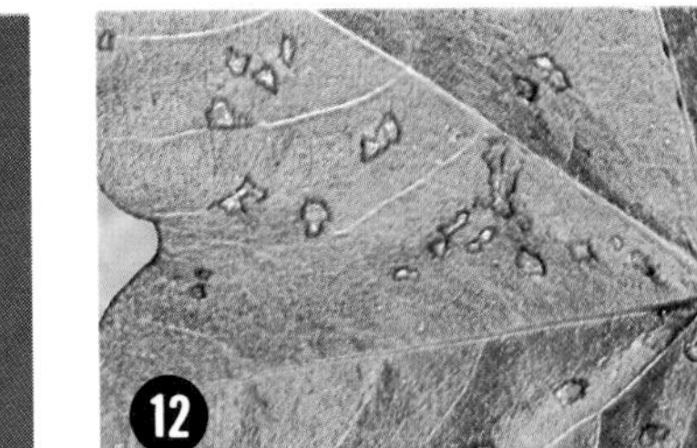

12 Angular Leaf Spot

13 Ascachyta Blight

14 Boll Rot

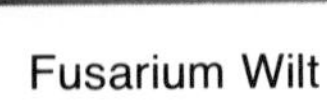

15 Fusarium Wilt

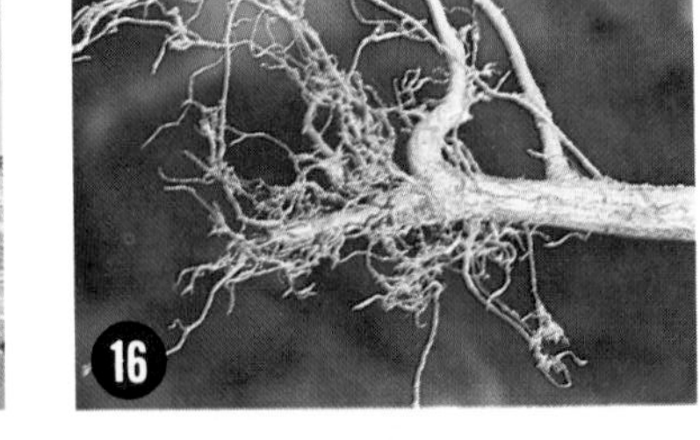
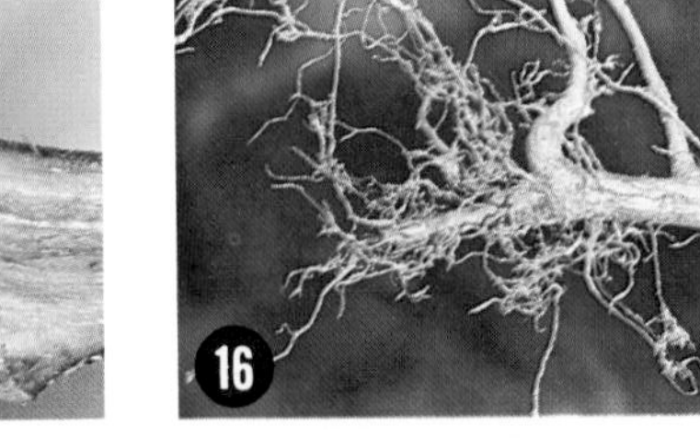

16 Nematode Damage to Roots

17 Seedling Blight

18 2-4,D Damage

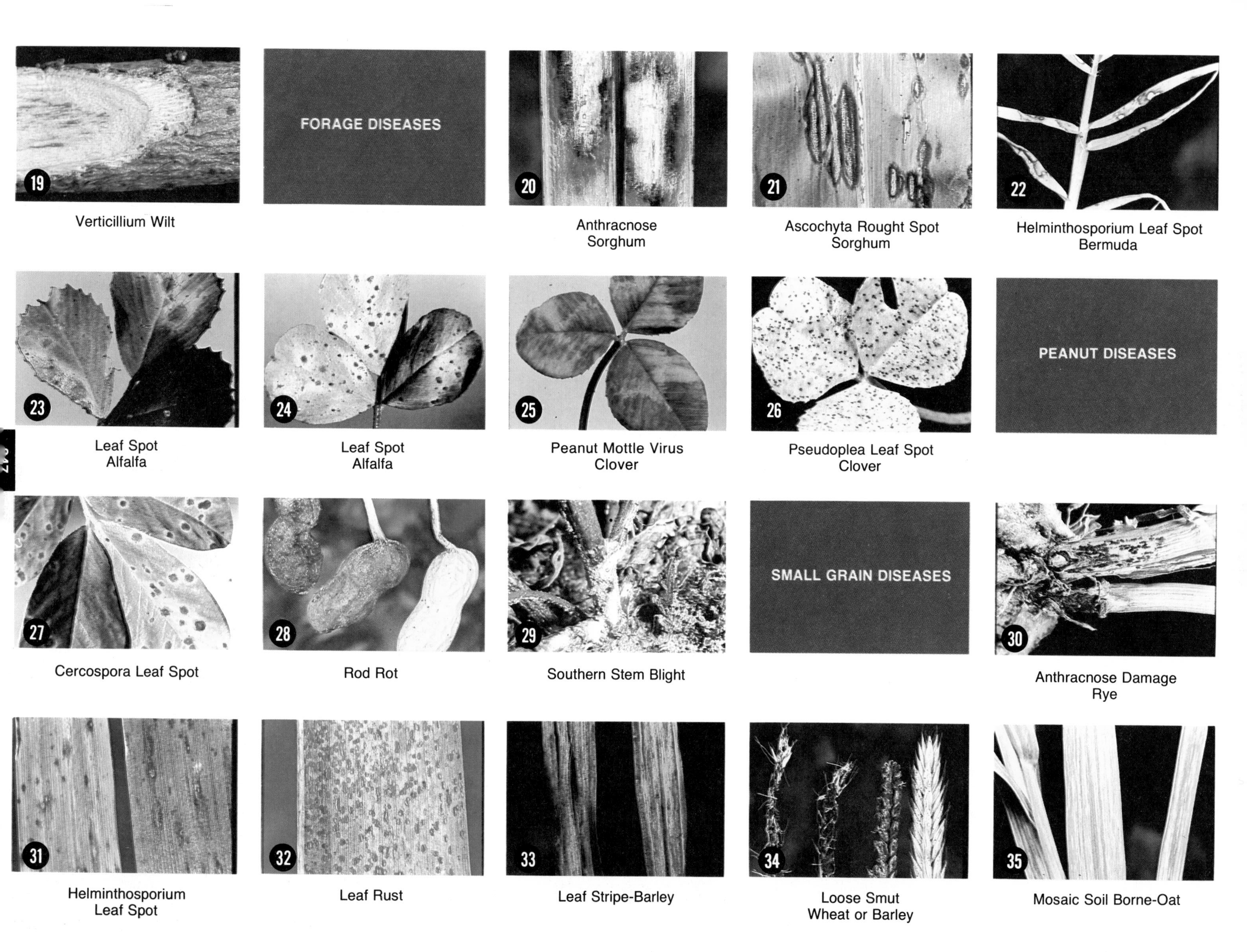

Verticillium Wilt

Anthracnose
Sorghum

Ascochyta Rought Spot
Sorghum

Helminthosporium Leaf Spot
Bermuda

Leaf Spot
Alfalfa

Leaf Spot
Alfalfa

Peanut Mottle Virus
Clover

Pseudoplea Leaf Spot
Clover

Cercospora Leaf Spot

Rod Rot

Southern Stem Blight

Anthracnose Damage
Rye

Helminthosporium
Leaf Spot

Leaf Rust

Leaf Stripe-Barley

Loose Smut
Wheat or Barley

Mosaic Soil Borne-Oat

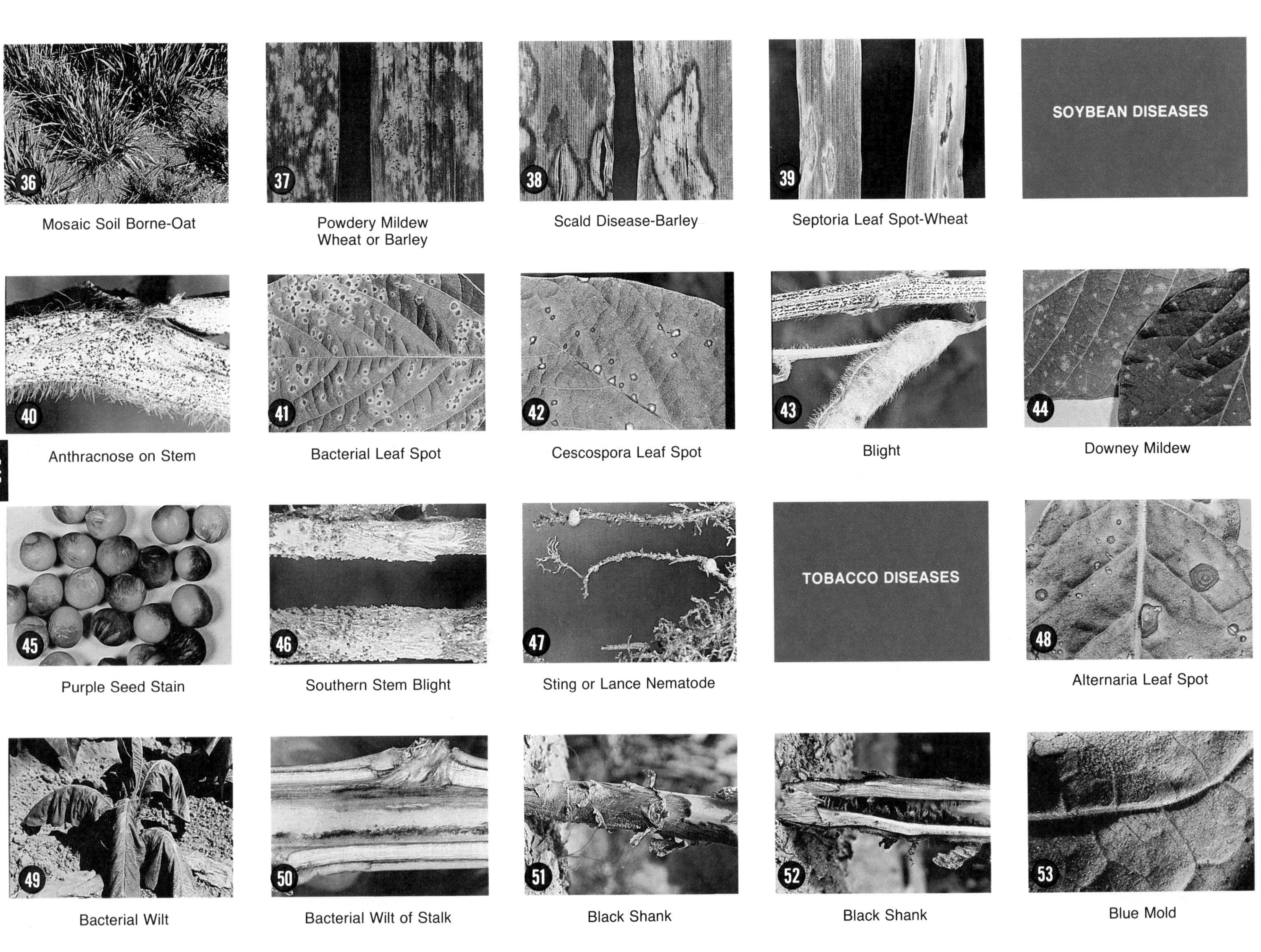

Mosaic Soil Borne-Oat

Powdery Mildew
Wheat or Barley

Scald Disease-Barley

Septoria Leaf Spot-Wheat

Anthracnose on Stem

Bacterial Leaf Spot

Cescospora Leaf Spot

Blight

Downey Mildew

Purple Seed Stain

Southern Stem Blight

Sting or Lance Nematode

Alternaria Leaf Spot

Bacterial Wilt

Bacterial Wilt of Stalk

Black Shank

Black Shank

Blue Mold

Tobacco Fleabeetle Damage

Fusarium Wilt

Hollow Stalk External

Hollow Stalk Internal

Lightning Damage

Lightning Damage

Mosaic

Rhizocronia and Pythium Seedling Damage

Root Knot Lesion Damage

Tobacco Budworm Damage

Wildfire

DISEASES OF HORTICULTURAL CROPS, TREES AND TURF AND WOOD ROTS

APPLE DISEASES

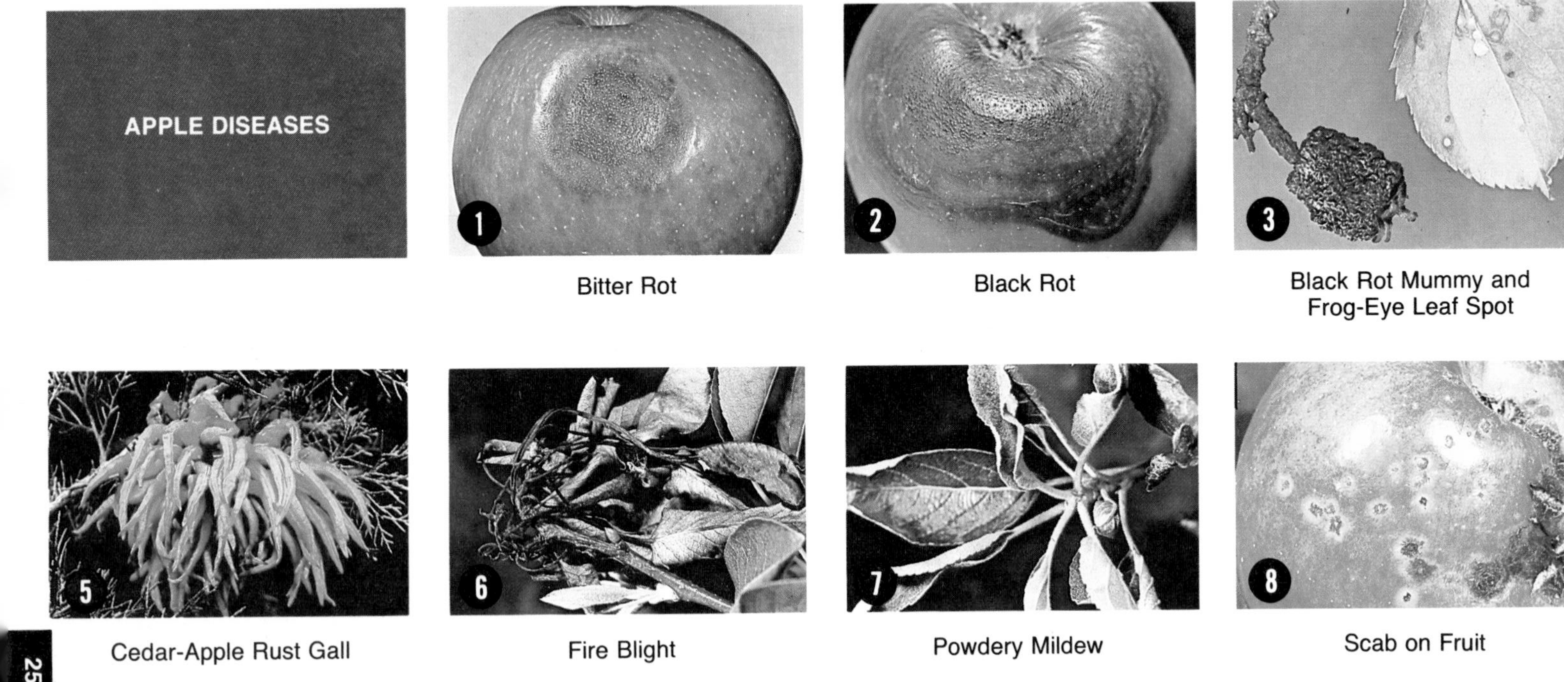

1 Bitter Rot

2 Black Rot

3 Black Rot Mummy and Frog-Eye Leaf Spot

4 Cedar-Apple Rust

5 Cedar-Apple Rust Gall

6 Fire Blight

7 Powdery Mildew

8 Scab on Fruit

9 Scab on Young Fruit

10 Sooty Blotch and Flyspeck

CRUCIFER DISEASES

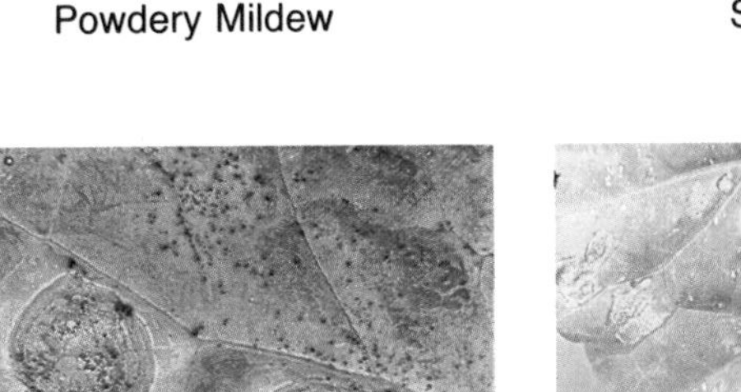

11 Alternaria and Downy Mildew-Cabbage

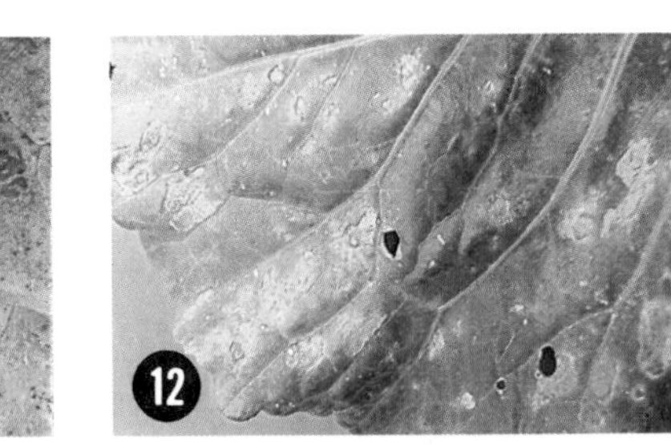

12 Downy Mildew Cabbage

13 White Spot-Turnip

CUCURBIT DISEASES

14 Alternaria Leaf Spot Cantaloupe

15 Anthracnose-Gourd

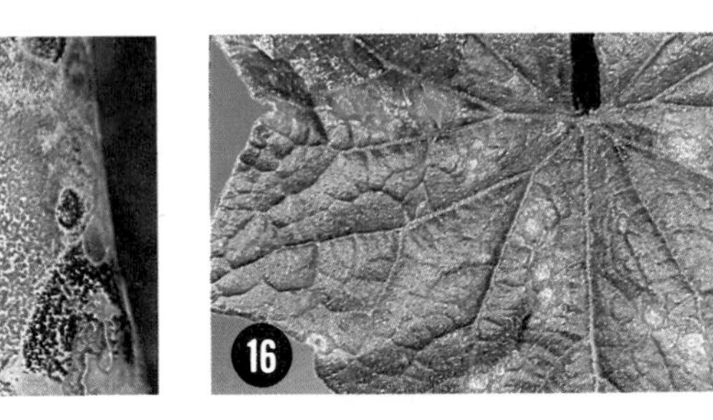

16 Anthracnose on Leaf Cucumber

17 Anthracnose-Fruit Watermelon

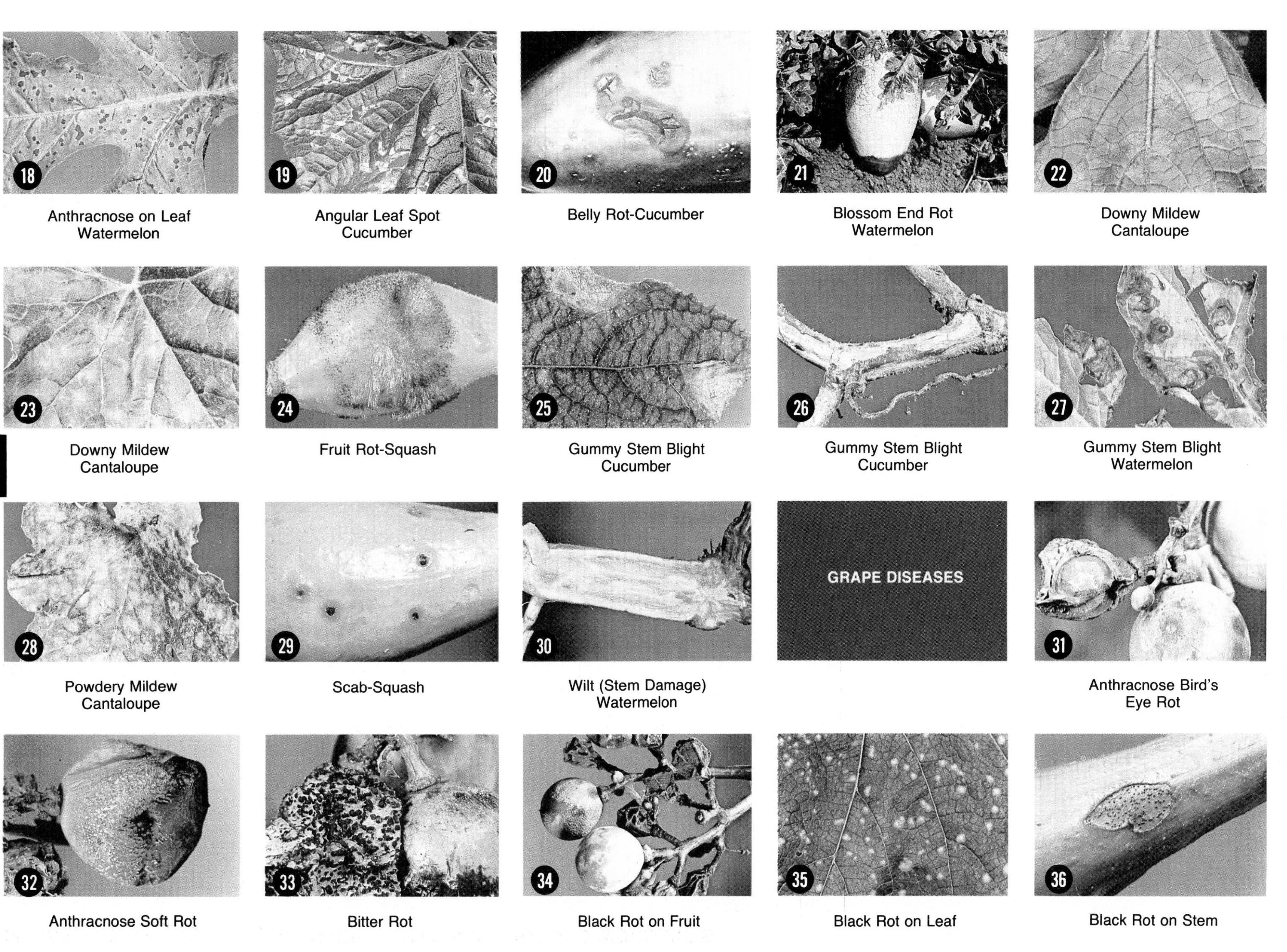

18 Anthracnose on Leaf Watermelon

19 Angular Leaf Spot Cucumber

20 Belly Rot-Cucumber

21 Blossom End Rot Watermelon

22 Downy Mildew Cantaloupe

23 Downy Mildew Cantaloupe

24 Fruit Rot-Squash

25 Gummy Stem Blight Cucumber

26 Gummy Stem Blight Cucumber

27 Gummy Stem Blight Watermelon

28 Powdery Mildew Cantaloupe

29 Scab-Squash

30 Wilt (Stem Damage) Watermelon

31 Anthracnose Bird's Eye Rot

32 Anthracnose Soft Rot

33 Bitter Rot

34 Black Rot on Fruit

35 Black Rot on Leaf

36 Black Rot on Stem

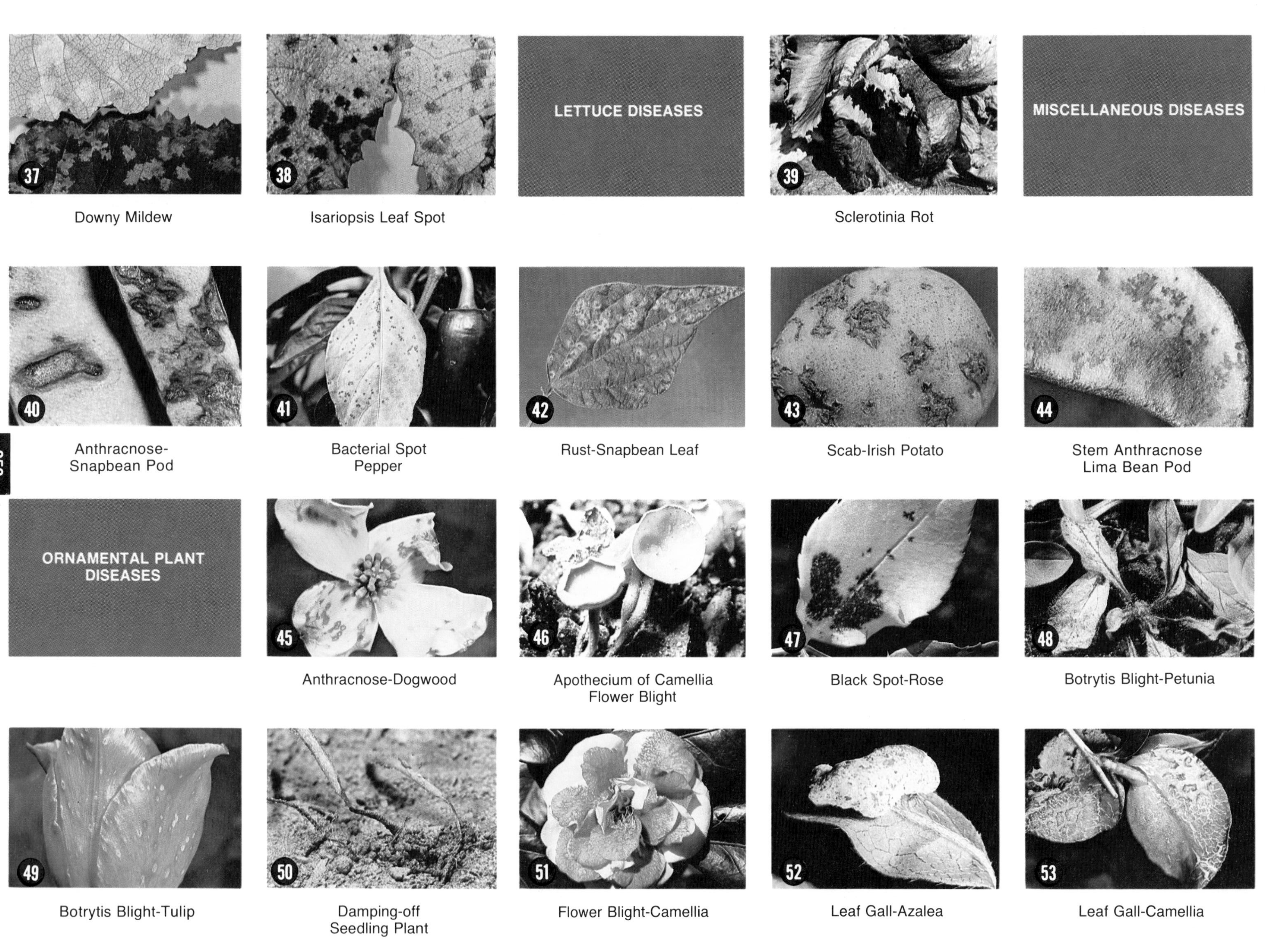

Downy Mildew

Isariopsis Leaf Spot

LETTUCE DISEASES

Sclerotinia Rot

MISCELLANEOUS DISEASES

Anthracnose-
Snapbean Pod

Bacterial Spot
Pepper

Rust-Snapbean Leaf

Scab-Irish Potato

Stem Anthracnose
Lima Bean Pod

ORNAMENTAL PLANT
DISEASES

Anthracnose-Dogwood

Apothecium of Camellia
Flower Blight

Black Spot-Rose

Botrytis Blight-Petunia

Botrytis Blight-Tulip

Damping-off
Seedling Plant

Flower Blight-Camellia

Leaf Gall-Azalea

Leaf Gall-Camellia

Leaf Spot Chrysanthemum

Leaf Spot-Iris

Lichen

Powdery Mildew Photinia

Septoria Leaf Spot Azalea

Twig Canker-Camellia

Bacterial Spot on Fruit Peach

Bacterial Spot on Leaf Peach

Blast Disease of Trunk Peach

Black Knot of Plum

Blossom Blight-Peach

Brown Rot on Fruit-Peach

Orchard Disease Problem Peach

Root Rot-Peach

Scab on Fruit-Peach

Scab on Twig-Peach

Crown Gall

Diseased Shuck and Young Nuts

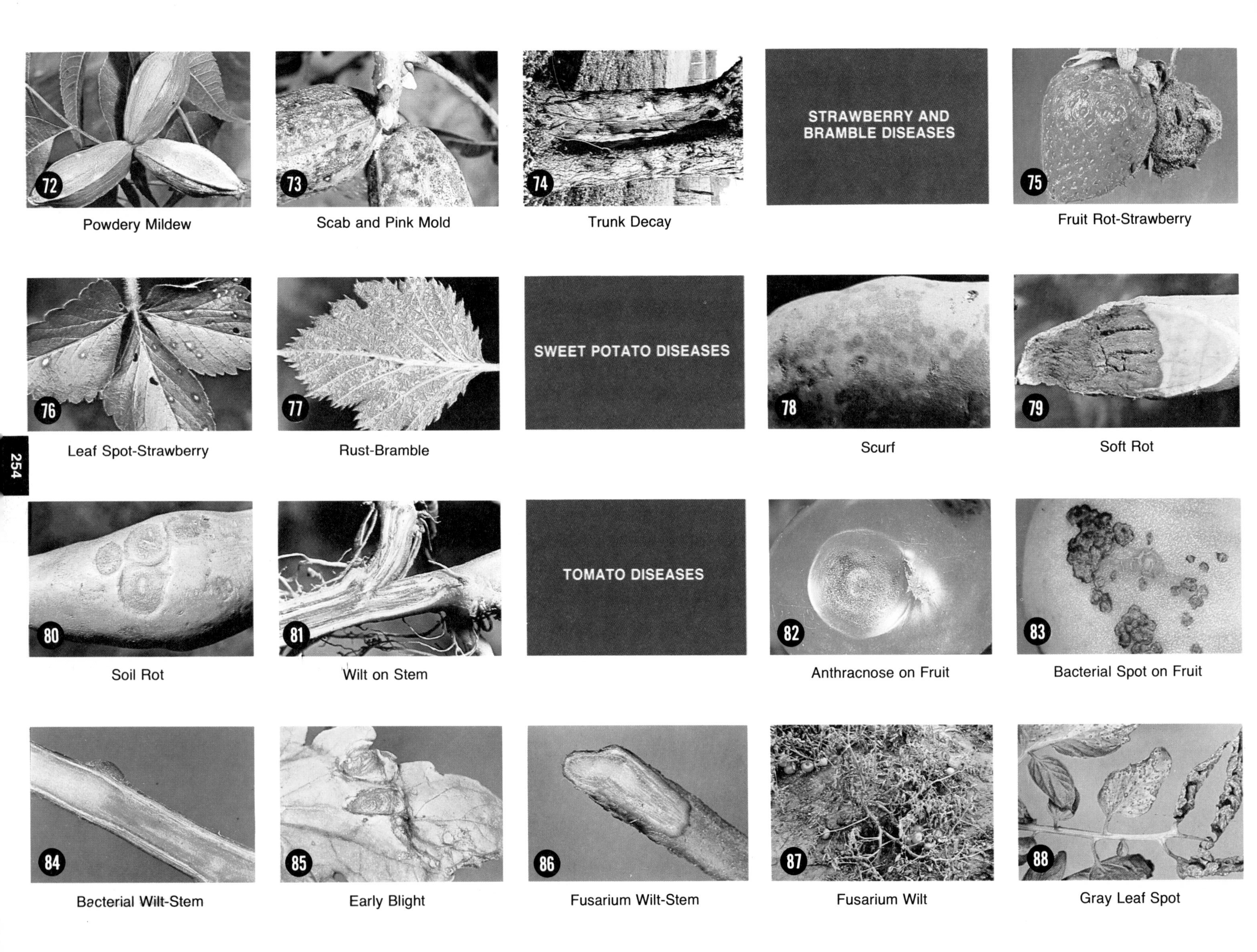

Powdery Mildew

Scab and Pink Mold

Trunk Decay

Fruit Rot-Strawberry

Leaf Spot-Strawberry

Rust-Bramble

Scurf

Soft Rot

Soil Rot

Wilt on Stem

Anthracnose on Fruit

Bacterial Spot on Fruit

Bacterial Wilt-Stem

Early Blight

Fusarium Wilt-Stem

Fusarium Wilt

Gray Leaf Spot

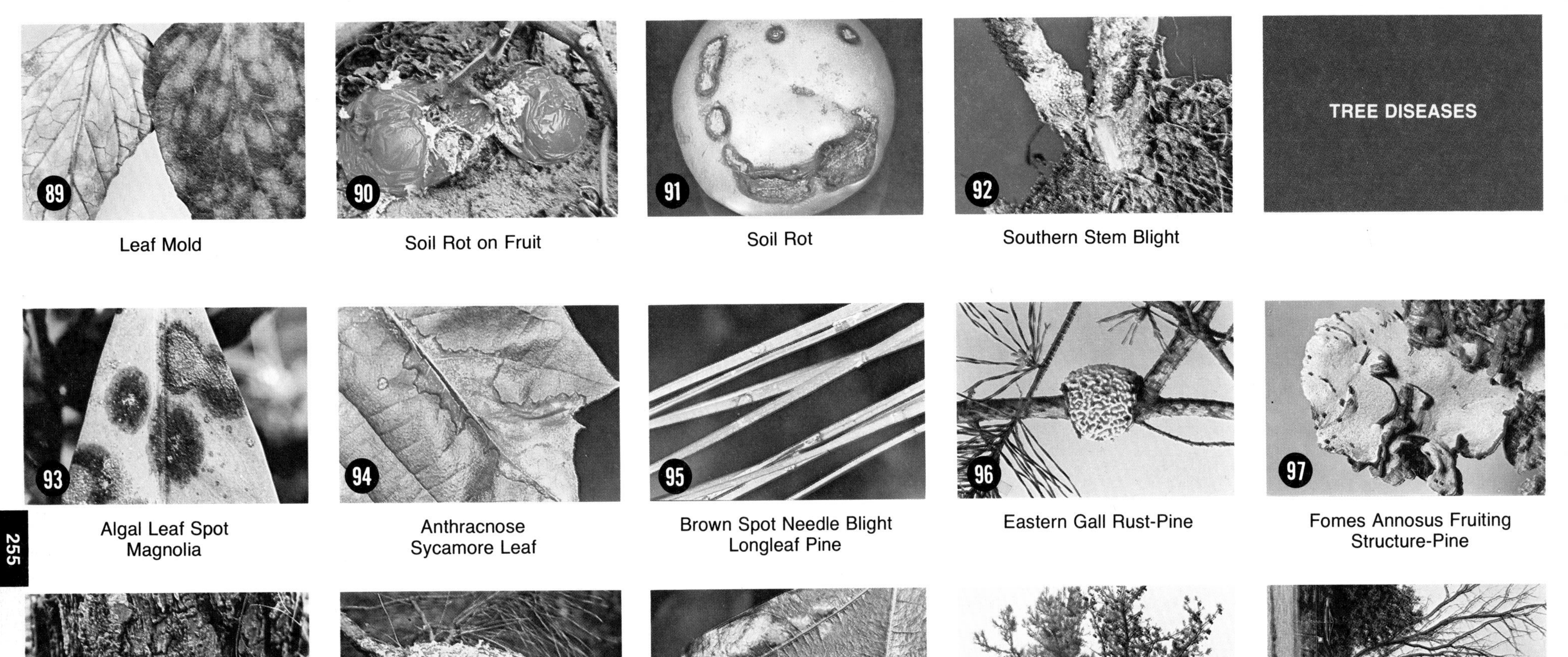

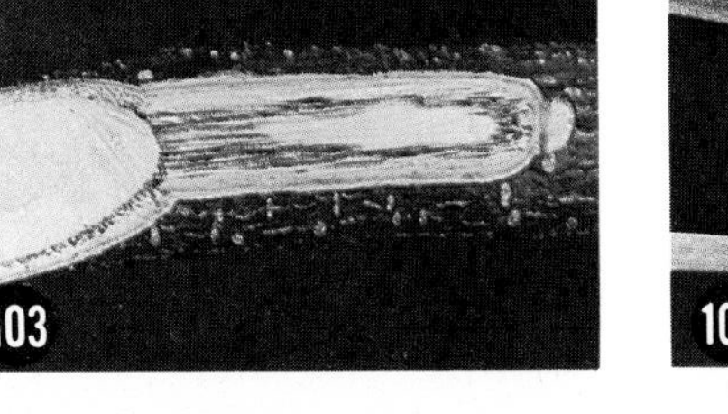
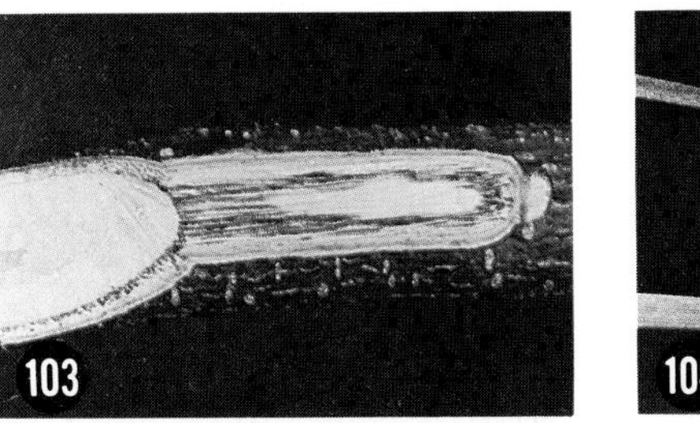
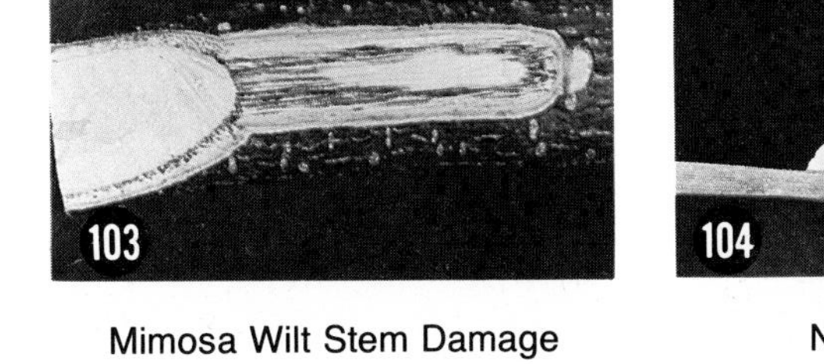
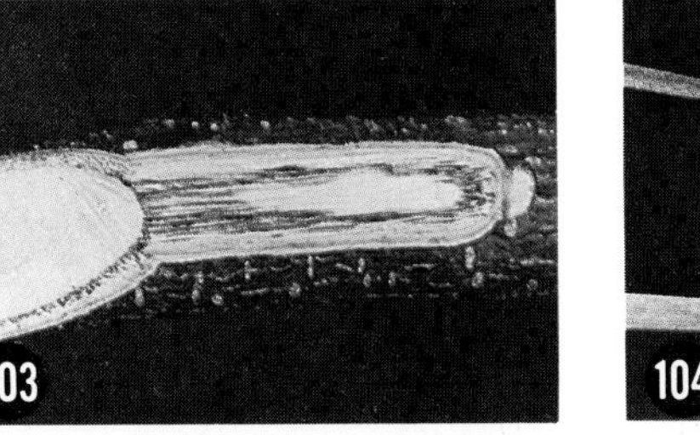

89 Leaf Mold

90 Soil Rot on Fruit

91 Soil Rot

92 Southern Stem Blight

TREE DISEASES

93 Algal Leaf Spot Magnolia

94 Anthracnose Sycamore Leaf

95 Brown Spot Needle Blight Longleaf Pine

96 Eastern Gall Rust-Pine

97 Fomes Annosus Fruiting Structure-Pine

98 Fomes Annosus on Trunk Pine

99 Fusiforme Rust-Pine

100 Leaf Blister-Oak

101 Little Leaf-Pine

102 Mimosa Wilt

103 Mimosa Wilt Stem Damage

104 Needle Rust-Pine

105 Powdery Mildew Oak Leaf

106 Tar Spot Disease Maple Leaf

TURF DISEASES

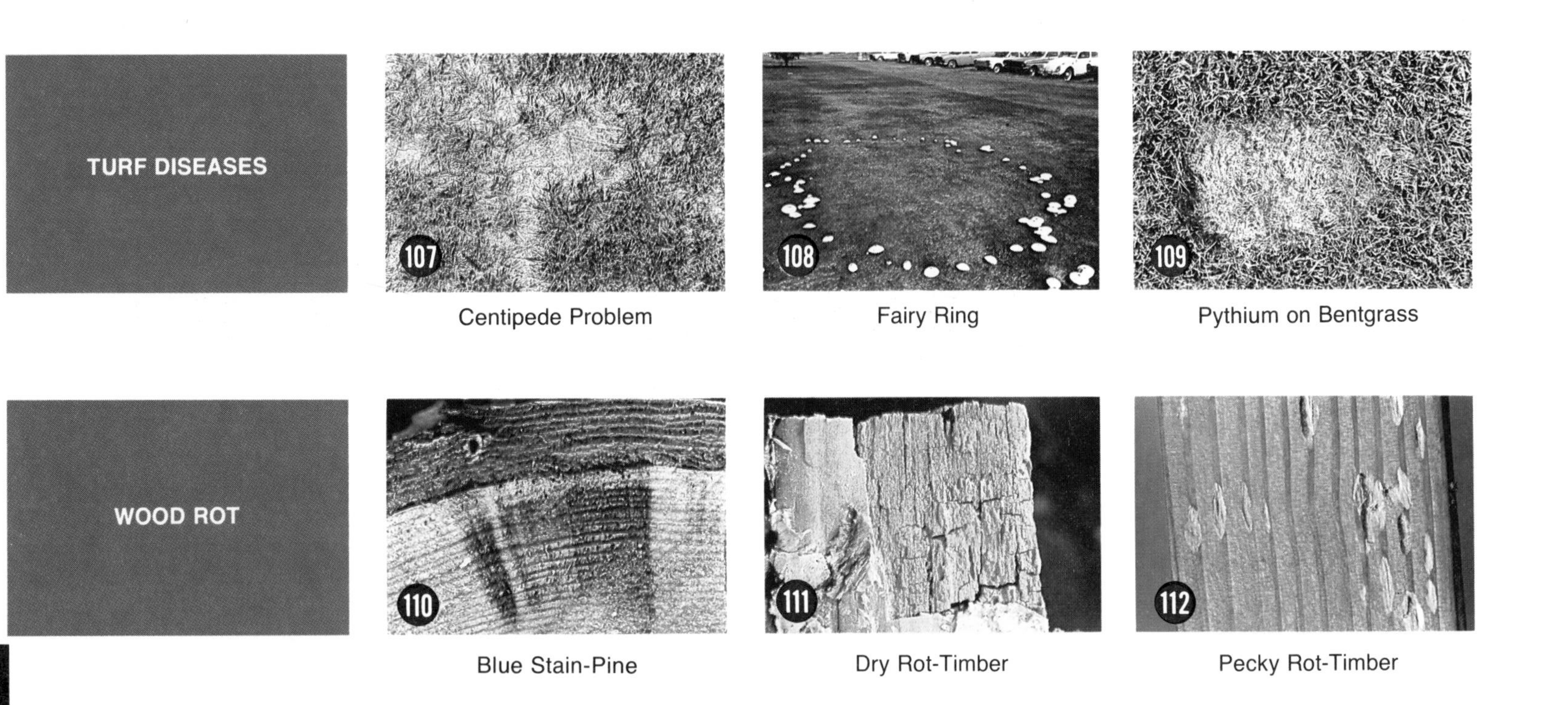

Centipede Problem

Fairy Ring

Pythium on Bentgrass

WOOD ROT

Blue Stain-Pine

Dry Rot-Timber

Pecky Rot-Timber

GLOSSARY

Definitions which follow apply specifically to subject matter in this book.

A

ABRASION—Any break in the skin, such as a cut, scratch, or sore.

ABRASIVE—Material that cuts or grinds.

ABSORPTION—Process of being "sucked up" or taken into; specifically, entry of a chemical into an organism such as a plant or insect.

ABSORPTIVE CLAY—A special type of clay which can take up liquids and is used to clean up pesticide spills.

ACARICIDE—Pesticide used to control mites and ticks.

ACCUMULATING PESTICIDE—Chemical which can build up in animals or the environment.

ACID-FORMING—Materials which leave an acid residue in the soil.

ACID SOIL—Soil with pH less than 7.0.

ACRE—An area of land equal to 43,560 sq. ft.

ACTIVE INGREDIENT—That part of a pesticide formulation directly responsible for the pesticidal effect.

ACTUAL DOSAGE—Amount of active ingredient (not formulated product) applied to a given area.

ACUTE TOXICITY—The intial rapid poisonous effect of a chemical as the result of one exposure.

ADHESIVE—An adjuvant that helps a pesticide stick to a surface.

ADJUVANT—Material added to a pesticide to improve the action of the active ingredient.

AEROSOL—Very fine particles of a pesticide suspended in air.

AEROSOL CAN—Container used to apply pesticide which is stored under pressure and driven through a fine nozzle by gas pressure.

AGITATE—Mixing a pesticide to keep it from settling or separating in the sprayer tank.

ALKALINE-FORMING—Materials which leave an alkaline residue in the soil.

ALKALINE SOIL—Soil with pH higher than 7.0.

AMMONIA—A gas made up of 82.25 percent nitrogen and 17.75 percent hydrogen.

ANALYSIS—The percentage composition of various nutrients in fertilizer.

ANNUAL—A plant that grows, matures, and produces seed in a single year and then dies.

ANTIDOTE—A treatment given to counteract the effects of a poison.

ANTISIPHONING DEVICE—Device (check valve) on the filling hose of a sprayer to prevent water from draining back to the source.

APIARY—A place where domestic bees are housed.

APPLICATION—Putting chemicals on or in soil or plants.

APPLICATOR—Device for applying chemicals, usually fertilizers or granular pesticides. A person who applies chemicals.

APPLICATION RATE—Amount of chemical applied per acre.

AQUATIC WEED—Weeds that grow in or near water.

ARTIFICIAL RESPIRATION—First aid for a person who has stopped breathing.

ATROPINE—Antidote for organophosphate and carbamate poisoning—full name is atropine sulfate.

ATTRACTANT—Substance used to attract pests.

ATTRACTIVE NUISANCE—Legal term for an object which might attract people, particularly children, and then cause them harm.

B

BAIT—Food or other material used to attract a pest to a trap or pesticide.

BAND APPLICATION—Application made in a narrow band, usually over or alongside a row.

BASAL APPLICATION—Application of a herbicide to the lower portion of the stems of woody plants.

BASE—Alkaline material; opposite of acid.

BIENNIAL—Plant that matures and produces seed during the second year of its life and dies after not more than two years of life.

BIOLOGICAL CONTROL—Use of biological agents, such as parasites and predators, to control pests.

BOOM—Pipe or tubing with several nozzles to apply chemicals over a wide area at one time.

BROADCAST APPLICATION—Application made nonselectively over a broad area.

BROADLEAF PLANTS—Plants with wide flat leaves.

BROAD-SPECTRUM PESTICIDE—Pesticide effective against a variety of pests.

C

CALCULATE—To figure with numbers.

CALIBRATE—To determine the application rate of a sprayer or other application equipment.

CANCEL—To discontinue registration of a pesticide.

CARBAMATES—A family of pesticides having similar chemical structures (Sevin, Furadan).

CARRIER—1. Material used to dilute the active ingredient in a chemical formulation. 2. Material used to carry a pesticide to its target. 3. Plant or animal carrying an infectious disease agent internally, but showing no marked symptoms.

CARBON DIOXIDE—A nonconbumstible gas formed by combining carbon and oxygen.

CARTRIDGE—The cylinder inserted in a respirator to absorb gases, fumes, or vapors.

CAUTION—Signal word used to alert users to slightly toxic pesticides.

CHEMICAL NAME—Scientific name for the active ingredient in a pesticide.

CHEMICALLY INACTIVE—Material which does not react with other materials.

CHLORINATED HYDROCARBONS—A family of pesticides with similar chemical structures (Chlordane, Lindane).

CHLOROSIS—Loss of green color (yellowing or whitening) in foliage.

CHRONIC POISONING—Poisoning as a result of small repeated doses of pesticide.

CHRONIC TOXICITY—Measure of the poisonous nature of a pesticide as a result of small repeated doses over a relatively long period of time.

CLAY—Soil with particles smaller than 0.002 mm in diameter; plastic when moist, hard when dry.

COMMON NAME—Simplified name for a pesticide. 2. Well-known or popular name for weeds and insects.

COMPATIBILITY—Adaptability of two or more pesticides to be mixed and used without adverse effect on the target or pesticidal action of the components.

COMPLETE FERTILIZER— A fertilizer which contains the three primary nutrients in sufficient quantity to be of value as a plant food.

CONCENTRATE—A pesticide, as sold, before being diluted for application.

CONCENTRATION—Amount of active ingredient in a formulation or mixture.

CONDEMNATION—Act of removing a crop or product from the market because residue of one or more pesticides exceeds the legal tolerance.

CONTACT—To touch or be touched by.

CONTAMINATE—To accidentally get a pesticide into other pesticides, seed, fertilizer, food, and feed.

CONTROL—To reduce the number or growth of pests in an area.

COPPER SULFATE—Most common source of copper in fertilizer.

CORM—A thick, rounded subsurface stem base with leaves and buds attached which acts as a vegetative reproductive structure.

CORROSION—Effect of metal being rusted, worn down, or eaten away by another substance.

CORROSIVE POISON—Poison which can severely burn the mouth or stomach.

D

DAYS TO HARVEST—Interval required by law between last pesticide application and harvest.

DECONTAMINATE—Remove or clean up a pesticide spill or residue to avoid or reduce danger or damage to desirable plants, insects, animals, people, or the environment.

DEFOLIANT—Chemical which causes foliage to die and fall off.

DEGRADE—Break down or decompose.

DEGREE OF EXPOSURE—Extent to which a person has been exposed to a toxic substance.

DEPOSIT—Pesticide left on leaves, stems, fruit, other plant parts, or skin after pesticide application.

DERMAL TOXICITY—Measure of how poisonous a pesticide is when absorbed through the skin.

DESSICANT—Chemical which draws moisture out of a plant causing it to wither and die.

DETERIORATE—Break down or decay.

DILUENT—Liquid or solid used to dilute a concentrated pesticide.

DILUTE—To reduce the concentration of a pesticide by adding water, oil, or other material.

DISEASE—Change in one or more physiological processes of plant or animal that impairs its ability to use energy.

DISINFECTANT—Pesticide or other chemical that kills or inactivates disease-causing bacteria.

DISPOSAL—Discarding a pesticide or pesticide container.

DIRECTED SPRAY—Application to a specific target area, such as a particular part of a plant.

DOMESTIC ANIMAL—Tame animal, such as a cow, sheep, or dog (livestock and pets).

DORMANT SPRAY—Pesticide application made while plants are not actively growing.

DOSE—Portion or amount of pesticide which reaches the target.

DOWNWIND—The direction toward which the prevailing wind blows.

DRIFT—Movement by wind of fine particles of spray or dust to an area not intended to be treated.

DUST—Finely-ground, dry mixture of an inert carrier and pesticide.

E

ECOLOGY—Study of the inter-relationships between plants, animals, and their surroundings.

EMETIC—Something which causes vomiting and is used as a first aid for some kinds of poisoning.

EMULSIFIABLE CONCENTRATE—A pesticide formulation with the active ingredient and an emulsifier suspended in a liquid.

EMULSIFIER—Chemical which helps one liquid form tiny droplets which remain suspended in another liquid—used to form a stable mixture of two liquids, such as oil and water, which would not ordinarily mix.

ENCAPSULATION—1. Disposing of pesticides and containers by sealing them in another safe container. 2. Method of formulation pesticides in capsules.

ENVIRONMENT—Surroundings or conditions, including water, soil, and air influencing the growth and well-being of organisms.

EPA—Environmental Protection Agency, a branch of the federal government which is responsible for formation and enforcement of rules and regulations concerning pesticides.

EPA REGISTRATION NUMBER—Number assigned by EPA to a product when it is registered and which must appear on all labels for the product.

ERADICANT FUNGICIDE—Fungicide which kills the fungus after the disease appears on or in a plant.

EXPOSE—To come into contact with a pesticide.

EXPOSURE—State of having been exposed.

F

FACE SHIELD—Transparent piece of protective equipment used to protect the face from exposure to pesticides.

FEED—Food for livestock and other domestic animals.

FILTER—1. To screen out unwanted material. 2. A device to strain out unwanted material.

FINAL TREATMENT—Last treatment before harvest.

FLOWABLE—Finely-ground solid material suspended in a liquid carrier.

FLUID—Liquid or gaseous substance that will flow.

FOAMING AGENT—Adjuvant which causes the pesticide mixture to form a thick foam which helps prevent drifting.

FOGGER—Aerosol generator.

FOLIAGE—Leaves, stems, and other aboveground plant parts.

FOLIAR SPRAY—Chemical applied to plant foliage.

FORMULA—Shorthand method of expressing the constituents of a compound with symbols and abbreviations.

FORMULATION—Mixture of one or more pesticides and other materials such as diluents and carriers as sold—does not include adjuvants or tank mixes added at the time of application.

FUME—Unpleasant or irritating smoke, vapor, or gas.

FUMIGANT—Volatile chemical that kills pests with a gas or vapor.

FUMIGATION—Pest control with fumigants.

FUNGI—Small plant organisms that lack chlorophyll and which cause rots, mildews and other diseases (singular: fungus).

FUNGICIDE—Pesticide used to kill or control fungi which cause plant diseases.

FERTILIZER—Material which contains one or more plant-food elements.

FERTILIZER FORMULA—Quantity and grade of materials in a fertilizer mixture.

FERTILIZER-PESTICIDE MIXTURE—Combination of fertilizer and pesticide.

FERTILIZER RATIO—Ratio of N, P, and K (nitrogen, phosphorus, and potassium) expressed in numbers.

FIXATION—Process of rendering available plant nutrients unavailable or fixed in the soil.

G

GAS MASK—Respirator which covers the face to protect the eyes, nose, and mouth.

GRANARY—Storage area for grain.

GRANULES—Pellets; a pesticide formulation of dry, ready-to-use, low-concentrate pesticide plus an inert carrier.

GROWTH REGULATORS—Chemicals which alter the normal growth or reproduction of a plant.

H

HAZARDS—Risk of danger to applicator, observers, livestock, and desirable plants.

HERBICIDE—Pesticide used to control unwanted plants.

HUMUS—Well-decomposed part of soil organic matter.

HYDRAULIC AGITATION—Mixing in a tank by means of liquid flow under pressure.

HYDRAULIC SPRAYER—Machine which applies pesticide by using water at high pressure and high volume.

HYDROGEN ION CONCENTRATION—pH; a measure of the acidity of alkalinity of a solution.

HYGROSCOPIC—Materials that attract and retain moisture.

I

ILLEGAL RESIDUE—Quantity of pesticide on the harvested crop which exceeds the legal limit.

INACTIVE—Material which will not react with other materials.

INCINERATOR—High-heat furnace or burner used to destroy pesticides.

INCOMPATIBLE—Not able to be mixed or used together.

INCORPORATE—To mix a pesticide into the soil.

INERT INGREDIENT—Inactive part of pesticide formulation.

INFESTATION—Pests in an area where they are not wanted.

INGEST—Eat or swallow.

INGREDIENT STATEMENT—That part of a pesticide-container label that lists the name and amount of each constituent in the mixture.

INHALATION—Taken into the lungs by breathing.

INHALATION TOXICITY—Poisonous effect of a pesticide breathed into the lungs.

INJECT—Force pesticide or fertilizer into a plant or the soil.

INSECTICIDE—Pesticide used to control or prevent damage by insects.

INSOLUBLE—Not soluble; will not dissolve.

INTEGRATED CONTROL—System using two or more pest-control methods in a planned program.

INTERVAL—A period of time—between pesticide applications, or between application and harvest or reentry to the treated area.

INVERT EMULSIFIER—Agent which allows water to remain suspended in an oil-based pesticide.

IRRITATING—Annoying; burning or inflaming skin or eyes.

J

JOINTLY LIABLE—Two or more individuals or companies share legal liability.

K

K—Symbol for potassium.

KILOGRAM (Kg)—Metric unit of weight approximately equal to 2.2 pounds.

L

LABEL—Printed information attached to a pesticide container.

LARVA—Immature or wormlike (grub) stage of an insect that does not resemble the adult form (plural: larvae).

LARVACIDE—Insecticide used to kill insect larvae.

LAY-BY—Crop development stage when final field operation, such as cultivation, is performed.

LD_{50}—Dose of a pesticide which kills half of a population of test animals if eaten or absorbed through the skin.

LEACHING—Removing nutrients from soil by passage of water through the soil profile.

LETHAL—Toxic or deadly.

LIABILITY—Legal responsibility.

LIABLE—Legally responsible.

LIME—Material with high calcium content used to neutralize soil acidity.

LIQUID FERTILIZER—Any fertilizer formulated as a liquid for handling and application.

LITER (L)—Metric unit of volume equal to approximately 1.06 quarts.

LOW VOLATILITY—Liquid or solid that does not evaporate readily.

LVC—Low-volume concentrate designed for ultra-low-volume applications.

M

MANURE—Animal waste used as fertilizer.

MARINE—Referring to animals and plants that live in the ocean.

MARL—Earth or soft rock deposits rich in calcium.

MATERIAL—A substance; a word often used here to refer to pesticides.

MAXIMUM DOSAGE—Largest amount of chemical that can safely be used without producing excessive residues or other ill effects.

MECHANICAL AGITATOR—Device which mixes spray materials in a tank by mechanical means.

MICRO-NUTRIENTS—Trace elements or minor nutrients—materials needed by plants in very small quantities.

MILLIGRAM (Mg)—Unit of weight in the metric system (1/1000 of a kilogram).

Mg/Kg—Amount of pesticide in milligrams per kilogram of body weight to produce a particular effect.

MISCIBLE—Capable of being mixed.

MITE—Tiny animal with eight legs, often mistaken for or referred to as an insect.

MITICIDE—Acaricide; pesticide used to control mites and ticks.

MOLD—Fungus with conspicuous, profuse, or woolly growth.

MULTIPURPOSE—Pesticide used for more than one purpose.

N

N—Symbol for nitrogen.

NATURAL ENEMIES—Predators and parasites (insects or diseases) that attack pests.

NEGLIGENCE—State of being neglectful.

NEMATICIDE—Pesticide used to control nematodes.

NEMATODE—Tiny, tubular, unsegmented, eellike worms that feed on plant roots.

NERVOUS SYSTEM—Brain, spinal cord, and nerves of humans and animals.

NEUTRALIZE—Counteract or destroy effectiveness of a chemical.

NITRATES—1. Materials carrying nitrogen in the nitrate form. 2. Salts or esters of nitric acid.

NONACCUMULATIVE—Pesticide that will not accumulate in the bodies of animals or the environment.

NONLABELED—Material or method which is not on a label and is illegal.

NONPERSISTENT—Having a short life or short period of effectiveness.

NONSELECTIVE PESTICIDE—Pesticide which controls a wide variety of pests, or which kills both desirable and undesirable organisms.

NOXIOUS WEED—A plant defined by law as particularly undesirable.

NYMPH—Immature stage of an insect which passes through three stages.

O

OIL SOLUTION—Chemicals dissolved in oil.

OPERATING SPEED—Steady ground speed of a machine (miles per hour).

ORAL—Through the mouth.

ORGANIC—Material largely composed of carbon.

ORGANIC MATTER—Decomposed, partially decomposed, or undecomposed plant or animal material.

ORGANIC SOIL—Soil high in organic matter.

ORGANISM—Any living thing: plants, animals, bacteria, insects, and fungi.

ORGANOPHOSPHATE—A group of similar insecticides, including malathion, parathion, and Diazinon.

ORIGINAL CONTAINER—The package the pesticide is in when sold.

ORNAMENTAL—Plant grown for beauty, accent, color, and screening.

OVICIDE—Insecticide used to kill insect eggs.

P

P—Symbol for phosphorus.

PARASITE—A plant, insect, or animal that lives and feeds on or in a living host plant, insect, or animal.

PATHOGEN—An organism or agent capable of causing disease.

PARTS PER MILLION (PPM)—A unit of measure for pesticide residues.

PENETRANT—An adjuvant that helps a pesticide enter through a plant surface.

PERENNIAL—A plant that normally lives more than two years.

PERSIST—To remain for a period of time.

PEST—An unwanted organism, such as a weed, insect, and virus.

PESTICIDE—A chemical or physical agent used to kill or control pests.

PESTICIDE KILL—Accidental death of nontarget organisms caused by a pesticide (may be due to careless or improper use).

PETROLEUM PRODUCT—Gasoline, kerosene, oil, and other products derived from crude oil.

pH—Measure of relative acidity of a solution.

PHOSPHORUS—A highly-reactive element that combines readily with other elements and is one of the three primary plant foods.

PHYTOTOXIC—Injurious to plants.

PLANT DISEASE—An abnormal plant condition caused by a pathogen, virus, or improper environmental condition.

POINT OF DRIP—Spray application just heavy enough to produce runoff from the leaves.

POISON—Chemical or other material which can cause illness or death.

POISON CONTROL CENTER— An agency (usually a hospital) which maintains an information file on first-aid procedures and antidotes for poisons.

POLLINATORS—Insects which spread pollen.

POLLUTE—To make something (air, water, soil) unsafe or impure because of carelessness or misuse.

PORT OF ENTRY—A place where foreign goods, animals, and plants enter a country.

POSTEMERGENT—Time period after young plants come through the soil surface (also: postemergence or postemerge).

POTASH—Potassium content of fertilizers expressed as potash (K_2O).

POTENCY—Strength, such as the degree of toxicity of a chemical.

PPM—Parts per million.

PRECAUTIONS—Safety measures taken in advance.

PRECIPITATE—A solid substance that forms in a liquid and settles out.

PREDATOR—An animal or insect that feeds on and destroys other animals or insects.

PRE-EMERGENT—Time period between planting of seeds and emergence of seedlings (also: pre-emergence or pre-emerge).

PREHARVEST—Time period immediately preceding harvest.

PREPLANT—Time period before a crop is planted.

PRESSURE—Force on an area, usually expressed as pounds per square inch; pressure causes a liquid to flow.

PRIMARY PLANT FOODS—The three basic nutrients required for plant growth (nitrogen, phosphorus, potassium).

PRODUCT—A term describing a material such as pesticide or fertilizer as sold to the user.

PROPERTIES—Characteristics or traits used to describe something, such as a chemical.

PROTECTANT—A chemical applied to prevent damage by pests or diseases.

PROTECTIVE GEAR—Clothes and other equipment used to guard workers against poisoning when working with toxic materials.

PUPA—Stage of insect metamorphosis between larva and adult (plural: pupae).

R

RATE—Amount of material applied to a plant or an area of land.

RECOMMENDED DOSAGE—Amount of pesticide to use to safely control a pest.

RE-ENTRY INTERVAL—Time span that must pass after application of a pesticide before it is safe to enter the treated area.

REGISTRATION—Acknowledgement of approval of a pesticide, by EPA, for the uses stated on the label.

REPELLENT OR REPELLANT—Chemical which discourages insects or other pests from entering a treated area.

RESIDUE—Pesticide remaining on or in a plant or other treated area following a time lapse after the application.

RESISTANCE—1. A characteristic that exists or is developed by natural selection that enables a pest population to survive normal application rates of pesticide. 2. Natural or developed characteristics of plants or animals which discourage attack by certain insects or disease organisms.

RESPIRATOR—A face mask equipped with filters to remove poisonous gases, spray, and dust particles from air to be breathed.

RESTRICTIONS—Limitations on use.

RESTRICTED-USE PESTICIDES—Pesticides which may be used only by licensed or certified applicators.

RODENTICIDE—Pesticide used to control rats, mice, and other rodents.

RUNAWAY PESTS—Pests which enter an area with no natural enemies and, as a result, reproduce in large numbers.

RUNOFF—Flow of surface liquid from land, plant, or animal surfaces.

S

SCIENTIFIC NAME—The name, usually Latin or Greek, used throughout the world to identify a specific plant or animal.

SECONDARY PLANT FOOD ELEMENTS—Calcium, magnesium and sulfur.

SEIZURE—Impounding of crop or animal products which contain more than allowable pesticide residues.

SELECTIVE PESTICIDE—A pesticide which controls some specific pest species without killing a crop or other desirable species.

SENSITIVE AREAS—Places particularly subject to damage from an improperly used pesticide.

SENSITIVE CROPS—Crops particularly susceptible to damage by a pesticide.

SHOCK—Severe reaction of a human or animal body to a serious injury.

SHORT-TERM PESTICIDE—A pesticide which breaks down into nontoxic substances relatively soon after application.

SIGNAL WORDS—The words on pesticide labels which indicate the toxicity of the pesticide: "Danger-Poison"; "Warning" or "Caution".

SLURRY—A thick paste-like mixture of wettable powder and water which is then diluted with additional water in the spray tank prior to application.

SOIL FUMIGANT—A pesticide applied as a vapor or gas to the soil surface or subsurface.

SOIL INJECTION—Placing pesticide below the surface of the soil.

SOIL STERILANT—The kill agent used to organisms in the soil with heat or chemicals.

SOIL TYPE—A term used in classifying different kinds of soil according to primary physical characteristics.

SOLUBLE POWDER—A finely-ground formulation of dry powder that dissolves readily in water or other specific liquids.

SOLUTION—Mixture of solid, liquid, or gas dissolved in a liquid.

SOLVENT—Liquid, such as water, kerosene, or oil, that will dissolve a pesticide and form a solution.

SPOT TREATMENT—Application to a small or limited area.

SPRAY—Application of a pesticide using water or another liquid as a carrier and applied in tiny droplets.

SPRAY CONCENTRATE—A formulation of pesticide that is diluted before application.

SPECIES—A group of living organisms with similar characteristics.

SPREADER-STICKER—An adjuvant which helps spray droplets spread and cover the surface more uniformly and to adhere longer and more tightly.

STERILIZE—Treatment, with chemicals or heat, to kill all organisms in an area.

STRUCTURAL PESTS—Insects and other pests which attack buildings.

SUCTION HOSE—Hose used to pull water from a stream or pond or other source into the sprayer tank, or to draw liquid from the tank to the sprayer pump.

SURFACE WATER—Any water located aboveground, such as lakes and streams.

SURFACTANT—Wetting agent that makes mixing easier and helps spread the mixture and wet the surface being treated.

SUSCEPTIBLE—Organism that can be killed or injured by a pesticide.

SUSPENDED—A pesticide use that is no longer approved—remaining stock of the material cannot be used.

SUSPENSION—Fine particles of a pesticide distributed throughout a liquid, solid, or gas.

SWATH—Width of the area treated in one pass by a sprayer or other applicator.

SYMPTOM—Evidence or indicator of disease; reaction of a plant to a pathogen; reaction of a person to pesticide overdose.

SYNERGISM—An increase in toxicity of chemicals mixed together, beyond their total potency as separate agent.

SYNTHETIC MATERIAL—Material manufactured, rather than being extracted from plants or animals.

SYSTEMIC—1. A pesticide taken up by one part of a plant and translocated to another part of the plant. 2. Disease wherein the pathogen spreads generally throughout the plant.

T

TANKAGE—Dried animal residues, usually from fat or gelatin.

TARGET—Area or plant intended to be treated with a pesticide.

TECHNICAL MATERIAL—Active, concentrated pesticide, before formulation and dilution.

TEST ANIMALS—Laboratory animals used in toxicity studies of pesticides.

TICK—A small, 8-legged, blood-sucking animal that superficially resembles an insect.

TILTH—Physical condition of the soil and its potential for cultivation of growing plants.

TOLERANCE—A term referring to the amount of pesticide that can remain in a plant or animal product to be eaten by humans or animals. Ability of a plant to develop even though attacked by disease or insect.

TOLERANT—Not susceptible to injury by a pesticide, parasite, or adverse conditions.

TOXIC—Poisonous.

TOXICANT—A poison.

TOXICITY—Degree to which a substance is poisonous.

TOXIN—"Poison" produced by a plant or animal.

TRACE ELEMENTS—Elements needed in very small quantities by plants.

TRADE NAME—Name given by a manufacturer to identify a product for advertising and sale.

TRANSPORT—Move from one place to another.

TREATED AREA—Place where a pesticide has been applied.

U

ULTRA-LOW-VOLUME (ULV)—Very low application rates; often of the pesticide as supplied by the manufacturer.

UNIFORMLY—The same over an entire area or each time.

UNINFORMED PERSONS—People not trained to handle and apply pesticides correctly and safely.

UNINTENTIONALLY—Accidentally; not meant to happen.

UREA—A synthetic organic-nitrogenous compound often used as a fertilizer.

USDA—The United States Department of Agriculture.

V

VAPOR—Gas, steam.

VAPORIZE—Evaporate or convert to a gas.

VERMIN—Noxious animal, pests, usually rodents or insects.

VICTIM—Someone who is injured or poisoned.

VOLATILITY—Tendency of a chemical to vaporize.

W

WEATHERING—Action of the wind, rain, and dust to remove pesticides from the surfaces to which they were applied.

WETTABLE POWDER—A pesticide formulation in the form of a powder which is easily wetted and mixed with water to form a suspension for application.

WIDE RANGE—Capability of a pesticide to control a variety of pests.

Z

ZERO TOLERANCE—A residue standard which does not allow any chemical residue to remain in or on a food product.

INDEX

A

B

B (continued)

C

D

L (continued)

M

N

N (continued)

O

P

P (continued)

Q

R

S

S (continued)

T (continued)

T

U

V

W

W (continued)

Y